FIRST-ORDER MODAL LOGIC

SYNTHESE LIBRARY

STUDIES IN EPISTEMOLOGY,

LOGIC, METHODOLOGY, AND PHILOSOPHY OF SCIENCE

VOLUME 277

MELVIN FITTING AND
RICHARD L. MENDELSOHN

*Lehman College and the Graduate Center,
CUNY, New York, U.S.A.*

FIRST-ORDER
MODAL LOGIC

KLUWER ACADEMIC PUBLISHERS
DORDRECHT / BOSTON / LONDON

A C.I.P. Catalogue record for this book is available from the Library of Congress.

ISBN 0-7923-5335-8 (PB)
ISBN 0-7923-5334-X (HB)

Published by Kluwer Academic Publishers,
P.O. Box 17, 3300 AA Dordrecht, The Netherlands.

Sold and distributed in North, Central and South America
by Kluwer Academic Publishers,
101 Philip Drive, Norwell, MA 02061, U.S.A.

In all other countries, sold and distributed
by Kluwer Academic Publishers,
P.O. Box 322, 3300 AH Dordrecht, The Netherlands.

Printed on acid-free paper

Printed in the Netherlands.

To Marsha, Robin, and Josh, with love
RLM
To Roma, who knows what I mean
MF

CONTENTS

PREFACE

After a rocky start in the first half of the twentieth century, modal logic hit its stride in the second half. The introduction of possible world semantics around the mid-century mark made the difference. Possible world semantics provided a technical device with intuitive appeal, and almost overnight the subject became something people felt they understood, rightly or wrongly. Today there are many books that deal with modal logic. But with very few exceptions, treatments are almost entirely of *propositional* modal logic. By now this is a well-worked area, and a standard part of philosophical training. *First-order* modal logic, on the other hand, is under-represented in the literature and under-developed in fact. This book aims at correcting the situation.

To make our book self-contained, we begin with propositional modal logic, and our presentation here is typical of the entire book. Our basic approach is semantic, using possible world, or Kripke, models. Proof-theoretically, our machinery is semantic tableaus, which are easy and intuitive to use. (We also include a treatment of propositional axiom systems, though axiom systems do not continue throughout the book.) Philosophically, we discuss the issues that motivate formal modal logics, and consider what bearing technical developments have on well-known philosophical problems. This three-pronged approach, Kripke semantics, tableaus, philosophical discussion, reappears as each new topic is introduced.

Classically, first-order issues like constant and function symbols, equality, quantification, and definite descriptions, have straightforward formal treatments that have been standard items for a long time. Modally, each of these items needs rethinking. First-order modal logic, most decidedly, is not just propositional modal logic plus classical quantifier machinery. The situation is much subtler than that. Indeed, many of the philosophical problems attending the development of modal logic have arisen because of the myopia of looking at it through classical lenses. For this reason we initially develop first-order modal logic without constants and functions. This arrangement separates the philosophical problems due to names and definite descriptions from those of quantifiers, and facilitates the development and understanding of first-order modal notions.

Successive chapters are arranged along the following lines. First, as we said, we introduce the quantifiers, and address the difference between actualist quantification (where quantifiers range over what actually exists) and possibilist quantification (where they range over what might exist). Semantically, this is the difference between varying domain models and constant domain models. (If you don't know what this means, you will.) Next, we introduce equality. We discuss Frege's well-known morning star/evening star puzzle in this modal context, while presenting a formal treatment of equality. We then return to the matter of actualist and possibilist quantification and show how to embed possibilist quantification in an actualist framework and conversely.

By this point in the exposition we are dealing both with a defined notion of existence as well as a primitive notion, and the time is ripe for a full-fledged discussion of the logical problems with existence and, in particular, of the idea that there are things that could exist but don't. Only after all of this has been examined are constants and function names introduced, along with the idea of predicate abstraction and rigidity.

Predicate abstraction is perhaps the most central notion of our treatment. This is a syntactic mechanism for abstracting a predicate from a formula, in effect providing a scoping mechanism for constants and function symbols similar to that provided for variables by quantifiers. Using predicate abstraction we are able to deepen the understanding of singular terms and reference, enabling distinctions that were either not visible or only dimly visible classically. Since the variable binding of predicate abstracts behaves similarly to the variable binding of quantifiers, analogues to some of the issues involving Barcan and converse Barcan formulas crop up again.

We continue our discussion of nonexistence by distinguishing it from nondesignation: we contrast terms that designate a nonexistent, e.g., "the golden mountain," and terms that fail to designate at all, e.g., "the round square." In the final chapter we present formal machinery for definite descriptions that borrows from both the Fregean and Russellian paradigms, and builds on the work in earlier chapters concerning these issues of existence and designation.

A primary goal of this text is to present first-order modal logic in an intuitively appealing way, so as to make more widely available versatile machinery for a new perspective on issues in philosophy of language and in metaphysics. Among other things, modal logic is a logic of change. It is our hope that this text on first-order modal logic will be an agent of change, sweeping through philosophical discussions, bringing a freshness, liveliness, and new vigor to ancient issues.

We wish to thank Graham Priest for very helpful comments on a draft of the manuscript for this book.

PROPOSITIONAL MODAL LOGIC

For analytic philosophy, formalization is a fundamental tool for clarifying language, leading to better understanding of thoughts expressed through language. Formalization involves abstraction and idealization. This is true in the sciences as well as in philosophy. Consider physics as a representative example. Newton's laws of motion formalize certain basic aspects of the physical universe. Mathematical abstractions are introduced that strip away irrelevant details of the real universe, but which lead to a better understanding of its "deep" structure. Later Einstein and others proposed better models than that of Newton, reflecting deeper understanding made possible by the experience gained through years of working with Newton's model.

"Model" is the key word here. A mathematical model of the universe is not the universe. It is an aid to the understanding—an approximately correct, though partial, description of the universe. As the Newton/Einstein progression shows, such models can improve with time, thought, and investigation.

What we present in this book is a mathematical model of the *modal* aspects of language. We apply formal techniques to clarify the interactions of modal issues with other important notions such as identity, names, and definite descriptions. We intersperse our formal mathematical presentation with discussions of the underlying philosophical problems that elucidate the basic issues.

In order to present a logic formally, several separate items are necessary. We need a rigorously defined *language* of course, but by itself this is pure, uninterpreted syntax. To give meaning to formulas of a language, we need a *semantics* that lets us single out those formulas we are primarily interested in, the *valid* ones. We also want a notion of *proof*, so that we can write down compact verifications that certain formulas are valid. And finally, we must connect the notion of proof with the semantics, so we can be sure it is exactly the valid formulas that have proofs.

We assume you are familiar with all this for classical propositional logic. There, *truth tables* provide the semantics—valid formulas are those for which every line evaluates to T (such formulas are generally called *tautologies*). You may have seen axiom systems for classical propositional logic, or tableau (tree) proof systems, or natural deduction systems. Since modal notions are more complex than propositional ones, one expects modal semantics and proof systems to be correspondingly more complex. Such expectations will be fulfilled shortly.

1

In this chapter we present syntax and semantics for several propositional modal logics. Proof systems follow in Chapters 2 and 3. Our treatment is relatively terse because propositional logics are not the main topic of this book; *first-order* modal logic is. Our treatment of this begins in Chapter 4. For fuller presentations of propositional modal logic, see (Hughes and Cresswell, 1968), (Hughes and Cresswell, 1984), (Hughes and Cresswell, 1996), (Chellas, 1980) or (Fitting, 1983).

With all this very general background, we still have not said what kinds of things modal notions are. It is time to begin.

1.1. WHAT IS A MODAL?

A modal qualifies the truth of a judgment. *Necessarily* and *possibly* are the most important and best known modal qualifiers. They are called "alethic" modalities, from the Greek word for *truth*. In traditional terminology, *Necessarily P* is an "apodeictic judgment," *Possibly P* a "problematic judgment," and *P*, by itself, an "assertoric judgment."

The most widely discussed modals apart from the alethic modalities are the *temporal* modalities, which deal with past, present, and future ("will be," "was," "has been," "will have been," etc.), the *deontic* modalities, which deal with obligation and permission ("may," "can," "must," etc.) and the *epistemic* or *doxastic* modalities, which deal with knowledge and belief ("certainly," "probably," "perhaps," "surely," etc.). Prior (1955), perhaps a bit tongue-in-cheek, quotes the 18th century logician Isaac Watts to show the variety of modals that are available:

We might also describe several *moral* or *civil Modes* of connecting two Ideas together (*viz.*) *Lawfulness* and *Unlawfulness, Conveniency* and *Inconveniency,* etc. whence we may form such *modal Propositions* as these: ... *It is lawful for Christians to eat Flesh in* Lent .... . There are several other *Modes* of speaking whereby a Predicate is connected with a Subject: Such as, *it is certain, it is doubtful, it is probable, it is improbable, it is agreed, it is granted, it is said by the ancients, it is written* etc. all which will form other Kinds of *modal Propositions.* (p. 215-216)

The modals capture what traditional linguists have categorized as *tense*, *mood* and *aspect*, but which, in more modern treatments of grammar, are lumped together as *adverbials*, i.e., as modifiers of adjectives or verbs, and, derivatively, of nouns. In the sentence-frame,

John is _____ happy

the blank space can be filled by any of the modals: *necessarily, possibly, contingently, known by me to be, believed by you to be, permitted to be, now, then*. Alternatively, these qualifiers can be inserted into the sentence-frame

It is ——— true that John is happy

and regarded as modifying the truth of the claim. Hence the two equivalent characterizations: the modal qualifies the predicate, or the modal qualifies the truth of a claim.

Not all modals give rise to modal *logics*, however. Usually there must be two modal operators that conform to the logical principles embodied in the Modal Square of Opposition (Section 1.4, Figure 1). But this is to presuppose that a logic of the modals is even possible, a view that has been much debated in the philosophical literature. Before we introduce the Square of Opposition, then, we will discuss these issues.

EXERCISES

EXERCISE 1.1.1.   Discuss the following.

1. Is 'probably' a modal?
2. Is 'truly' a modal?
3. Is 'needless to say' a modal?
4. Is 'not' a modal?

1.2. CAN THERE BE A MODAL LOGIC?

Aristotle set the tone for the older tradition of logic—a tradition that stretched from antiquity, through the medieval period, as far as Leibniz—by including the study of modality in his book on the syllogism:

Since there is a difference according as something belongs, necessarily belongs, or may belong to something else (for many things belong indeed, but not necessarily, others neither necessarily nor indeed at all, but it is possible for them to belong), it is clear that there will be different syllogisms to prove each of these relations, and syllogisms with differently related terms, one syllogism concluding from what is necessary, another from what is, a third from what is possible. *Prior Analytics* i. 8 ($29^b29 - 35$) in (Ross, 1928)

The modal syllogism never attained the near universal acceptance of Aristotle's more familiar logic of categorical statements, partly because of serious confusions in his treatment (documented by Kneale and Kneale (1962)). The same problems infected the discussions of many who followed him. Indeed, Kneale (1962) quotes a medieval logician who warns students away

from the study of modality if they are to retain their sanity! Modality was
closely linked with theological speculation and the metaphysics of the school-
men, both of which were largely repudiated by the Enlightenment. The combination of interminable confusions and discredited metaphysics has given
sustenance to those who have urged that there cannot be a modal logic.

The primary claim of the opposition is that modality has no place in the
content of a judgment, so that the modal is inert with regard to the logical
connections between that judgment and others. Kant (1781) is representative.
Here is what he says about the category of the modals presented in his *Table
of Judgments* in Book I of the *Transcendental Analytic*:

> The *modality* of judgments is a quite peculiar function. Its distinguishing character-
> istic is that it contributes nothing to the content of the judgment (for, besides quantity,
> quality, and relation, there is nothing that constitutes the content of a judgment), but
> concerns only the value of the copula in relation to thought in general. (p. 10)

And Kant's assessment was echoed by Frege (1879):

> What distinguishes the apodeictic from the assertoric judgment is that it indicates
> the existence of general judgments from which the proposition may be inferred—an
> indication that is absent in the assertoric judgment. If I term a proposition 'necessary,'
> then I am giving a hint as to my grounds for judgment. *But this does not affect the
> conceptual content of the judgment; and therefore the apodeictic form of a judgment
> has not for our purposes any significance.* (p. 4)

Frege (1879) laid the groundwork for the modern treatment of logic by providing a notation and a set of axioms and rules of inference for propositional,
first-order, and higher-order logic. His decision to omit modality was an influential factor in delaying the contemporary development of modal logic.

Frege's remarks above are found in a section where he explains which
distinctions he will attempt to capture in his logical symbolism, and which
he will not. His *Begriffsschrift* (literally, *Concept Script*) was designed to capture
conceptual content, "that part of judgments which affects the *possible inferences* .... Whatever is needed for a valid inference is fully expressed; what
is not needed is for the most part not indicated either ... ." (p. 3) Compare,
for example, "The Greeks defeated the Persians at Plataea" and "The Persians
were defeated by the Greeks at Plataea." The active and passive forms of a
judgment, he argues, each have exactly the same inferential connections—any
judgment that entails (or is entailed by) the first entails (or is entailed by) the
second. The grammatical distinction may therefore be of psychological significance, but it is of no logical significance, so the active/passive distinction is
not formally marked. By contrast, there is a clear logical distinction between
"The Greeks defeated the Persians at Plataea *and* Thermopylae," on the one
hand, and "The Greeks defeated the Persians at Plataea *or* Thermopylae," on

the other; and both conjunction and disjunction are therefore represented in his *Begriffsschrift*.

It is true, as Frege says, that the content represented by P remains the same whether preceded by "necessarily," "possibly," or nothing at all. But it does not follow that the apodeictic, problematic and assertoric judgments all have the same conceptual content. Contrary to what Frege says, modal distinctions do affect possible inferences. Let us avail ourselves of modern notation: $\Box P$ for *It is necessary that P* and $\Diamond P$ for *It is possible that P*. Now,

$$P \supset \Diamond P$$

(i.e., *It's actual, so it's possible*) is usually considered to be valid—Hughes and Cresswell (1968) call it the "Axiom of Possibility"—but its converse,

$$\Diamond P \supset P$$

(i.e., *It's possible, so it's actual*) is not. By Frege's own criterion, P and $\Diamond P$ differ in conceptual content. Frege's logic does not capture these formal inferences, but they are inferences nonetheless, and there is no compelling reason why logic should not be widened to accommodate them. Can there be a modal logic, then? There is no *a priori* reason why not. The most effective way of showing that there can be a modal logic is to actually construct one. We will construct several!

1.3. WHAT ARE THE FORMULAS?

Now it is time to be more specific about the formal language we will be using. The language of propositional modal logic extends that of classical propositional logic. Classically, formulas are built up from *propositional letters* (also called *propositional variables*). We use P, Q, R, etc. for these—they stand for unanalyzed propositions. (Sometimes one also allows *propositional constants*, $\top$ and $\bot$, intended to represent truth and falsehood respectively. We generally will not. Propositional letters are combined into more elaborate formulas using *propositional connectives* such as $\neg$ (not), $\wedge$ (and), $\vee$ (or), $\supset$ (if, then), $\equiv$ (if, and only if), etc. The first, $\neg$, is *unary*, while the others are *binary*. There are other binary connectives besides these, so the list can be longer. On the other hand, it is possible to define some of these from others—both proof-theoretically and semantically—so the list can be shorter. We will sometimes take one approach, sometimes the other, depending on convenience.

What is added to this syntax of classical logic are two new unary operators, $\Box$ (necessarily) and $\Diamond$ (possibly). Again, it will turn out these are interdefinable, so the list can be shorter. Also there are various *multi-modal* logics, so the list can be longer. We will say more about such variations later.

DEFINITION 1.3.1. [Propositional Modal Formulas] The set of (propositional modal) *formulas* is specified by the following rules.

1. Every propositional letter is a formula.
2. If X is a formula, so is $\neg X$.
3. If X and Y are formulas, and $\circ$ is a binary connective, $(X \circ Y)$ is a formula.
4. If X is a formula, so are $\Box X$ and $\Diamond X$.

Using this definition, for example, $((\Box P \wedge \Diamond Q) \supset \Diamond (P \wedge Q))$ is a formula. It can be read, "If P is necessary, and Q is possible, then it is possible that both P and Q." Informally, we will generally omit the outer parentheses of a formula. Thus the formula we just gave may appear as $(\Box P \wedge \Diamond Q) \supset \Diamond (P \wedge Q)$.

We are using $\Box$ and $\Diamond$ to symbolize modal operators. Some books, following a different tradition, use L and M instead—(Hughes and Cresswell, 1996) is an example. In addition, when considering issues of knowledge, $\mathcal{K}$ is often used in place of $\Box$. If more than one person's knowledge is concerned, $\mathcal{K}_a$, $\mathcal{K}_b$, etc. may be used—this is an example of a multi-modal logic. Likewise, modal operators corresponding to actions can be considered. For example, the action might be that of going from New York to San Francisco. If $\Box$ is intended to denote this action, then if P is the proposition, "It is 1pm," and Q is the proposition "It is after 2pm," then $P \supset \Box Q$ would be read, "If it is 1pm, then after I go from New York to San Francisco (no matter how I do so), it will be after 2pm." A modal logic intended for consideration of actions would also naturally be a multi-modal logic, since there are many different actions—generally the notation involves $\Box$ with action-denoting letters placed inside.

In this book we confine our presentation to just $\Box$ and $\Diamond$ since, if the formal aspects of these two are firmly understood, generalizations are fairly straightforward.

EXERCISES

EXERCISE 1.3.1.   Using the definition verify that, in fact, $((\Box P \wedge \Diamond Q) \supset \Diamond (P \wedge Q))$ is a formula.

1.4. ARISTOTLE'S MODAL SQUARE

A rough minimal requirement for a modal logic is that $\Box$ and $\Diamond$ satisfy the Modal Square of Opposition given below in Figure 1. (Additional requirements have been suggested, but there is no generally accepted standard. See (Lukasiewicz, 1953; Bull and Segerberg, 1984; Koslow, 1992) for alternatives.)

The Modal Square is due to Aristotle, who worked out the basic logical connections between necessity and possibility in his *De Interpretatione*. In Chapter 12, he argued that the negation of

(1.1)   It is possible that P

is not "It is possible that *not* P," but rather "It is *not* possible that P." Similarly, he argued that the negation of "It is necessary that P" is not "It is necessary that *not* P," but rather "It is *not* necessary that P."

In Chapter 13, he connected up the two modalities by identifying

(1.2)   It is *not* necessary that *not* P.

as the equivalent of (1.1). Aristotle's claim that (1.1) is equivalent to (1.2) reflects the standard (first half) of the interdefinability of the two modal operators:

$$\Diamond P \equiv \neg\Box\neg P$$
$$\Box P \equiv \neg\Diamond\neg P.$$

And we thereby get the Modal Square of Opposition given in Figure 1.

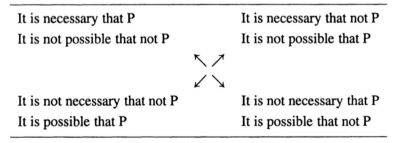

| It is necessary that P | It is necessary that not P |
| It is not possible that not P | It is not possible that P |

| It is not necessary that not P | It is not necessary that P |
| It is possible that P | It is possible that not P |

Figure 1.   The Modal Square of Opposition

The Modal Square of Opposition is analogous to the more familiar Square of Opposition for Categorical Statements given in Figure 2. In both cases the schemata across the top row are *contraries* (cannot both be true), the schemata across the bottom row are *subcontraries* (cannot both be false), the

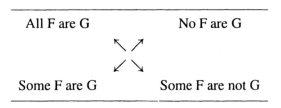

Figure 2.   The Square of Opposition for Categorical Statements

schemata in each column are *subalternatives* (the top implies the bottom), and the schemata across the diagonals are *contradictories* (cannot have the same truth value).[1]

1.5. INFORMAL INTERPRETATIONS

In Section 1.1, we characterized a modal as an adverbial that fills the blank in a sentence frame like

John is _____ happy.

The situation is actually somewhat more complicated. Adverbials differ in their favored placement. *Necessarily* must precede the adjective; *now* and *then* can occur either before or after the adjective; *here* and *there* must come after the adjective. These are peculiarities of English, and are independent of the underlying logical sense. In what follows, we will find it helpful to standardize the frame, with the modifier coming after the adjective, like this:

John is happy _____,

The modal we insert qualifies John's happiness, opening up a spectrum along which modifications of his happiness can be compared. Let us flesh out the adverbial position as

John is happy <u>under interpretation *i*</u>.

Think of an interpretation *i* as a particular assignment of truth values to the nonlogical particles in the proposition, with the logical particles understood as the usual boolean functions. On the simplest reading of $\Box$ and $\Diamond$, a proposition is necessarily true if it comes out true for every interpretation—i.e., if it

---

[1] This does assume the modal logic is sufficiently strong—at least **D** (to be introduced soon). Some of these properties do not hold for weaker logics.

is a logical truth—and a proposition is possibly true if it comes out true for at least one interpretation—i.e., if it is consistent. Our modal propositions

$\Box$ John is happy

$\Diamond$ John is happy

therefore correspond respectively to the quantified propositions

($\forall i$) (John is happy under interpretation i)

($\exists i$) (John is happy under interpretation i)

"John is happy," without a modal operator, is understood as "John is happy under interpretation i_0," where i_0 is the interpretation corresponding to the real world.

It is more usual in modal logic to speak of a possible interpretation as a *possible world* or a *possible state*. The terminology and the idea that a necessary truth is true in all possible worlds is derived from Leibniz.

Let us see how this reading works for $\Box P \supset P$. If P is true in every possible world, then it is true in the actual world. So, this is intuitively a valid claim. $P \supset \Box P$, on the other hand, is an invalid claim. Just because P is true in the actual world, it need not be true in *every* possible world.

Modern modal logic employs a modification of Leibniz's notion by taking into consideration the way in which the modal relates the possible worlds. This is called an *accessibility* relation. We will have much to say about accessibility, so for now one example must suffice: temporal worlds are ordered by their place in time. Take a world Δ to be accessible from a world Γ if Γ occurs no later in the temporal order than Δ. A proposition is *necessarily* true if it is true at every *accessible* possible world; a proposition is *possibly* true if it is true at some *accessible* possible world. Let w_i be a possible world, and let w_0 be the actual world. Let $\mathcal{R}$ be the accessibility relation: $x\mathcal{R}y$ is read y *is accessible from* x. Then

$\Box$ John is happy

corresponds to

($\forall w_i$)$_{w_0\mathcal{R}w_i}$ (John is happy at w_i)

where we read the quantifier as, "for all w_i in the relation $\mathcal{R}$ to w_0," or less formally, as, "for all w_i accessible from w_0."

Consider $\Box P \supset P$ again. Once accessibility relations are brought in, this says that if P is true at every accessible possible world, then it is true at the real world. For this to remain valid, the real world must be accessible to itself—which it is, according to the temporal definition just given. But we could have defined the accessibility relation differently, so that the real world is not accessible to itself. In such a setting this formula is not valid.

The notion of accessibility, introduced by Kripke (1963a), makes it possible to model a rich variety of modal logics. If we consider interpretations or worlds connected by an *accessibility relation*, then various algebraic constraints on the relation turn out to correspond to specific modal principles. The formal connection between these algebraic constraints and modal principles is discussed in Section 1.8.

Let us see informally how the accessibility relation enables us to extend our understanding of the modal operators. Consider, for example,

(1.3) $\Diamond P \supset \Box \Diamond P$

Let $\Diamond P$ mean *It is conceivable that P*. Then (1.3) says that if there is a possible world we can conceive of from the actual one—i.e., *an accessible (conceivable) world*—in which P is true, then from every accessible (conceivable) world there is an accessible (conceivable) world in which P is true. Is this so? If P is conceivable, is it inconceivable that it is inconceivable that P? It is reasonable to suppose that from the standpoint of a Greek disciple of Euclid, it is inconceivable that parallel lines meet; but this is certainly not inconceivable to us. So, from our perspective, it is conceivable that parallel lines meet, making the antecedent of (1.3) true; but there is at least one conceivable situation (putting ourselves in the sandals of an ancient Greek) from which it is inconceivable that parallel lines meet, making the consequent of (1.3) false. Presumably we can put ourselves into the Greek's sandals but he cannot put himself into our shoes, because we know something about the past, but he knows nothing about the future. This means that, in the case of *conceivability*, the accessibility relation is not symmetric and so (1.3) does not hold.

Once accessibility among worlds has been highlighted as the preeminent characteristic of modal logics, we can interpret the modal operators far afield from their originally intended meanings. Harel (1984) gives the flavor of *Dynamic Logic*. Suppose we are speaking about all possible execution states relevant to some computer program α, and we define an accessibility relation $s\mathcal{R}_\alpha t$ such that t is a possible next state of α with current state s. We obtain a multi-modal logic with $\boxdot P$ read as, "Every possible execution of α leads to a situation in which P is true," and $\Diamond\!\!\!\!\cdot P$ read as, "It is possible to execute α reaching a situation in which P is true." Hence one can generalize the subject called modal logic: it is an attempt to characterize talk of deformations or transitions or alternatives.

EXERCISE 1.5.1. There are four truth-table definable functions of a single input.

1. Show that we can define a truth-functional operator $\Box$ for which $\Box P \supset P$ is a thesis but for which $P \supset \Box P$ is not.
2. Show that if we also require that $\neg \Box P$ not be a thesis, no truth-functional operator will do.

EXERCISE 1.5.2. Suppose every world is accessible to every other. Would (1.3) then be valid? Discuss informally.

1.6. WHAT ARE THE MODELS?

Although several formal semantics have been created for modal logics, *possible world* semantics has become the standard. It was introduced in the 1960's, by Kripke, in (Kripke, 1963b; Kripke, 1963a; Kripke, 1965), and independently by Hintikka, in (Hintikka, 1961; Hintikka, 1962). It provides an elegant mathematical treatment that is intuitively meaningful. Indeed, we have already used it informally in our discussions of the previous section.

The notion of possible world is of a piece with standard model-theoretic interpretations of classical logic. As Kripke (1980) says:

The main and original motivation for the 'possible world analysis'—and the way it clarified modal logic—was that it enabled modal logic to be treated by the same set theoretic techniques of model theory that proved so successful when applied to extensional logic. (p 19)

A valid formula (i.e., logical truth) of classical propositional logic is one that comes out true for every *possible* assignment of truth values to the statement letters. This notion of alternative interpretations should not pose any more of a philosophical problem in modal logic than it does in classical logic. Put a slightly different way, in so far as there are philosophical difficulties with the notion of possible world, these are difficulties that are already to be found in understanding multiple interpretations for statements of classical logic.

The difficulty is exacerbated by the fact that we say in modal logic, "There is a possible world such that ... ," and so we quantify over possible worlds. According to a criterion of ontological commitment put forward by Quine (1948), this means that we are committed to the existence of such things. But some of these are intended to be mere possibilities, not actualities, and so we appear to be committed to the existence of things we would not otherwise want to be committed to. It looks as though we have all of a sudden increased our store of existents. Not only do we have the way things are, but also the way things might be. Once again, however, we must underscore that the very same difficulty crops up in the classical case, for we speak there of alternative interpretations, and we consider *all* possibilities—all truth-table lines—and these are possibilities that cannot all hold simultaneously. Since the notion of possible world is useful and coherent, it is quite clear that the problem is not with the notion of possible world but with the philosophical theories about them.

DEFINITION 1.6.1. [Frame] A *frame* consists of a non-empty set, $\mathcal{G}$, whose members are generally called *possible worlds*, and a binary relation, $\mathcal{R}$, on $\mathcal{G}$, generally called the *accessibility relation*. Since a frame consists of two parts, we follow general mathematical practice and say a frame is a pair, $\langle \mathcal{G}, \mathcal{R} \rangle$, so we have a single object to talk about.

Understand that, while *possible world* is suggestive terminology, it commits us to nothing. In the mathematical treatment of frames, possible worlds are any objects whatsoever—numbers, sets, goldfish, etc.

We will generally use Γ, Δ, etc. to denote possible worlds. If Γ and Δ are in the relation $\mathcal{R}$, we will write $\Gamma \mathcal{R} \Delta$, and read this as Δ is *accessible from* Γ, or Δ is an *alternative world* to Γ.

DEFINITION 1.6.2. [Model] A frame is turned into a modal model by specifying which propositional letters are true at which worlds. We use the following notation for this. A *propositional modal model*, or *model* for short, is a triple $\langle \mathcal{G}, \mathcal{R}, \Vdash \rangle$, where $\langle \mathcal{G}, \mathcal{R} \rangle$ is a frame and $\Vdash$ is a relation between possible worlds and propositional letters. If $\Gamma \Vdash P$ holds, we say P is *true at the world* Γ. If $\Gamma \Vdash P$ does not hold we symbolize this by $\Gamma \nVdash P$, and say P is *false at* Γ.

Incidentally, $\Vdash$ is often read as "forces," a usage which comes from proving the independence results of set theory. In (Smullyan and Fitting, 1996) modal logic is, in fact, used quite directly for this purpose, illustrating how close the connection is.

Keep classical propositional logic in mind as partial motivation. There, each line of a truth table is, in effect, a model. And a truth table line is specified completely by saying which propositional letters are assigned T and which are assigned F. Modal models are more complex in that a *family* of such truth assignments is involved, one for each member of $\mathcal{G}$, and there is some relationship between these assignments (represented by $\mathcal{R}$, though we have not seen its uses yet). Now in the classical case, once truth values have been specified at the propositional letter level, they can be calculated for more complicated formulas. The modal analog of this is to calculate the truth value of non-atomic formulas *at each possible world*. The notational mechanism we use is to extend the relation $\Vdash$ beyond the propositional letter level.

DEFINITION 1.6.3. [Truth in a Model] Let $\langle \mathcal{G}, \mathcal{R}, \Vdash \rangle$ be a model. The relation $\Vdash$ is extended to arbitrary formulas as follows. For each $\Gamma \in \mathcal{G}$:

1. $\Gamma \Vdash \neg X \Longleftrightarrow \Gamma \nVdash X$.
2. $\Gamma \Vdash (X \wedge Y) \Longleftrightarrow \Gamma \Vdash X$ and $\Gamma \Vdash Y$.
3. $\Gamma \Vdash \Box X \Longleftrightarrow$ for every $\Delta \in \mathcal{G}$, if $\Gamma \mathcal{R} \Delta$ then $\Delta \Vdash X$.
4. $\Gamma \Vdash \Diamond X \Longleftrightarrow$ for some $\Delta \in \mathcal{G}$, $\Gamma \mathcal{R} \Delta$ and $\Delta \Vdash X$.

We will give concrete examples in the next section, but first a few remarks. The first two clauses of Definition 1.6.3 say that, at each world, the propositional connectives $\wedge$ and $\neg$ behave in the usual truth table way. The last two clauses are what is special. The first says necessary truth is equivalent to truth in all possible worlds, but limited to those that are accessible from the one you are dealing with. The second says possible truth is equivalent to truth in some accessible possible world.

Suppose we think of $\vee$ as a defined connective in the usual way: $(X \vee Y)$ abbreviates $\neg(\neg X \wedge \neg Y)$. Then, using various of the conditions above,

$$\Gamma \Vdash (X \vee Y) \Longleftrightarrow \Gamma \Vdash \neg(\neg X \wedge \neg Y)$$
$$\Longleftrightarrow \Gamma \nVdash (\neg X \wedge \neg Y)$$
$$\Longleftrightarrow \Gamma \nVdash \neg X \text{ or } \Gamma \nVdash \neg Y$$
$$\Longleftrightarrow \Gamma \Vdash X \text{ or } \Gamma \Vdash Y$$

Alternatively we could take $\vee$ as primitive, and add to the definition of truth in a model the additional condition:

$$\Gamma \Vdash (X \vee Y) \Longleftrightarrow \Gamma \Vdash X \text{ or } \Gamma \Vdash Y$$

Either way leads to the same thing. Thus there are two possible approaches to the various binary connectives not mentioned in Definition 1.6.3. Whether defined or primitive, we want the condition for $\vee$ above, as well as the following.

$$\Gamma \Vdash (X \supset Y) \Longleftrightarrow \Gamma \nVdash X \text{ or } \Gamma \Vdash Y$$
$$\Longleftrightarrow \text{ if } \Gamma \Vdash X \text{ then } \Gamma \Vdash Y$$
$$\Gamma \Vdash (X \equiv Y) \Longleftrightarrow \Gamma \Vdash X \text{ if and only if } \Gamma \Vdash Y$$

Incidentally, just as we could take various propositional connectives as either primitive or defined, the same is the case with the modal operators. Check for yourselves that if we took $\Diamond X$ as an abbreviation for $\neg\Box\neg X$, the result would be equivalent to the characterization in Definition 1.6.3.

The notation we are using, $\Gamma \Vdash X$, does not mention the model, only a possible world and a formula. This is fine if it is understood which model we are working with, but if more than one model is involved, the notation can be ambiguous. In such cases we will write $\mathcal{M}, \Gamma \Vdash X$, where $\mathcal{M} = \langle \mathcal{G}, \mathcal{R}, \Vdash \rangle$ is a model. We take this to mean: X is true at the world Γ of the collection $\mathcal{G}$ of possible worlds of the model $\mathcal{M}$. We will suppress explicit mention of $\mathcal{M}$ whenever possible, though, in the interests of a simpler notation.

EXERCISES

EXERCISE 1.6.1.   Show that, at each possible world Γ of a modal model, $\Gamma \Vdash (\Box X \equiv \neg\Diamond\neg X)$ and $\Gamma \Vdash (\Diamond X \equiv \neg\Box\neg X)$.

EXERCISE 1.6.2.   Show that if a world Γ of a model has no worlds accessible to it, then at Γ every formula is necessary, but none are possible.

1.7. EXAMPLES

In order to give a feeling for the behavior of modal models, we give several examples, and a few exercises. The pattern of presentation we follow is this. Models are given using diagrams, with boxes representing possible worlds. For two worlds of such a model, Γ and Δ, if $\Gamma\mathcal{R}\Delta$, we indicate this by drawing an arrow from Γ to Δ. We say explicitly which propositional letters are true at particular worlds. If we do not mention some propositional letter, it is taken to be false.

We begin with examples showing certain formulas *are not* true at worlds in a particular model, then move on to examples showing certain other formulas *are* true at worlds of models, under very broad assumptions about the models.

EXAMPLE 1.7.1.
    Here is the first of our specific models.

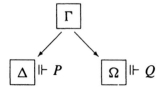

In this model, $\Delta \Vdash P \vee Q$ since $\Delta \Vdash P$. Likewise, $\Omega \Vdash P \vee Q$. Since Δ and Ω are the only possible worlds in this model that are accessible from Γ, and $P \vee Q$ is true at both of them, $\Gamma \Vdash \Box(P \vee Q)$. On the other hand, we do not have $\Gamma \Vdash \Box P$, for if we did, since Ω is accessible from Γ, we would have $\Omega \Vdash P$, and we do not. Similarly we do not have $\Gamma \Vdash \Box Q$ either, and so we do not have $\Gamma \Vdash \Box P \vee \Box Q$. Consequently $\Box(P \vee Q) \supset (\Box P \vee \Box Q)$ is not true at Γ.

You might try showing $(\Diamond P \wedge \Diamond Q) \supset \Diamond(P \wedge Q)$ is also not true at Γ.

EXAMPLE 1.7.2.
    This time, $\Gamma \nVdash \Box P \supset \Box\Box P$.

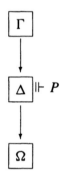

$\Gamma \Vdash \Box P$ since $\Delta \Vdash P$, and Δ is the only world accessible from Γ. If we had $\Gamma \Vdash \Box\Box P$, it would follow that $\Delta \Vdash \Box P$, from which it would follow that $\Omega \Vdash P$, which is not the case. Consequently $\Gamma \nVdash \Box P \supset \Box\Box P$.

EXAMPLE 1.7.3.
    Show for yourself that, in the following model, $\Gamma \nVdash \Diamond P \supset \Box\Diamond P$.

EXAMPLE 1.7.4. Again, show for yourself that in the following model, $\Gamma \not\Vdash$ $(\Diamond\Box P \wedge \Diamond\Box Q) \supset \Diamond\Box(P \wedge Q)$.

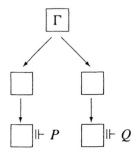

EXAMPLE 1.7.5. This is a counter-example to lots of interesting formulas.

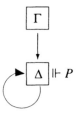

Since P is true at Δ, and Δ is the only world accessible from Δ, $\Box P$ is true at Δ. But then again, since $\Box P$ is true at Δ, it follows that $\Box\Box P$ is also true there, and likewise $\Box\Box\Box P$, and so on.

Now, since Δ is the only world accessible from Γ, and P is true at Δ, $\Box P$ is true at Γ. Likewise, since $\Box P$ is true at Δ, $\Box\Box P$ is true at Γ, as is $\Box\Box\Box P$, and so on. On the other hand, P itself is not true at Γ. Thus at Γ, all of the following formulas are false: $\Box P \supset P$, $\Box\Box P \supset P$, $\Box\Box\Box P \supset P$, and so on.

Next we turn to examples showing certain formulas must be true at worlds, under very general circumstances.

EXAMPLE 1.7.6. In Example 1.7.1 we showed $\Box(P \vee Q) \supset (\Box P \vee \Box Q)$ could be false at a world under the right circumstances, so we cannot assume $\Box$ distributes over $\vee$. Now we show $\Box$ does distribute over $\wedge$, that is, $\Box(P \wedge Q) \supset (\Box P \wedge \Box Q)$ is true at every possible world of every model. This time we draw no pictures, since it is not a *particular* model we are dealing with, but the class of *all* models.

Suppose $\langle \mathcal{G}, \mathcal{R}, \Vdash \rangle$ is a model, and $\Gamma \in \mathcal{G}$. Also suppose $\Gamma \Vdash \Box(P \wedge Q)$; we show it follows that $\Gamma \Vdash \Box P \wedge \Box Q$.

Let Δ be an arbitrary member of $\mathcal{G}$ that is accessible from Γ, that is, $\Gamma \mathcal{R} \Delta$. Since $\Gamma \Vdash \Box(P \wedge Q)$, it follows that $\Delta \Vdash P \wedge Q$, and hence $\Delta \Vdash P$ and also $\Delta \Vdash Q$. Since Δ was arbitrary, at *every* possible world accessible from Γ, both P and Q are true, hence at Γ both $\Box P$ and $\Box Q$ are true. But then, $\Gamma \Vdash \Box P \wedge \Box Q$.

In a similar way, it can be shown that the converse, $(\Box P \wedge \Box Q) \supset \Box(P \wedge Q)$ is true at every world, as is $\Box(P \supset Q) \supset (\Box P \supset \Box Q)$. We leave these to you.

Sometimes binary relations satisfy special conditions, and many of these have standard names. Here are a few of them. A relation $\mathcal{R}$ on $\mathcal{G}$ is *reflexive* if $\Gamma \mathcal{R} \Gamma$ holds for every $\Gamma \in \mathcal{G}$, that is, if every world is accessible from itself. Also $\mathcal{R}$ is *transitive* if $\Gamma \mathcal{R} \Delta$ and $\Delta \mathcal{R} \Omega$ always imply $\Gamma \mathcal{R} \Omega$; $\mathcal{R}$ is *symmetric* if $\Gamma \mathcal{R} \Delta$ always implies $\Delta \mathcal{R} \Gamma$.

EXAMPLE 1.7.7. Suppose $\langle \mathcal{G}, \mathcal{R}, \Vdash \rangle$ is a model in which $\mathcal{R}$ is transitive. Then $\Gamma \Vdash \Box P \supset \Box\Box P$, for each $\Gamma \in \mathcal{G}$. We show this as follows.

Suppose $\Gamma \Vdash \Box P$; we must show $\Gamma \Vdash \Box\Box P$. And to show $\Gamma \Vdash \Box\Box P$, let Δ be any member of $\mathcal{G}$ such that $\Gamma \mathcal{R} \Delta$ and show $\Delta \Vdash \Box P$. And to show this, let Ω be any member of $\mathcal{G}$ such that $\Delta \mathcal{R} \Omega$ and show $\Omega \Vdash P$. But, $\Gamma \mathcal{R} \Delta$ and $\Delta \mathcal{R} \Omega$, and $\mathcal{R}$ is transitive, so $\Gamma \mathcal{R} \Omega$. Also, we began by supposing that $\Gamma \Vdash \Box P$, so it follows that $\Omega \Vdash P$, which is what we needed.

EXAMPLE 1.7.8. Suppose $\langle \mathcal{G}, \mathcal{R}, \Vdash \rangle$ is a model.

1. If $\mathcal{R}$ is reflexive, $\Box P \supset P$ is true at every member of $\mathcal{G}$.
2. If $\mathcal{R}$ is symmetric, $P \supset \Box\Diamond P$ is true at every member of $\mathcal{G}$.
3. If $\mathcal{R}$ is both symmetric and transitive, $\Diamond P \supset \Box\Diamond P$ is true at every member of $\mathcal{G}$.

We leave the verification of these items to you as exercises.

EXERCISES

EXERCISE 1.7.1.   Continue Example 1.7.1 and show that $(\Diamond P \wedge \Diamond Q) \supset \Diamond(P \wedge Q)$ is not true at Γ.

EXERCISE 1.7.2.   Complete Example 1.7.3 by showing that $\Gamma \nVdash \Diamond P \supset \Box\Diamond P$.

EXERCISE 1.7.3.   Complete Example 1.7.4 by showing that $\Gamma \nVdash (\Diamond\Box P \wedge \Diamond\Box Q) \supset \Diamond\Box(P \wedge Q)$.

EXERCISE 1.7.4. Show the following are true at every possible world of every model.

1. $(\Box P \wedge \Box Q) \supset \Box(P \wedge Q)$
2. $\Box(P \supset Q) \supset (\Box P \supset \Box Q)$

EXERCISE 1.7.5. Show the three parts of Example 1.7.8.

1.8. SOME IMPORTANT LOGICS

During the course of the twentieth century, a large number of modal logics have been created and investigated. Partly this was because of different intended applications, and partly it was because our intuitions about necessary truth are not fully developed. While everyone (probably) agrees that, whatever "necessity" might mean, necessary truths have necessary consequences, opinions differ about whether necessary truths are so by necessity, or contingently. And probably few indeed have thought much about whether a proposition whose necessity is possibly necessary is necessarily true, or even possible. Once iterated modalities arise, intuition and experience begin to be inadequate. In part, different formal modal logics have arisen in an attempt to capture differences in the behavior of iterated modalities. There is no point here in trying to choose between them. We simply present a few of the more famous systems that have arisen, and leave it at that. The main purpose of this book is the treatment of *first-order* modal issues and for this, variations between modal logics at the propositional level have minor significance.

Originally, different modal logics were characterized by giving different sets of axioms for them. One of the early successes of possible world semantics was the characterization of most standard modal logics by placing simple mathematical conditions on frames. We want to emphasize this: conditions are placed on *frames*. Although models are what we deal with most often, frames play a central role. We will see the importance of frames throughout this book.

In this section we discuss a handful of the best-known modal logics: **K**, **D**, **T**, **K4**, **B**, **S4**, and **S5**, and we use possible world semantics for their definition. Think of these logics as just a sampling of what can be obtained using this approach.

DEFINITION 1.8.1. [**L**-Valid] We say the model $\langle \mathcal{G}, \mathcal{R}, \Vdash \rangle$ is *based on* the frame $\langle \mathcal{G}, \mathcal{R} \rangle$. A formula X is *valid in a model* $\langle \mathcal{G}, \mathcal{R}, \Vdash \rangle$ if it is true at every world of $\mathcal{G}$. A formula X is *valid in a frame* if it is valid in every model based on that frame. Finally, if **L** is a collection of frames, X is **L**-valid if X is valid in every frame in **L**.

Different modal logics are characterized semantically as the **L**-valid formulas, for particular classes **L** of frames. To give a simple example, the logic called **T** is characterized by the class of frames having the property that each world is accessible from itself. To make things easier to state, we recall some standard terminology that was briefly introduced earlier.

DEFINITION 1.8.2.   Let ⟨$\mathcal{G}$, $\mathcal{R}$⟩ be a frame. We say it is:

1. *reflexive* if $\Gamma \mathcal{R} \Gamma$, for every $\Gamma \in \mathcal{G}$;
2. *symmetric* if $\Gamma \mathcal{R} \Delta$ implies $\Delta \mathcal{R} \Gamma$, for all Γ, $\Delta \in \mathcal{G}$;
3. *transitive* if $\Gamma \mathcal{R} \Delta$ and $\Delta \mathcal{R} \Omega$ together imply $\Gamma \mathcal{R} \Omega$, for all Γ, Δ, $\Omega \in \mathcal{G}$;
4. *serial* if, for each $\Gamma \in \mathcal{G}$ there is some $\Delta \in \mathcal{G}$ such that $\Gamma \mathcal{R} \Delta$.

The logic **T**, mentioned above, is characterized by the class of reflexive frames. From now on we, ambiguously, use **T** to denote *both* the class of reflexive frames, and the logic determined by that class. This should cause no real difficulties—context will always determine which is meant when it matters. The chart in Figure 3 supplies standard names for several more frame collections, and their corresponding logics. We use these names in what follows.

Logic	Frame Conditions
K	no conditions
D	serial
T	reflexive
B	reflexive, symmetric
K4	transitive
S4	reflexive, transitive
S5	reflexive, symmetric, transitive

Figure 3.   Some Standard Modal Logics

We saw in the previous section, in Example 1.7.6, that $\square(P \wedge Q) \supset (\square P \wedge \square Q)$ is true at every world of every model. In other words, it is **K**-valid. In that same Example, $(\square P \wedge \square Q) \supset \square(P \wedge Q)$ and $\square(P \supset Q) \supset (\square P \supset \square Q)$ were also said to be **K**-valid, and verification was left as an exercise. Likewise in Example 1.7.7 it was shown that $\square P \supset \square \square P$ is valid in any model whose frame is transitive. It follows that this formula is **K4**-valid, and also **S4**- and **S5**-valid as well, since these include transitivity among their frame conditions. Example 1.7.8 shows that $\square P \supset P$ is **T**-valid, that $P \supset \square \diamond P$ is **B**-valid, and that $\diamond P \supset \square \diamond P$ is **S5**-valid.

If a formula X is **K4**-valid it is true in all transitive frames, and so in all transitive and reflexive frames, and hence it is also **S4**-valid. Thus **K4** is a *sublogic* of **S4**. In a similar way, **B** is a sublogic of **S5**. Trivially, **K** is a sublogic of all the others. Finally, every reflexive relation is automatically serial (why?) so **D** is a sublogic of **T**. All this is summarized in the following diagram.

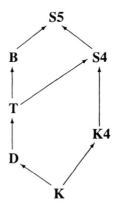

Every frame in **T** is also a frame in **D**, but the converse is not true. Still it is conceivable that, *as logics*, **D** and **T** are the same. That is, it is conceivable that the same formulas are valid in the two classes of frames. (In fact, there are well-known examples of two distinct classes of frames that determine the same logic.) However, this is not the case here. Example 1.7.8 shows that $\Box P \supset P$ is **T**-valid. But the model in Example 1.7.5 is based on a **D**-frame, and $\Box P \supset P$ is not valid in it. Thus as logics **D** and **T** are different. As a matter of fact, the inclusions among logics that are shown in the diagram above are the *only* ones that hold among these logics.

EXERCISES

EXERCISE 1.8.1.   Show $\Box P \supset \Diamond P$ is valid in all serial models.

EXERCISE 1.8.2.   Prove that a frame $\langle \mathcal{G}, \mathcal{R} \rangle$ is transitive *if and only if* every formula of the form $\Box P \supset \Box \Box P$ is valid in it.

EXERCISE 1.8.3.   There is a modal logic, sometimes called **S4.3** (due to Dummett and Lemmon) which is characterized semantically by the class of frames $\langle \mathcal{G}, \mathcal{R} \rangle$ for which $\mathcal{R}$ is reflexive, transitive, and *linear*, that is, for all $\Gamma, \Delta \in \mathcal{G}$, either $\Gamma \mathcal{R} \Delta$ or $\Delta \mathcal{R} \Gamma$.

   1. Show that $\Box(\Box P \supset \Box Q) \vee \Box(\Box Q \supset \Box P)$ is valid in all **S4.3** frames.

2. Show that $P \supset \Box \Diamond P$ can be falsified in some **S4.3** frame.

3. Show that $\Diamond \Box (P \supset Q) \supset (\Diamond \Box P \supset \Diamond \Box Q)$ is valid in all **S4.3** frames.

EXERCISE 1.8.4. This exercise has many parts, but each one is not difficult. Show the inclusions between logics that are given in the diagram above are the only ones that hold.

1.9. LOGICAL CONSEQUENCE

In classical logic the notion of *follows from* is fundamental. Loosely speaking, one says a formula X follows from a set of formulas S provided X must be true whenever the members of S are true. More formally, one says X is a *logical consequence* of S (in classical propositional logic) provided that for every assignment of truth values to propositional letters, if every member of S evaluates to T, then X also evaluates to T. This is often symbolized by $S \models X$. Many important facts about classical propositional logic are quite easily stated using the notion of logical consequence. For instance the *compactness* property says, if $S \models X$ then for some finite subset $S_0 \subseteq S$ we have $S_0 \models X$. Tarski instituted the general study of consequence relations, and their fundamental role has been accepted ever since.

Not surprisingly, things are more complex for modal logics. We want to keep the intuition that X is a logical consequence of S if X is true whenever the members of S are true. But a modal model can contain many possible worlds. Do we want to take X to be a consequence of S if X is true at each *world* at which the members of S are true, or do we want to take it as meaning X is valid in every *model* in which the members of S are valid? No such dichotomy is available classically, but the choice presents itself modally, and the two notions are most decidedly *not* equivalent. This has caused a certain amount of confusion from time to time. As it happens, both versions are important, so we briefly sketch a treatment of them using notation from (Fitting, 1983). We omit proofs of basic facts here; it is enough if one understands the kind of complications modal notions bring. Proofs can be found in (Fitting, 1983) and in (Fitting, 1993).

DEFINITION 1.9.1. [Consequence] Let **L** be one of the frame collections given in Figure 3 of Section 1.8. Also let S and U be sets of formulas, and let X be a single formula. We say X is a consequence in **L** of S as *global* and U as *local* assumptions, and write $S \models_L U \rightarrow X$, provided: for every frame $\langle \mathcal{G}, \mathcal{R} \rangle$ in the collection **L**, for every model $\langle \mathcal{G}, \mathcal{R}, \Vdash \rangle$ based on this frame in which all members of S are valid, and for every world $\Gamma \in \mathcal{G}$ at which all members of U are true, we have $\Gamma \Vdash X$.

Thus $S \models_L U \to X$ means X is true at those worlds of **L** models where the members of U are true, provided the members of S are true throughout the model, i.e., at every world. Admittedly, this is a more complex notion than classical consequence. Local and global assumptions play quite different roles, as the following examples make clear.

EXAMPLE 1.9.2.   We take **L** to be **K**, the collection of all frames, S to be empty, so there are no global assumptions, U to contain only $\Box P \supset P$, and X to be $\Box\Box P \supset \Box P$. The following model shows $\emptyset \models_K \{\Box P \supset P\} \to \Box\Box P \supset \Box P$ *does not* hold.

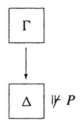

In this model, $\Gamma \not\Vdash \Box P$ since $\Gamma \mathcal{R} \Delta$ and $\Delta \not\Vdash P$. Consequently $\Gamma \Vdash (\Box P \supset P)$, so the only member of $\{\Box P \supset P\}$ is true at Γ. On the other hand, $\Delta \Vdash \Box P$ since there are no worlds alternate to Δ at which P fails. Since Δ is the only world accessible from Γ, it follows that $\Gamma \Vdash \Box\Box P$. We have already seen that $\Gamma \not\Vdash \Box P$, so $\Gamma \not\Vdash (\Box\Box P \supset \Box P)$.

We thus have a counterexample to $\emptyset \models_K \{\Box P \supset P\} \to \Box\Box P \supset \Box P$ fails. Every global assumption is valid in the model since there are none. But at Γ all local assumptions are true and $\Box\Box P \supset \Box P$ is not.

Comparison of this example with the following will make clear the difference in behavior between local and global assumptions.

EXAMPLE 1.9.3.   This time we show that $\{\Box P \supset P\} \models_K \emptyset \to \Box\Box P \supset \Box P$ *does* hold.

Let $\langle \mathcal{G}, \mathcal{R}, \Vdash \rangle$ be a **K** model in which $\Box P \supset P$ is valid. Let Γ be an arbitrary member of $\mathcal{G}$ (at which all the local assumptions hold, since there are none). We show $\Gamma \Vdash \Box\Box P \supset \Box P$.

Suppose $\Gamma \Vdash \Box\Box P$. Let Δ be an arbitrary member of $\mathcal{G}$ such that $\Gamma \mathcal{R} \Delta$. Then $\Delta \Vdash \Box P$. Since $\Box P \supset P$ is valid in the model, $\Delta \Vdash (\Box P \supset P)$, and so also $\Delta \Vdash P$. Since Δ was arbitrary, $\Gamma \Vdash \Box P$.

Many of the general properties of classical consequence carry over to modal consequence. For instance, classical consequence obeys a *monotonicity* condition: if X is a consequence of a set, it is also a consequence of any extension of that set. A similar result holds modally.

PROPOSITION 1.9.4. Suppose $S \models_L U \rightarrow X$, and $S \subseteq S'$ and $U \subseteq U'$. Then $S' \models_L U' \rightarrow X$.

Classical consequence has the *compactness* property, as we noted at the beginning of this section. This too carries over to a modal setting.

PROPOSITION 1.9.5. Suppose $S \models_L U \rightarrow X$. Then there are finite sets $S_0 \subseteq S$ and $U_0 \subseteq U$ such that $S_0 \models_L U_0 \rightarrow X$.

One of the most used features of classical consequence is that it obeys the deduction theorem. Classically, this says that $S \cup \{Y\} \models X$ if and only if $S \models (Y \supset X)$. It is often used to replace the proof of an implication, $Y \supset X$, by a derivation of a simpler formula, X, from a premise, Y. But modal consequence differs significantly here. More precisely, there are two versions of the deduction theorem, depending on whether Y is taken as a local or as a global assumption.

PROPOSITION 1.9.6. [Local Deduction] $S \models_L U \cup \{Y\} \rightarrow X$ if and only if $S \models_L U \rightarrow (Y \supset X)$.

PROPOSITION 1.9.7. [Global Deduction] $S \cup \{Y\} \models_L U \rightarrow X$ if and only if
$S \models_L U \cup \{Y, \square Y, \square\square Y, \square\square\square Y, \ldots\} \rightarrow X$.

The Local Deduction Proposition is essentially the same as the classical version. In a sense, it is the peculiar form of the Global Deduction Proposition that is at the heart of the complexities of modal consequence. In subsequent chapters we will need the notion of modal consequence from time to time, but it will almost always be with an empty set of local assumptions. Different books on modal logic often make special assumptions about whether local or global assumptions are fundamental, and so the very definition of modal deduction differs significantly from book to book.

DEFINITION 1.9.8. We abbreviate $S \models_L \emptyset \rightarrow X$ by $S \models_L X$.

EXERCISES

EXERCISE 1.9.1. Show that $\emptyset \models_K \{\square P \supset P\} \rightarrow \square\square P \supset P$ does not hold, but $\{\square P \supset P\} \models_K \emptyset \rightarrow \square\square P \supset P$ does.

EXERCISE 1.9.2. Prove Proposition 1.9.4.

EXERCISE 1.9.3.   Let $\Box^n Y$ denote $\Box\Box\ldots\Box Y$, where we have written a string of n occurrences of $\Box$. Use the various facts given above concerning modal consequence and show the following version of the deduction theorem: $S \cup \{Y\} \models_L U \to X$ if and only if, for some n, $S \models_L U \to (\Box^0 Y \wedge \Box^1 Y \wedge \ldots \wedge \Box^n Y) \supset X$.

1.10. TEMPORAL LOGIC

The temporal interpretations of $\Box$ and $\Diamond$ are as old as the study of modality itself. In the present century, it was the temporal semantics developed over a number of years by A. N. Prior that provided the clearest understanding of modal logic, until general possible world semantics was created. We have stronger intuitions about temporal worlds than we do about the more abstract and austere alethically possible worlds, and frequently we will find ourselves using the temporal interpretation to aid in the understanding of modal concepts.

For Frege, a complete thought, a *Gedanke*, is an eternal, nonlinguistic object whose truth value never changes. Quine has a corresponding notion of an *eternal sentence*. The sentence

(1.4)   John is happy

is not an eternal sentence; it does not, *by itself*, express a complete thought. For, unless John is very unusual, there are periods in history in which it is true and periods in history in which it is false: it is not, by itself, timelessly true or false. If we were to add a time determination to the sentence, we would stabilize its truth value. Let us, then, qualify *when* John is happy. The sentence frame is effectively filled with a time index t,

(1.5)   John is happy <u>at time t,</u>

where t occupies a position that can be bound by a quantifier. We suppose there to be a range of times at some of which, perhaps, John is happy, and at others, not. Suppose that John bought the latest model computer in November of 1994. Then, it is reasonable to suppose that the sentence,

(1.6)   John is happy <u>in November 1994</u>

is true. Suppose, as always happens, that a more powerful computer is introduced some months later and the one he bought has dropped significantly in price. Then, it is reasonable to suppose that the sentence,

(1.7)   John is happy <u>in February 1995</u>

is false. We can existentially generalize on the true (1.6) to get the true

> ($\exists t$) John is happy <u>at time t</u>.

Given the falsity of (1.7), we have the falsity also of its universal generalization

> ($\forall t$) John is happy <u>at time t.</u>

Treating (1.4) as (1.5), in effect, is to regard *that John is happy* not as a proposition, but as a property of times. On this reading, (1.5) is an open sentence. When t is inside the scope of a quantifier, it is bound by the quantifier; when outside the scope of a quantifier, it acts as an indexical *now* denoting the designated (actual) time.

   This gives us our first example of a modal logic with a temporal interpretation.

EXAMPLE 1.10.1.   This simplest temporal interpretation of $\Box P$ and $\Diamond P$, which parallels our alethic treatment of *necessarily* and *possibly*, takes

> $\Box$ John is happy
>
> $\Diamond$ John is happy

to correspond, respectively, to

> ($\forall t$) John is happy at time t
>
> ($\exists t$) John is happy at time t

As with the alethic interpretation, $P \supset \Diamond P$ turns out true and $\Diamond P \supset P$ turns out false; similarly, $\Box P \supset P$ turns out true and $P \supset \Box P$ turns out false.

   Let us complicate our temporal interpretation by introducing the following future and past tense operators:

> FP :   It will sometime be the case that P
>
> PP :   It was sometime the case that P

where the verb in P is now untensed. It is common to define two more tense operators, GP ($\equiv \neg F \neg P$) and HP ($\equiv \neg P \neg P$):

> GP :   It will always be the case that P
>
> HP :   It has always been the case that P

   We can, using these tense operators, capture a considerable number of natural language tenses. Let P be "John wins the election." Then we have the following ways of interpreting natural language tenses:

(a)	P	John wins the election
(b)	$\mathbf{F}P$	John will win the election
(c)	$\mathbf{P}P$	John won the election
(d)	$\mathbf{PP}P$	John had won the election
(e)	$\mathbf{FP}P$	John will have won the election
(f)	$\mathbf{PF}P$	John would win the election

We will explain the iterated operators in (d) – (f) shortly.

We can develop two different modal logics from these operators by taking $\Box P$ and $\Diamond P$, respectively, to be $\mathbf{G}P$ and $\mathbf{F}P$, or equally well, to be $\mathbf{H}P$ and $\mathbf{P}P$. (The reader can verify that the Modal Square of Opposition is satisfied either way.)

These modal operators can be nested, and the complications thus introduced are nontrivial. We have, to this point, made no provisions for iterations of temporal operators. This is clear from our quantificational representation. If $\Diamond P$ says "There will be a time t such that John wins the election at t," then putting another temporal operator in front will appear to have no effect since there is no free variable left to bind. Yet, from the intuitive reading we give to the modal operators above, there are obvious distinctions: (e) is distinct from (f), for example, and (c) is distinct from (d). How are we to handle these cases?

Let the present time be t. Then corresponding to the modal claims

$\mathbf{F}$ John wins the election,

$\mathbf{P}$ John wins the election,

$\mathbf{G}$ John wins the election,

$\mathbf{H}$ John wins the election,

we have, respectively,

$(\exists t')(t' > t \wedge$ John wins the election at $t')$,

$(\exists t')(t' < t \wedge$ John wins the election at $t')$,

$(\forall t')(t' > t \wedge$ John wins the election at $t')$,

$(\forall t')(t' < t \wedge$ John wins the election at $t')$.

Then (e) and (f) become, respectively,

$(\exists t'')(t'' > t \wedge (\exists t')(t' < t'' \wedge$ John wins the election at $t'))$

and

$(\exists t'')(t'' < t \wedge (\exists t')(t' > t'' \wedge$ John wins the election at $t'))$

(We leave (*d*) as an exercise for the reader.) So, (*e*) says that there will be a time in the future before which John wins the election: he will have won the election; and (*f*) says that there is a time in the past after which John wins the election: he would win the election.

Thus we have our second, more elaborate, temporal interpretation of the modal operators.

EXAMPLE 1.10.2. Taking $\Box$ and $\Diamond$ to be **G** and **F**, respectively, gives us a different logic from the one in Example 1.10.1 even though the Modal Square of Opposition is still satisfied. For example, the thesis $\Box P \supset P$, fails. Just because it *will* always be true that P, it does not follow that P is true *now*. On the other hand, $\Box P \supset \Diamond P$ holds: if it will always be true that P, then there will be at least one (future) occasion at which P is true (assuming that time does not end). Other differences emerge when we consider iterations of the modal operators. Consider: $P \supset \Box \Diamond P$. This fails because from the fact that P is true, it does not follow that it will always be the case that there will be some further time at which it is true. On the other hand, $\Box P \supset \Box \Box P$, holds because the temporal order is clearly transitive: if P holds at every future time, then at every future time P will hold at every (yet) future time.

EXAMPLE 1.10.3. We can get yet another modal logic by defining $\Box$ and $\Diamond$ as follows:

$$\Diamond P = P \vee \mathbf{F} P$$
$$\Box P = P \wedge \mathbf{G} P$$

Thus $\Box P$ means P is and will remain the case; $\Diamond P$ means P is now or will at sometime be the case. (Prior (1957) claims that *Possibly* and *Necessarily* were used in this sense by the Megarian logician Diodorus.) In this system, $\Diamond P \supset \Box \Diamond P$ will fail. Suppose it either is or will be the case that P. Must this now and forever in the future be so? No. We might go too far into the future, to the point when P is a completed process. For instance, let P be "Scientists discover a cure for the common cold." We hope that $\Diamond P$ is the case—someday a cure will be discovered. But once discovered, it cannot be discovered again, so $\Box \Diamond P$ is not the case.

There is an ordering relation that is imposed by time in temporal logic: *later than*. The temporal ordering is transitive, but it is neither reflexive nor symmetric. So, if we take $\Box P$ to be $\mathbf{G} P$, a proposition is necessarily true if it holds at all future worlds: a world, or state, is accessible if it is later in the temporal ordering. If, on the other hand, we take $\Box P$ to be $P \wedge \mathbf{G} P$, then a proposition is necessarily true if it holds now and forever later. $\Box P \supset P$ fails for the former reading, but it holds for the latter one.

EXERCISE 1.10.1.   What is the corresponding quantifier translation of item
(d) **PP**P? Use our treatment of items (e) and (f) as models.

1.11. EPISTEMIC LOGIC

We can also interpret the usual modal symbols to create a logic of knowledge.
Hintikka (1962) introduces the following readings for $\square$ and $\lozenge$ respectively:

$\mathcal{K}_a P$: a knows that P
$\mathcal{P}_a P$: It is possible, for all that a knows, that P

This is a multi-modal logic: each subscript identifies an individual. There is a
similar vocabulary for belief:

$\mathcal{B}_a P$: a believes that P
$\mathcal{C}_a P$: It is compatible with everything a believes that P

In each case, we have the usual interdefinability of the modal operators:

$$\mathcal{K}_a P \equiv \neg \mathcal{P}_a \neg P$$
$$\mathcal{B}_a P \equiv \neg \mathcal{C}_a \neg P$$

The theorems of (Hintikka, 1962) include all tautologies plus

(1.8) $\mathcal{K}_a(P \supset Q) \supset (\mathcal{K}_a P \supset \mathcal{K}_a Q)$

(1.9) $\mathcal{K}_a P \supset P$

(1.10) $\mathcal{K}_a P \supset \mathcal{K}_a \mathcal{K}_a P$

and the two rules of inference, *Modus Ponens* (Y follows from X and $X \supset Y$)
and *Necessitation* ($\mathcal{K}_a X$ follows from X). Formula (1.8) will be our **K** axiom
in Chapter 3, corresponding to frame condition **K** in Figure 3. Likewise (1.9)
will be our **T** axiom; and (1.10) will be our **4** axiom. With **T** and **4** on top of
K, we have what is called an **S4** system.
    $\mathcal{K}_a P$ is perhaps better read as *It follows from what a knows that P*, or
perhaps as *a is entitled to know that P*. For, in this system, one knows all
the logical consequences of anything one knows: if $P \supset Q$ is logically true,
and if $K_a P$, then $\mathcal{K}_a Q$. In this logic, then, the knower is treated as *logically
omniscient*. This feature of the system is made even more puzzling because
of (1.10), which is called the *Principle of Positive Introspection*: one cannot
know something and yet fail to know that one knows it. Suppose, for example,
that Goldbach's Conjecture is provable (although we have no proof as yet).

Then, anyone who knows that the Peano Axioms are true—most likely, any reader of this book—knows, by logical omniscience, that Goldbach's Conjecture is true, and knows, by (1.10), that he knows that Goldbach's Conjecture is true. Yet it is unlikely that any reader of this book is in possession of a proof of Goldbach's conjecture, so it is unlikely that any reader of this book knows that Goldbach's Conjecture is true, and even more unlikely that any reader of this book knows that he knows that Goldbach's Conjecture is true. Formula (1.10) is justified, despite its title, not because knowing that P carries with it a certain mental feeling which one knows one has by introspection. Hintikka is working with the view that knowledge is justified, true belief.[2] So (1.10) is justified because, if one knows that P, one must be justified in believing that P, and one cannot be justified without being aware of being justified. It is not just that if one knows that P, there are reasons justifying P; rather, if one knows that P, these justifying reasons must be *your* reasons, and so you cannot help but be aware of them. Logical Omniscience simply tells us that if Q follows from P on logical grounds, every knower is aware of these grounds. This is a very strong condition on knowledge, and it is why we prefer reading $\mathcal{K}_a P$ as *a is entitled* to know that P, rather than as *a knows that P*.

Of course, there are some constraints on knowledge. It is interesting, in this regard, to note that by contrast with (1.10), the *Principle of Negative Introspection*,

(1.11) $\neg \mathcal{K}_a P \supset \mathcal{K}_a \neg \mathcal{K}_a P$

is *not* a theorem of this system. Again, this has nothing to do with introspection. The reasons are a bit more mundane: you might not know that P and yet *believe* that you do know it. (1.11), however, requires that if you don't know that P, then you *know*, and so *believe*, you don't know it. The following also fails to be a theorem:

(1.12) $P \supset \mathcal{K}_a \mathcal{P}_a P$

And this seems appropriate. For, a statement P might be true and yet one need not know that it is compatible with everything one believes. The following all do hold in Hintikka's system:

(1.13) $\mathcal{K}_a \mathcal{K}_b P \supset \mathcal{K}_a P$

(1.14) $\mathcal{K}_a P \supset \mathcal{P}_a P$

(1.15) $\mathcal{K}_a P \equiv \mathcal{K}_a \mathcal{K}_a P$

---

[2] "Believe it, for you know it." *Merry Wives of Windsor*, William Shakespeare.

(1.13) expresses the transmissibility of *knowledge*.[3] (1.14) essentially tells us that every world has an alternative, which we know from the reflexivity expressed in axiom (1.9). It can also be shown that $\mathcal{K}_a P \supset \mathcal{K}_a Q$ follows from $P \supset Q$. This is a form of logical omniscience: one knows all the logical consequences of anything one knows. It is a controversial feature of the system. In some sense, we are talking about an idealized version of knowledge, and not knowledge as people generally use the term.

A logic of belief is more problematic than a logic of knowledge because we are not entirely consistent in our belief sets. In any event, Hintikka (1962) suggests that

(1.16) $\mathcal{B}_a P \supset \mathcal{B}_a \mathcal{B}_a P$

is true. In defense of (1.16), Hintikka argues that the two formulas, $\mathcal{B}_a P$ and $\mathcal{B}_a (P \supset \neg \mathcal{B}_a P)$ are inconsistent. On the other hand, he suggests that none of the following are true:

$$\mathcal{B}_a \mathcal{B}_a P \supset \mathcal{B}_a P$$
$$\mathcal{B}_a \mathcal{B}_b P \supset \mathcal{B}_a P$$
$$\mathcal{B}_a P \supset \mathcal{K}_a \mathcal{B}_a P$$

The semantics of Hintikka (1962) is in terms of what he calls *model sets*, and his proof procedure is a precursor of what has since become known as *tableau rules*. On the usual semantics for classical logic, a model consists of an assignment of a truth value to every propositional symbol and an interpretation of the logical connectives as boolean functions on these truth values, so that each complex sentence of the language is assured a unique truth value. Hintikka, by contrast, takes a model to be a *model set*, that is, a collection of sentences (deemed to be true) which might be thought of as a partial description of the world. A model set μ is governed by the following rules:

(C.¬) If $P \in \mu$ then not $\neg P \in \mu$

(C.∧) If $P \wedge Q \in \mu$ then $P \in \mu$ and $Q \in \mu$

(C.∨) If $P \vee Q \in \mu$ then $P \in \mu$ or $Q \in \mu$ (or both)

(C.¬¬) If $\neg\neg P \in \mu$ then $P \in \mu$

(C.¬∧) If $\neg(P \wedge Q) \in \mu$ then $\neg P \in \mu$ or $\neg Q \in \mu$ (or both)

(C.¬∨) If $\neg(P \vee Q) \in \mu$ then $\neg P \in \mu$ and $\neg Q \in \mu$

---

[3] We do not have the corresponding principle for *belief*, i.e. $\mathcal{B}_a \mathcal{B}_b P \supset \mathcal{B}_a P$.

These are just the classical conditions. For modal logic we need the notion of a *model system*—a collection of model sets plus an alternativeness relation (actually, one for each knower a). The following additional rules govern the modal operators:

(C.P*) If $\mathcal{P}_a P \in \mu$ and if μ belongs to a model system Ω, then there is in Ω at least one alternative μ^* to μ (with respect to a) such that $P \in \mu^*$

(C.KK*) If $\mathcal{K}_a Q \in \mu$ and if μ^* is any alternative to μ (with respect to a) in model system Ω, then $\mathcal{K}_a Q \in \mu^*$

(C.K) If $\mathcal{K}_a P \in \mu$ then $P \in \mu$

(C.$\neg$ K) If $\neg \mathcal{K}_a P \in \mu$, then $\mathcal{P}_a \neg P \in \mu$

(C.$\neg$ P) If $\neg \mathcal{P}_a P \in \mu$, then $\mathcal{K}_a \neg P \in \mu$

The proof procedure is by *reductio*: if the assumption that $\neg X$ belongs to a model set leads to an inconsistency, then X must belong to every model set, i.e., it must be true in all models, which is to say that it is valid.

EXAMPLE 1.11.1.   Show: $(\mathcal{K}_a P \wedge \mathcal{K}_a Q) \supset \mathcal{K}_a(P \wedge Q)$

1	$\mathcal{K}_a P \wedge \mathcal{K}_a Q \in \mu$	Assumption
2	$\neg \mathcal{K}_a(P \wedge Q) \in \mu$	Assumption
3	$\mathcal{P}_a \neg(P \wedge Q) \in \mu$	(C.$\neg$ K), 2
4	$\neg(P \wedge Q) \in \mu^*$	(C.P*), 3
5	$\mathcal{K}_a P \in \mu$	(C.$\wedge$), 1
6	$\mathcal{K}_a Q \in \mu$	(C.$\wedge$), 1
7	$\mathcal{K}_a P \in \mu^*$	(C.KK*), 5
8	$\mathcal{K}_a Q \in \mu^*$	(C.KK*), 6
9	$P \in \mu^*$	(C.K), 7
10	$Q \in \mu^*$	(C.K), 8

$$11 \quad \neg P \in \mu^* \;\Big|\; 12 \quad \neg Q \in \mu^*$$
$$\times \qquad\qquad \times$$

In line 4, μ^* is an alternative model set to μ. Also lines 11 and 12 are the two alternatives deriving from line 4. One alternative is impossible, because of 9 and 11; the other is impossible, because of 10 and 12.

Epistemic logics continue to be of interest to philosophers. Nozick (1981) has raised some interesting problems with (1.8), $\mathcal{K}_a(P \supset Q) \supset (\mathcal{K}_a P \supset \mathcal{K}_a Q)$, in regard to skepticism. Let E be some experiential statement, e.g., "I see a human hand before me." Let S be the crucial skeptical statement that I am unable to disprove, e.g., "I am dreaming" (or, alternatively, "I am a brain in a vat"). Now, it is quite clear that if E is true, i.e., if I *do* see a human hand before me (and not just something that appears to be a human hand), then S must be false, i.e., it cannot be the case that I am now dreaming. So, we have

(1.17) $E \supset \neg S$

and also, as just pointed out, we know this to be the case.

(1.18) $\mathcal{K}_a(E \supset \neg S)$

Now, the skeptical position is just that I am unable to rule out the possibility that I am dreaming or that I am a brain in a vat. That is,

(1.19) $\neg \mathcal{K}_a \neg S$

An instance of (1.8) is

(1.20) $\mathcal{K}_a(E \supset \neg S) \supset (\mathcal{K}_a E \supset \mathcal{K}_a \neg S)$

From this, contraposing the consequent, we have the following.

(1.21) $\mathcal{K}_a(E \supset \neg S) \supset (\neg \mathcal{K}_a \neg S \supset \neg \mathcal{K}_a E)$

From (1.18) and (1.19), applying *Modus Ponens* twice, we derive

(1.22) $\neg \mathcal{K}_a E$.

In effect, then, I have no experiential knowledge. Nozick (1981) suggests that we should reject (1.8). On the other hand, (1.22) could be correct. This interplay of philosophy and modal logic is a good example of how the sharpening provided by logic formalization enables us to grasp a philosophical puzzle.

EXERCISES

EXERCISE 1.11.1.   Prove each of (1.13), (1.14) and (1.15) using Hintikka's model sets.

1.12. HISTORICAL HIGHLIGHTS

Now, armed with some formal machinery, it is time for a (highly selective) glance back in history at some well-known problems involving modal reasoning.

Aristotle's Development of the Square

Aristotle had to fight hard to work out The Modal Square of Opposition we presented earlier in Figure 1; confusions constantly threatened to obliterate the gains. Aristotle opens *De Interpretatione* 13 with a somewhat different square of opposition, and which we reproduce as Figure 4. In this figure, the

It is necessary that P	It is necessary that not P
It is not possible that not P	It is not possible that P

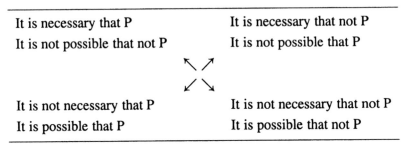

It is not necessary that P	It is not necessary that not P
It is possible that P	It is possible that not P

Figure 4. The Modal Square of Opposition in *De. Int.* 13

apodeictic forms have been transposed on the bottom row: *It is possible that P* is alternately expressed as *It is not necessary that P*, and *It is possible that not P* is alternately expressed as *It is not necessary that not P*. Aristotle had not clearly separated two readings of "possible" in *De Interpretatione* 13: on the one hand, he takes it to mean *not impossible*; on the other hand, he takes it to mean *contingent*. As Ackrill (1963) points out, the top half of the table appears to be driven by the first notion of "possible" while the bottom half appears to be driven by the second notion, i.e., of "contingent."[4]

The intention in a Square of Opposition is that the forms in the left column are to be *subalternatives*, with the top implying the bottom. But while *It is necessary that P* implies *It is possible that P*, it cannot imply the supposedly equivalent formulation *It is not necessary that P*. Aristotle saw the problem and sorted things out, eventually winning his way through to the square we gave in Figure 1. He broke apart the two forms in the lower left hand corner. But his argument reveals a typical modal confusion that spills over to infect the relatively good grasp he had of nonmodal distinctions. At *De Interpretatione* 22[b]10, he says:

For the necessary to be is possible to be. (Otherwise the negation will follow, since it is necessary either to affirm or to deny it; and then, if it is not possible to be, it is impossible to be; so the necessary to be is impossible to be—which is absurd.) (Ackrill, 1963), p. 63

---

[4] Not quite, however. *It is contingent that P* is equivalent to the conjunction *It is possible that P but not necessary that P*, not to *It is not necessary that P* alone.

Aristotle's conclusion, which we might express so,

(1.23) $\Box P \supset \Diamond P$

is readily acceptable in any of the modal logics discussed that are at least as powerful as **D**. But the reasoning Aristotle put forward to justify (1.23) is faulty. To be sure, as he says, "it is necessary either to affirm or to deny it," that is

Either (1.23) or not-(1.23).

But *not-(1.23)* ought to be

$\neg(\Box P \supset \Diamond P)$

Aristotle, however, concludes:

$\Box P \supset \neg \Diamond P.$

Ackrill (1963) regards this as a misapplication of the Law of Excluded Middle; but it is perhaps better viewed as a *scope* error. Aristotle confused the negation of the conditional with the negation of the consequent.

EXERCISES

EXERCISE 1.12.1.   Can a Square of Opposition be constructed when $\Diamond$ is interpreted to mean "It is contingent that," where *contingent* means true but not necessary?

EXERCISE 1.12.2.   Informally, we take $\Box P \supset \Diamond P$ to be valid for the following reason: If P is necessary, then P is true; and if P is true, then P is certainly possible. **D**, however, is usually thought of as a *Deontic* logic (which is why it is called **D**); it is a logic of *obligations* and *permissions*. $\Box P \supset P$ is not valid in **D**. One can readily understand why: just because an action is obligatory, it does not follow that it is performed. The exercise comes in three parts:

  1. Verify that $\Box P \supset P$ is not valid in **D**.
  2. Verify that $\Box P \supset \Diamond P$ is valid in **D**.
  3. Verify that $\Box P \supset \Diamond P$ is not valid in **K**.

Aristotle's Future Sea Battle

The ancient Greek philosophers were much concerned with the relation between *necessity* and *determinism*. Events do not appear to occur haphazardly, but regularly, with earlier events affecting the course of later events. If an event B has its roots in an earlier event A, then it would seem that, given A, B could not help but occur; and if the earlier event A could not be altered, then it would be necessary, so we have not merely the *relative necessity* of B given A, but the *absolute necessity* of B. Everything, then, would seem to be occurring of necessity.

Aristotle meant by *necessity* "the unalterability of whatever has already happened." (Ackrill, 1963, p. 113) Now, *unalterability* appears to identify a fundamental asymmetry between past and future: the past is unalterable in a way the future is not. In one of the most famous arguments from antiquity, found in *De Interpretation 9*, Aristotle suppressed issues of causation and faced the problem of determinism in terms of *truth, truth at a given time* and *necessity*. One cannot undo what has already been done. But if it is true *now* that there will be a sea battle tomorrow, the future also appears determined and unalterable, and the asymmetry collapses:

Again, if it is white now it was true to say earlier that it would be white; so that it was always true to say of anything that has happened that it would be so. But if it was always true to say that it was so, or would be so, it could not not be so, or not be going to be so. But if something cannot not happen it is impossible for it not to happen; and if it is impossible for something not to happen it is necessary for it to happen. Everything that will be, therefore, happens necessarily. So nothing will come about as chance has it or by chance; for if by chance, not of necessity. (18^b9 ff) in (Ackrill, 1963), pp 50–51.

This is the famous problem of future contingents.

Aristotle clearly rejects the *fatalistic* or *deterministic* conclusion. "Absurdities," he calls them: if "everything is and happens of necessity ... there would be no need to deliberate ... ." (18^b26 ff) And he offers the following solution beginning at 19^a23:

What is, necessarily is, when it is; and what is not, necessarily is not, when it is not. But not everything that is, necessarily is; and not everything that is not, necessarily is not. For to say that everything that is, is of necessity, when it is, is not the same as saying unconditionally that it is of necessity. Similarly with what is not. And the same account holds for contradictories: everything necessarily is or is not, and will be or will not be; but one cannot divide and say that one or the other is necessary. I mean, for example: it is necessary for there to be or not to be a sea-battle tomorrow; but it is not necessary for a sea-battle to take place tomorrow, nor for one not to take place—though it is necessary for one to take place or not to take place. So, since statements are true according to how the actual things are, it is clear that wherever these are such as to allow of contraries as chance has it, the same necessarily holds for the contradictories also. This happens with things that are not always so or are not always not so. With these it is necessary for one or the other of the contradictories to be true or false—not however, this one or that one, but as chance has it; or for one to be true *rather* than the other, yet not *already* true or false. (Ackrill, 1963), pp 52–53.

Aristotle denies that events in the future are already determined. His response to the argument is to deny that statements about the future are in some sense already true or false. But there is considerable disagreement among commentators about what he meant by this. The text has subtle complexities that provide an interesting and instructive modal story.

Aristotle's claim, "Everything necessarily is or is not," introduces a possible source of confusion. For simplicity, we will take this to be a variant of the Law of Excluded Middle: $P \vee \neg P$. Depending upon the scope we assign to "necessarily," we get two different claims. On the one hand, it might mean that the claim that P is true or $\neg P$ is true is necessarily true:

(1.24) $\Box(P \vee \neg P)$

On the other hand, it might mean that P is necessarily true or $\neg P$ is necessarily true:

(1.25) $\Box P \vee \Box \neg P$

Clearly, (1.24) is correct and (1.25) is problematic. If we maintain (1.25), then, together with *Excluded Middle*, the modality collapses, i.e., $P \equiv \Box P$: *Whatever is true is necessarily true; whatever is false is necessarily false.*

To be more precise, the modality collapses in **T**. It does not collapse in **K**. (We leave this as an exercise.) It collapses in **T** by the following reasoning. (1.25) says $\Box P \vee \Box \neg P$. By classical logic, $X \vee Y$ is equivalent to $\neg Y \supset X$. So, (1.25) gives us $\neg \Box \neg P \supset \Box P$. But in **T**, $P \supset \neg \Box \neg P$, so we have $P \supset \Box P$. And since $\Box P \supset P$ is valid in **T**, we have $P \equiv \Box P$.

It is not clear which of (1.24) or (1.25) Aristotle holds. In the passage quoted above, Aristotle appears to be clearly endorsing (1.24) but it is not as clear that he is rejecting (1.25). For, as Ackrill (1963) notes, the claim that "one cannot divide" might be understood to rule out either of the following inferences:

$$\frac{\Box(P \vee \neg P)}{\text{So, } \Box P} \qquad \frac{\Box(P \vee \neg P)}{\text{So, } \Box \neg P}$$

This is not yet (1.25). Nonetheless, passages like this sustain one interpretation of the chapter, on which the deterministic conclusion is presumed to follow because of scope confusion. This interpretation, which Ackrill (1963) calls "the preferred interpretation," sees Aristotle as arguing that the clearly and unproblematically true (1.24) is confused with the dubious (1.25), leading the unsuspecting modal novice to believe that since he is committed to (1.24), he is thereby committed to the *fatalistic* position.

This interpretation, however, does not place any special premium on statements about the future, and this appears to be the point Aristotle stresses in his solution. On a second interpretation, as Ackrill (1963) puts it, "Aristotle holds that a statement with a truth-value automatically has a necessity-value (if true, necessary; if false, impossible), but he claims that a statement may lack a truth-value and acquire one later." (p. 140)

There are two ways of understanding Aristotle's solution on this second interpretation. On the one hand, it could be that Aristotle accepts the following argument.

$$(a) \quad P \vee \neg P$$
$$(b) \quad P \supset \Box P$$
$$(c) \quad \neg P \supset \Box \neg P$$
$$\text{SO} \quad \Box P \vee \Box \neg P$$

But what he objects to is substituting $\mathbf{F}P$ (i.e., *It will be the case that P*—see the discussion in Section 1.5) for P in step (b). Unfortunately this interpretation is clouded by Aristotle's remarks in the quotation above in which he appears to sanction the substitution of $\mathbf{F}P$ for P in $\Box(P \vee \neg P)$. An alternative view is that the argument requires the intermediate steps of ascriptions of truth to statements. Let **True** P mean *It is true that P*. Then, on this view, the argument goes

$$(a) \quad P \vee \neg P$$
$$(b) \quad P \supset \textbf{True } P$$
$$(c) \quad \neg P \supset \textbf{True } \neg P$$
$$(d) \quad \textbf{True } P \supset \Box P$$
$$(e) \quad \textbf{True } \neg P \supset \Box \neg P$$
$$\text{SO} \quad \Box P \vee \Box \neg P$$

Kneale and Kneale (1962) favor this latter reading, and they see Aristotle as rejecting steps (b) and (c): they distinguish the *Law of Excluded Middle*, i.e., $P \vee \neg P$, from the *Principle of Bivalence*, i. e., **True** $P \vee$ **True** $\neg P$, and they assert that Aristotle rejected the latter but not the former. Statements about future contingents lack truth values. More precisely, for Kneale and Kneale (1962), Aristotle holds that although *It is true that* there either will be or won't be a sea battle tomorrow, one cannot infer that either *it is true that* there will be a sea battle tomorrow or *it is true that* there won't be a sea battle tomorrow.

A related view was held by some of the ancients. They took Aristotle's *true* to be *determinately* or *definitely true*. It is now definitely or determinately true that there either will be a sea battle tomorrow or not. But it is not now definitely true or determinately true that there will be a sea battle tomorrow nor is it now definitely true or determinately true that there won't be a sea battle tomorrow. The future contingent is not now determinately or definitely true (or false), so that the future is still open.[5]

EXERCISES

EXERCISE 1.12.3.   Show that $\Diamond P \supset \Box P$ is not valid in **K**.

EXERCISE 1.12.4.   We showed that the modality collapses in **T** if $\Box P \vee \Box \neg P$. Show that the argument must break down for **K**.

The Master Argument of Diodorus Cronus

Two schools of logic were distinguished in antiquity: the Peripatetic School, which followed the teaching of Aristotle, and the Stoic School, which was founded by Zeno. The most famous logical practitioner among the Stoics was Chrysippus, who was deeply influenced by the Megarians Diodorus Cronus and Philo (Kneale and Kneale, 1962; Mates, 1961).

Diodorus is responsible for one of the most famous of the Stoic arguments, the so-called "Master Argument." It is a matter of speculation as to *who* is the Master referred to—Diodorus or Fate. (Kneale and Kneale, 1962, p. 119) records the description from Epictetus:

The Master Argument seems to have been formulated with some such starting points as these. There is an incompatibility between the three following propositions, "Everything that is past and true is necessary," "The impossible does not follow from the possible," and "What neither is nor will be is possible." Seeing this incompatibility, Diodorus used the convincingness of the first two propositions to establish the thesis that nothing is possible which neither is nor will be true.

Sorabji (1980) notes that the first two premises can both be found in Aristotle, but not the conclusion: Aristotle believed that something could happen even though it never does—this cloak could be cut up even though it never is. There is good reason to think that Diodorus's argument is a sharpened version of the fatalistic argument Aristotle had considered in *De Interpretatione* 9. We will now try to reconstruct Diodorus's argument using the tools of modal logic.

---

[5] This notion of *definite truth* has a formal representation using the notion of *supervaluations*. For a good discussion of supervaluations, see (Fraassen, 1966).

Let us symbolize Diodorus's first premise, *Everything that is past and true is necessary*, as

(1.26) $PP \supset \Box PP$,

where PP is the tense-logical representation introduced in Section 1.5 for *It was the case that P*. There are a number of ways in which to interpret his second premise, *The impossible does not follow from the possible*. We shall take it as the Rule of Inference:[6]

(1.27) $$\frac{P \supset Q}{\Diamond P \supset \Diamond Q}$$

We shall also need one more assumption that is not explicit in the argument, namely, that time is discrete.[7] Then we can show that the specifically Diodorian claim follows: *What neither is nor will be is impossible*, i.e.

(1.28) $(\neg P \wedge \neg FP) \supset \neg \Diamond P$.

(*FP* is the tense-logical representation for *It will be the case that P*).

To be concrete, let *P* be "There is a sea battle taking place." Now, suppose there is no sea battle taking place and never will be, i.e.,

(1.29) $\neg P \wedge \neg FP$

If no sea battle *is* taking place and none *will*, then—and right here is where the assumption of the discreteness of time comes in—there was a time (for example, the moment just before the present) after which no sea battle would take place, i.e.,

(1.30) $(\neg P \wedge \neg FP) \supset P \neg FP$

It follows by modus ponens from (1.29) and (1.30) that there was a time in the past when it was true that no sea battle would later take place, i.e.,

(1.31) $P \neg FP$,

So, by modus ponens and the first of Diodorus's premises, (1.26),

(1.32) $\Box P \neg FP$

---

[6] Note that (1.27) is quite different from $(P \supset Q) \supset (\Diamond P \supset \Diamond Q)$: to infer $\Diamond P \supset \Diamond Q$ by the rule, we require not merely that $P \supset Q$ be true, but that it be logically true.

[7] Prior (1967), whose reconstruction is similar to ours, also makes this very same assumption.

which is to say that it is not possible that it not have been the case that it would be true that no sea battle will ever take place,[8] i.e.,

(1.33) $\neg\Diamond\neg\mathbf{P}\neg\mathbf{F}P$.

So, $\neg\mathbf{P}\neg\mathbf{F}P$ is impossible. But if the sea battle were taking place, it would not have been the case that no sea battle would ever take place, i.e.,

(1.34) $P \supset \neg\mathbf{P}\neg\mathbf{F}P$,

This is a logical truth based on the meaning of the tense operators. So, by our Rule of Inference (1.27),

(1.35) $\Diamond P \supset \Diamond\neg\mathbf{P}\neg\mathbf{F}P$,

And since we have the negation of the consequent in (1.33), an application of *Modus Tollens* yields

(1.36) $\neg\Diamond P$.

On the assumption of (1.29) we have derived (1.36) using the Diodorian premises (1.26) and (1.27). Thus we have the Diodorian conclusion (1.28).

Diodorus's definition *The possible is that which either is or will be [true]* was apparently widely known throughout the ancient world (Mates, 1961; Prior, 1957).[9] And widely debated. It is none other than the *Principle of Plenitude*: every possibility is realized (at some time).

---

[8] This is a good example of the virtues of symbolization: it is highly unlikely that we would have been able to come up with this English rendering so readily without the symbolic formulation.

[9] His definitions of the remaining modals are entirely consistent:

The impossible is that which, being false, will not be true
The necessary is that which, being true, will not be false
The nonnecessary is that which either is or will be false

Formally, these turn out as follows:

$$\Diamond P \equiv P \vee \mathbf{F}P$$
$$\neg\Diamond P \equiv \neg P \wedge \mathbf{G}\neg P$$
$$\Box P \equiv P \wedge \mathbf{G}P$$
$$\neg\Box P \equiv \neg P \vee \mathbf{F}\neg P$$

The Once and Future Conditional

Some modern logicians read *If p, then q* as the *material conditional*: i.e., it is false if the antecedent is true and consequent false; otherwise it is true. Some logicians have also called it the *philonean conditional*, after the Stoic logician Philo, who argued for this reading. Just as *necessity* was a source of heated dispute among the ancients, so too was the conditional. Mates (1961) reports that Diodorus disputed Philo's reading of the conditional, preferring to regard a conditional as true only when it is *impossible* for the antecedent to be true and consequent false—of course, in his special temporal sense of impossible, viz., that at no time is the antecedent true and consequent false. So, as Mates (1961) (p. 45) puts it, "A conditional holds in the Diodorean sense if and only if it holds *at all times* in the Philonean sense." How appropriate that the modern reintroduction of modal logic should turn on this very same issue.

Lewis (1918) is largely responsible for bringing modal logic back into the mainstream of logic.[10] As with Diodorus, Lewis found the material conditional wanting as a formalization of *If p, then q*, especially insofar as it entailed the so-called Paradoxes of the Material Conditional. These are, first, that a false statement (materially) implies anything, i.e., $\neg P \supset (P \supset Q)$, and, second, that a true statement is (materially) implied by anything, i.e., $Q \supset (P \supset Q)$. Lewis sought to tighten up the connection between antecedent and consequent, and he introduced a new connective, symbolized by $\dashv$ (known as "the fishhook"). $P \dashv Q$, expresses that P *strictly implies* Q, and is definable in terms of $\Box$ as follows:

$$P \dashv Q =_{def} \Box(P \supset Q).$$

We might very well term this the *diodorean conditional*, for to say that P *strictly implies* Q is to say that the material (i.e., philonean) conditional holds necessarily. It is equivalent to $\neg\Diamond(P \wedge \neg Q)$: it is impossible for P to be true and Q to be false.

Lewis's introduction of $\dashv$ was, however, flawed by use/mention confusion. In *P strictly implies Q*, we appear to be speaking about propositions, not using them, and so *necessity* appears to be a predicate of sentences, not a propositional operator, as we have treated it in our formulation of modal logic. Lewis, unfortunately, was not particularly careful in this regard. For example, he says in (Lewis and Langford, 1959)

... the relation of strict implication expresses precisely that relation which holds when valid deduction is possible, and fails to hold when valid deduction is not possible. In that sense, the system of Strict Implication may be said to provide that canon

---

[10] Lewis eventually introduced five distinct formulations for modal logic, of which S4 and S5 are the most well known.

and critique of deductive inference which is the desideratum of logical investigation. (p. 247)

Note the two-fold confusion in this passage. First, the expression "valid deduction" combines semantical and syntactical notions without clearly distinguishing them. Second, whether syntactical or semantical, the relation is one that holds between sentences: that the relation holds is a fact about sentences, and so the variables on either side of $\dashv$ are to be replaced by names of sentences (since one is speaking about them) rather than by sentences themselves. These confusions were seized upon by foes of formalized modal logic, and Lewis's achievements were overshadowed in the subsequent controversy, a controversy which has its roots deep in the history of the subject, as we shall now see.

The Reality of Necessity

In order to understand the continuing controversy about modal logic, we must return to our discussion in Section 1.2 about whether the modal is part of the content of a judgment.

The cognitive content of a sentence, the proposition it expresses, is commonly thought of as encapsulated by its *truth conditions*, i.e., the features of the reality it purports to describe that make it true. The proposition that *John is a man*, for example, is true if and only if the individual *John* has the property of *being a man*. But the proposition that *John is necessarily a man* will require more for its truth than just that the individual *John* has the property of *being a man*—there will also have to be something in reality corresponding to his *necessarily* being a man. The something more is usually thought of as a *real essence*—a property an individual has necessarily, i.e., a property that is part of his essential nature so that he could not exist if he lacked the property. *Real essences*, *natural kinds*, *natural laws*, and related notions were characteristic of the scholastic thinking that derived from Aristotelian metaphysics. This prevalent way of thinking admitted necessary connections in reality, between an individual and some of its properties (real essences), between a group of individuals and some of their common properties (natural kinds), between an individual or group doing something and another (possibly the same) individual or group's doing something (natural laws). From the Realist standpoint, a modal is comfortably conceived of in much the same way as any other part of a proposition, namely, as representing a feature of reality that determines the truth of the proposition.

One can readily understand, then, how a healthy skepticism about these necessities in nature will nudge one towards the Kant/Frege position described in Section 1.2 which denies that the modals belong to the content of a claim. The treatment of causality in (Hume, 1888) captures this view.

A proposition saying that events of type A cause events of type B is widely thought to make a claim about the way things must be. Whenever events of type A occur, events of type B *must* occur. Hume, however, denies that there are necessary connections in nature. And if there are no necessary connections in nature, then no part of a proposition can represent them as anything that determines the truth of the proposition. So the "must" in the causal claim is not part of the proposition at all. Its role must be assigned elsewhere. Hume (1888) locates the seeming necessary connection between cause and effect in the habitual connection of our *ideas* of these causes and effects: the invocation of the idea of the cause immediately invokes the idea of the effect.

Despite their widely divergent views, Kant and Frege appear, like Hume, to have located necessity in the realm of ideas. Kneale and Kneale (1962) say Kant holds "that a modal adverb such as "possibly" represents only a way of thinking the thought enunciated by the rest of the sentence in which it occurs." (p. 36) Frege (1879), in a similar vein, says: "If a proposition is presented as possible, then either the speaker is refraining from judgment . . . or else he is saying that in general the negation of the proposition is false." (p. 5) Both believe that the apodeictic and problematic indicators form no part of the content of the judgment. But the role they are supposed to play is a bit uncertain. One reading is that the modal introduces the speaker's attitude toward the content, clearly locating it in non-cognitive psychology; another reading is that the modal connects the content with other judgments, moving it back into the realm of logic.

On the first reading, a judgment of possibility is a possible judgment. Appending "possibly" to a content indicates that one is merely entertaining it but not committing oneself to it. Or, it might have the force of "perhaps." But this view seems correct only if one looks at assertions of modal claims. For, if a modal claim is not itself asserted, but forms part of a larger claim, occurring, for instance, as antecedent or consequent of a conditional like $(P \supset \Diamond P)$, this view no longer looks plausible.

It is quite possible that Frege (1879) came to believe that there is no difference between the assertoric and apodeictic judgments because he fell prey to a well-known modal error. Given the modal inference rule of *Necessitation* characteristic of *normal* modal logics,

(1.37) If $\vdash P$ then $\vdash \Box P$

($\vdash$ symbolizes *it is asserted that*) and the "Axiom of Necessity," as Hughes and Cresswell (1996) call it,

(1.38) $\Box P \supset P$

we obtain the derived rule:[11]

(1.39) $\vdash P$ if, and only if $\vdash \Box P$

And (1.39) makes it look as though there really is no logical difference in content between the assertoric and the apodeictic judgments. But, of course, this is not so. $\Box P$ and P differ in content. The true (1.39) must be distinguished from the (usually) false

(1.40) $P \equiv \Box P$

The collapse of the modality effected by (1.40) is avoided: although (1.38) (i.e., *If necessarily true, then true*) is true, its converse,

(1.41) $P \supset \Box P,$

(i.e., *If true, then necessarily true*), is (usually) not.

On the second reading, the modal relates this content to other contents. As Frege said, to claim that a content is necessary is to say that "it follows only from universal judgments." The necessity of a content appears to be a function of its justification. And this brings us back to the position of Lewis.

It should be quite clear to the reader how the philosophical issues concerning modality have polarized. The Realist about necessity has no problem taking a modal as part of the proposition. The Anti-Realist about necessity takes the modal out of the proposition and raises it up a level, either as a comment about the propositional representation or, noncognitively, as an expression of one's attitude about the propositional representation. The Anti-Realist about necessity in the twentieth century has been most comfortable regarding modal claims, if cognitive at all, as explicitly about sentences, and so as *metalinguistic* claims, a level at which the development of any interesting modal logics is seriously restricted. In a very famous comment, Marcus (1992) characterized the most serious contemporary foe of quantified modal logic, W. V. O. Quine, as believing "that it was conceived in sin: the sin of confusing use and mention."(p. 5) But the real evil of quantified modal logic for Quine is its apparent commitment to "the metaphysical jungle of Aristotelian essentialism." Quine, and his fellow critics of quantified modal logic, instead of arguing the merits, even the coherence of essentialism, have drawn battle lines around the coherence of modal logic itself. And this criticism has invariably been that quantified modal logic rests on logical errors, like use/mention confusion, or illegal substitutions, errors that arise because meaningful elements of a sentence are taken to represent reality when no

---

[11] (1.38), as we mentioned earlier, holds in **T**, but not in **K**. Even so, (1.39) holds in **K**. We leave this as an exercise.

such representation can be taking place. The nature of the criticisms stems from underlying philosophical and metaphysical outlooks, from which point of view modal elements really don't belong in the content of a proposition (for they represent nothing real), and any logic that treats them as part of the content of a proposition is flirting with logical disaster. We will be revisiting these arguments throughout this book.

EXERCISES

EXERCISE 1.12.5. Show that, in **K**, a formula P is valid if and only if $\Box P$ is valid. (Note (1.39).) This is hard if one uses validity directly, but becomes easier when tableaus are available. See Exercise 2.2.3.

CHAPTER TWO

TABLEAU PROOF SYSTEMS

2.1. WHAT IS A PROOF

Informally, a proof is an argument that convinces. Formally, a proof is of a formula, and is a finite object constructed according to fixed syntactic rules that refer only to the structure of formulas and not to their intended meaning. The syntactic rules that define proofs are said to specify a *proof procedure*. A proof procedure is *sound* for a particular logic if any formula that has a proof must be a valid formula of the logic. A proof procedure is *complete* for a logic if every valid formula has a proof. Then a sound and complete proof procedure allows us to produce "witnesses," namely proofs, that formulas are valid.

There are many kinds of proof procedures. Loosely they divide themselves into two broad categories: *synthetic* and *analytic*. The terms are suggestive, but not terribly precise. An analytic proof procedure decomposes the formula being proved into simpler and simpler parts. A synthetic proof procedure, by contrast, builds its way up to the formula being proved. Analytic proof procedures tend to be easier to use, since the field of play is sharply limited: one never looks outside the formula being proved. Synthetic proof procedures, on the other hand, are more common, and often produce especially elegant proofs.

The most common example of a synthetic proof procedure is an *axiom system*. Certain formulas are taken as axioms. A proof starts with these axioms and, using rules of inference that produce new formulas from old, builds up a sequence of formulas that finally ends with the formula being proved.

Tableau systems are one of the more common analytic proof procedures. These are *refutation systems*. To prove a formula we begin by negating it, analyze the consequences of doing so, using a kind of tree structure, and, if the consequences turn out to be impossible, we conclude the original formula has been proved. Tableau systems come in several versions—see (Fitting, 1998) for an account of their history and variety. We will be presenting *prefixed tableaus*, which originated in (Fitting, 1972b), and took on their present form in (Massacci, 1994; Massacci, 1998)—see also (Goré, 1998).

We discuss tableau systems first, devoting this chapter to them, because their structure is more obviously related to that of the possible world models we have been using, and also because it is easier to discover proofs in tableau systems than in axiom systems. Axiom systems are also presented, in

Chapter 3, because they are almost universally present in the literature, and because they give additional insight into modal logics.

2.2. TABLEAUS

There are many varieties of tableaus. The kind we present uses *prefixed formulas*, and proofs are particular *trees* of such things. So, let us begin at the beginning. A *tree* is a kind of graph. Rather than giving a formal definition, we give a representative example which should suffice to convey the general idea.

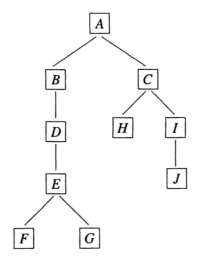

The lettered square boxes are called *nodes*. (In tableaus, the nodes will be prefixed formulas, which are defined below.) Node A is the *root node* and nodes F, G, H, and J are *leaves*. (Thus botanically, our trees are upside-down.) Node A has two *children*, B and C, with B the *left child* and C the *right child*. Nodes C and E also have two children; all other nodes that are not leaves have just one child each. (Our nodes will always have at most two children—more complex trees are possible, but we will not need them.) If one node is the child of another, the second is the *parent* of the first—for example, C is the parent of both H and I. The sequence A, B, D, E is an example of a *path*. The sequence A, B, D, E, F is another path, and it is also a *branch*, which means it is a maximal path. Trees can be infinite, but such trees will not come up for some time, so we can ignore them for now.

DEFINITION 2.2.1. [Prefix] A *prefix* is a finite sequence of positive integers. A *prefixed formula* is an expression of the form $\sigma\, X$, where σ is a prefix and X is a formula.

We will write prefixes using periods to separate integers, 1.2.3.2.1 for instance. Also, if σ is a prefix and n is a positive integer, $\sigma.n$ is σ followed by a period followed by n.

The intuitive idea is that a prefix, σ, names a possible world in some model, and $\sigma\, X$ tells us that X is true at the world σ names. Our intention is that $\sigma.n$ should always name a world that is accessible from the one that σ names. Other facts about prefixes depend on which modal logic we are considering, but this is enough to get started.

An attempt to construct a tableau proof of a formula Z begins by creating a tree with $1\,\neg Z$ at its root (and with no other nodes). Intuitively, the formula at the root says Z is false (that is, $\neg Z$ is true) at some world of some model (the world named by 1). Next, branches are "grown" according to certain *Branch Extension Rules*, to be given shortly. This yields a succession of *tableaus for* $1\,\neg Z$. Finally, we need each branch to be *closed*, something we define after we give the rules for growing the branches.

Suppose we have $\sigma\, X \wedge Y$ at a node. Intuitively this says $X \wedge Y$ is true at the world that σ names, hence both X and Y are true there, and so we should have both $\sigma\, X$ and $\sigma\, Y$. Then any branch going through the node with $\sigma\, X\wedge Y$ can be lengthened by adding a new node with $\sigma\, X$ to the end, then another after that, with $\sigma\, Y$. This is an example of a Branch Extension Rule. On the other hand, if we have a node with $\sigma\, X \vee Y$, intuitively $X \vee Y$ is true at the world that σ names, so one of X or Y is true there, and so we should have one of $\sigma\, X$ or $\sigma\, Y$. Then any branch going through this node can be lengthened by taking the final node on it (a leaf), and giving it *two* children, $\sigma\, X$ as left child and $\sigma\, Y$ as right child. This is another Branch Extension Rule.

All the binary propositional connectives fit one or the other of these patterns. We give them in two groups, beginning with the conjunctive cases (like $\sigma\, X \wedge Y$), followed by the disjunctive cases (like $\sigma\, X \vee Y$). Incidentally, $\equiv$ is something of an anomaly, but with a little stretching, it too fits the general patterns.

DEFINITION 2.2.2. [Conjunctive Rules] For any prefix σ,

$$\frac{\sigma\, X \wedge Y}{\begin{array}{c}\sigma\, X\\ \sigma\, Y\end{array}} \qquad \frac{\sigma\, \neg(X \vee Y)}{\begin{array}{c}\sigma\, \neg X\\ \sigma\, \neg Y\end{array}} \qquad \frac{\sigma\, \neg(X \supset Y)}{\begin{array}{c}\sigma\, X\\ \sigma\, \neg Y\end{array}} \qquad \frac{\sigma\, X \equiv Y}{\begin{array}{c}\sigma\, X \supset Y\\ \sigma\, Y \supset X\end{array}}$$

DEFINITION 2.2.3. [Disjunctive Rules] For any prefix σ,

$$\frac{\sigma\ X \vee Y}{\sigma\ X \mid \sigma\ Y} \qquad \frac{\sigma\ \neg(X \wedge Y)}{\sigma\ \neg X \mid \sigma\ \neg Y}$$

$$\frac{\sigma\ X \supset Y}{\sigma\ \neg X \mid \sigma\ Y} \qquad \frac{\sigma\ \neg(X \equiv Y)}{\sigma\ \neg(X \supset Y) \mid \sigma\ \neg(Y \supset X)}$$

Incidentally, Smullyan's *uniform notation* can be used to group similar cases together in a convenient fashion—see (Smullyan, 1968) or (Fitting, 1996a).

There is one more classical case before we turn to the modal operators.

DEFINITION 2.2.4. [Double Negation Rule] For any prefix σ,

$$\frac{\sigma\ \neg\neg X}{\sigma\ X}$$

Now for the modal rules. Suppose we have $\sigma\ \Diamond X$, so at the world that σ names, $\Diamond X$ is true. Then there must be some world accessible to that one at which X is true. By our naming convention, such a world should have a name of the form $\sigma.n$, but we shouldn't use a prefix of this form if it already has a meaning, since it might name the wrong world. The solution is to use $\sigma.n$, *where this prefix has never been used before*. Then we are free to make it a name for a world at which X is true. This simple idea gives us the following rules.

DEFINITION 2.2.5. [Possibility Rules] If the prefix $\sigma.n$ is new to the branch,

$$\frac{\sigma\ \Diamond X}{\sigma.n\ X} \qquad \frac{\sigma\ \neg\Box X}{\sigma.n\ \neg X}$$

Finally, consider $\sigma\ \Box X$. At the world that σ names, $\Box X$ is true, so at every world accessible from that one, X is true. It is possible that the prefix $\sigma.n$ doesn't name anything, because we might not have used it, but if it does name, it names a world accessible from the one that σ names, so we should have $\sigma.n\ X$. This gives us the following rules.

DEFINITION 2.2.6. [Basic Necessity Rules] If the prefix $\sigma.n$ already occurs on the branch,

$$\frac{\sigma\ \Box X}{\sigma.n\ X} \qquad \frac{\sigma\ \neg\Diamond X}{\sigma.n\ \neg X}$$

As with the propositional connectives, these cases too can be condensed using uniform notation—see (Fitting, 1983). There is a strong similarity between these modal tableau rules and tableau rules for quantifiers. This is no coincidence.

DEFINITION 2.2.7. [Closure] A tableau branch is *closed* if it contains both $\sigma\, X$ and $\sigma\, \neg X$ for some formula X. A branch that is not closed is *open*. A tableau is *closed* if every branch is closed.

A closed branch obviously represents an impossible state of affairs—there is no model in which the prefixed formulas on it could all be realized. A closed tableau for $1\, \neg Z$, in effect, says all the consequences of assuming that Z could be false in some world of some model are contradictory.

DEFINITION 2.2.8. [Tableau Proof] A closed tableau for $1\, \neg Z$ is a *tableau proof* of Z, and Z is a *theorem* if it has a tableau proof.

We have seen semantic characterizations of several propositional modal logics, but (so far) only one tableau system. It is, in fact, a proof procedure for **K**. In the next section we give additional rules to produce tableau systems for the other modal logics we have discussed. But first an example.

EXAMPLE 2.2.9. According to Example 1.7.6, $\Box(P \wedge Q) \supset (\Box P \wedge \Box Q)$ is **K**-valid. Here is a tableau proof of it. (The numbers to the right of prefixed formulas are for reference only, and are not part of the tableau structure.)

$$
\begin{array}{ll}
1 & \neg[\Box(P \wedge Q) \supset (\Box P \wedge \Box Q)] \quad 1. \\
1 & \Box(P \wedge Q) \quad 2. \\
1 & \neg(\Box P \wedge \Box Q) \quad 3.
\end{array}
$$

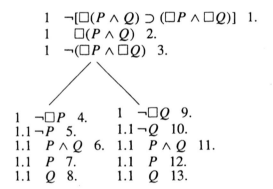

In this tableau, item 1 is, of course, how the proof begins. Items 2 and 3 are from 1 by a Conjunction Rule; 4 and 9 are from 3 by a Disjunction Rule; 5 is from 4 by a Possibility Rule (note, the prefix 1.1 is new on the branch at this point); 6 is from 2 by a Basic Necessity Rule (note, the prefix 1.1 already occurs on the branch at this point); 7 and 8 are from 6 by a Conjunction Rule. The left branch is closed because of 5 and 7. The right branch has a similar explanation.

EXERCISES

EXERCISE 2.2.1.   In the tableau system given above, for the modal logic **K**, prove the following.

1. $(\Box P \land \Box Q) \supset \Box(P \land Q)$
2. $\Box(P \supset Q) \supset (\Box P \supset \Box Q)$
3. $(\Box P \land \Diamond Q) \supset \Diamond(P \land Q)$
4. $(\Box \Diamond P \land \Box \Box Q) \supset \Box \Diamond(P \land Q)$

EXERCISE 2.2.2.   Give a tableau proof of $(\Box P \lor \Box Q) \supset \Box(P \lor Q)$. On the other hand, the converse, $\Box(P \lor Q) \supset (\Box P \lor \Box Q)$, is not provable. Explain why a tableau proof is impossible. Discuss the intuitive reasons why one would not want this formula to be provable.

EXERCISE 2.2.3.   Show that if $\Box X$ has a tableau proof, so does X. (See Exercise 1.12.5.)

2.3. MORE TABLEAU SYSTEMS

In the previous section we gave a single tableau system, which we will eventually show proves exactly the valid formulas of **K**. Now we add various rules to that system, to produce proof procedures for the other modal logics we have considered. First we state various rules abstractly, then we say which combinations of them give which logics. The rules in these forms are from (Massacci, 1994; Massacci, 1998; Goré, 1998).

DEFINITION 2.3.1. [Special Necessity Rules] For prefixes σ and $\sigma.n$ already occurring on the tableau branch:

$$T \qquad \frac{\sigma\ \Box X}{\sigma\ X} \qquad \frac{\sigma\ \neg\Diamond X}{\sigma\ \neg X}$$

$$D \qquad \frac{\sigma\ \Box X}{\sigma\ \Diamond X} \qquad \frac{\sigma\ \neg\Diamond X}{\sigma\ \neg\Box X}$$

$$B \qquad \frac{\sigma.n\ \Box X}{\sigma\ X} \qquad \frac{\sigma.n\ \neg\Diamond X}{\sigma\ \neg X}$$

$$4 \qquad \frac{\sigma\ \Box X}{\sigma.n\ \Box X} \qquad \frac{\sigma\ \neg\Diamond X}{\sigma.n\ \neg\Diamond X}$$

$$4r \qquad \frac{\sigma.n\ \Box X}{\sigma\ \Box X} \qquad \frac{\sigma.n\ \neg\Diamond X}{\sigma\ \neg\Diamond X}$$

Now, for each of the logics of Section 1.8, we get a tableau system by adding to the system for **K** additional rules, according to the following chart.

Logic	Rules
D	D
T	T
K4	4
B	$B, 4$
S4	$T, 4$
S5	$T, 4, 4r$

We begin with a few very simple examples of proofs using these rules, then a somewhat more complicated one.

EXAMPLE 2.3.2. Here is a proof, using the **B** rules, of $\Diamond\Box X \supset X$. As usual, numbers to the right of each item are for reference only, and are not an official part of the tableau proof.

$$
\begin{array}{lll}
1 & \neg(\Diamond\Box X \supset X) & 1. \\
1 & \Diamond\Box X & 2. \\
1 & \neg X & 3. \\
1.1 & \Box X & 4. \\
1 & X & 5.
\end{array}
$$

Formula 1 is the usual starting point. Formulas 2 and 3 are from 1 by a Conjunctive Rule; formula 4 is from 2 by a Possibility Rule; then 5 is from 4 by Special Necessity Rule B. Closure is by 3 and 5.

EXAMPLE 2.3.3.   The following is a proof, using the **T** rules, of $[\Box(X \vee Y) \wedge \neg X] \supset Y$.

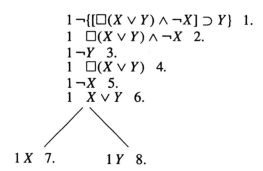

$$1 \neg\{[\Box(X \vee Y) \wedge \neg X] \supset Y\} \quad 1.$$
$$1 \quad \Box(X \vee Y) \wedge \neg X \quad 2.$$
$$1 \neg Y \quad 3.$$
$$1 \quad \Box(X \vee Y) \quad 4.$$
$$1 \neg X \quad 5.$$
$$1 \quad X \vee Y \quad 6.$$

$$1 \, X \quad 7. \qquad 1 \, Y \quad 8.$$

Formulas 2 and 3 are from 1 by a Conjunctive Rule; 4 and 5 are from 2 the same way; formula 6 is from 4 by Special Necessity Rule T; 7 and 8 are from 6 by a Disjunctive Rule. Closure is by 5 and 7 on the left branch, and 3 and 8 on the right.

EXAMPLE 2.3.4.   Here is a proof of greater complexity, using the **S4** rules, of the formula

$$\Box\Diamond(\Box X \supset \Box\Diamond Y) \supset (\Box X \supset \Box\Diamond Y)$$

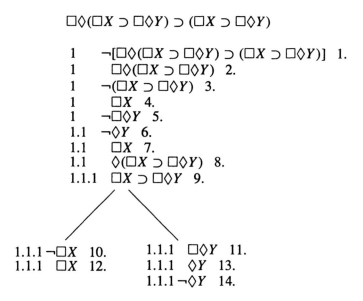

$$1 \quad \neg[\Box\Diamond(\Box X \supset \Box\Diamond Y) \supset (\Box X \supset \Box\Diamond Y)] \quad 1.$$
$$1 \quad \Box\Diamond(\Box X \supset \Box\Diamond Y) \quad 2.$$
$$1 \quad \neg(\Box X \supset \Box\Diamond Y) \quad 3.$$
$$1 \quad \Box X \quad 4.$$
$$1 \quad \neg\Box\Diamond Y \quad 5.$$
$$1.1 \quad \neg\Diamond Y \quad 6.$$
$$1.1 \quad \Box X \quad 7.$$
$$1.1 \quad \Diamond(\Box X \supset \Box\Diamond Y) \quad 8.$$
$$1.1.1 \quad \Box X \supset \Box\Diamond Y \quad 9.$$

$$1.1.1 \neg\Box X \quad 10. \qquad 1.1.1 \quad \Box\Diamond Y \quad 11.$$
$$1.1.1 \quad \Box X \quad 12. \qquad 1.1.1 \quad \Diamond Y \quad 13.$$
$$1.1.1 \neg\Diamond Y \quad 14.$$

Formula 1 is the usual starting point. Formulas 2 and 3 are from 1 by a Conjunctive Rule; 4 and 5 are from 3 also by a Conjunctive Rule; 6 is from 5 by a Possibility Rule; 7 is from 4 by Special Necessity Rule 4; 8 is from 2 by a Basic Necessity Rule; 9 is from 8 by a Possibility Rule; 10 and 11 are from 9 by a Disjunctive Rule; 12 is from 7 by Special Necessity Rule 4; 13 is from 11 by Special Necessity Rule T; and 14 is from 6 by Special Necessity Rule 4. The left branch is closed because of 10 and 12; the right branch because of 13 and 14.

Finally we consider the special case of **S5**, which actually has two prefixed tableau systems. We gave one system above—here is a simple example that uses it.

EXAMPLE 2.3.5.   Using the **S5** rules, here is a proof of $\Diamond \Box P \supset \Box P$.

$$
\begin{array}{lll}
1 & \neg(\Diamond \Box P \supset \Box P) & 1. \\
1 & \Diamond \Box P & 2. \\
1 & \neg \Box P & 3. \\
1.1 & \Box P & 4. \\
1.2 & \neg P & 5. \\
1 & \Box P & 6. \\
1.2 & P & 7.
\end{array}
$$

In this, 2 and 3 are from 1 by a Conjunctive Rule; 4 is from 2 by a Possibility Rule, as is 5 from 3; 6 is from 4 by Special Necessity Rule $4r$; 7 is from 6 by a Basic Necessity Rule.

While there is certainly nothing wrong with this **S5** tableau system, there is a greatly simpler version. Take as prefixes just *positive integers*, not sequences of them. And replace all the modal rules by the following two.

DEFINITION 2.3.6.   [**S5** Possibility Rule] If the integer k is new to the branch,

$$
\frac{n \Diamond X}{k \, X} \qquad \frac{n \, \neg \Box X}{k \, \neg X}
$$

DEFINITION 2.3.7.   [**S5** Necessity Rule] If the integer k already occurs on the branch,

$$
\frac{n \Box X}{k \, X} \qquad \frac{n \, \neg \Diamond X}{k \, \neg X}
$$

We conclude the section with an example using this simplified **S5** system.

EXAMPLE 2.3.8.   We give a proof of $\Box P \lor \Box\neg\Box P$.

$$1 \; \neg(\Box P \lor \Box\neg\Box P) \quad 1.$$
$$1 \; \neg\Box P \quad 2.$$
$$1 \; \neg\Box\neg\Box P \quad 3.$$
$$2 \; \neg P \quad 4.$$
$$3 \; \neg\neg\Box P \quad 5.$$
$$3 \; \Box P \quad 6.$$
$$2 \; P \quad 7.$$

Formulas 2 and 3 are from 1 by a Conjunctive Rule; 4 is from 2 by the Possibility Rule; 5 is from 3 similarly; 6 is from 5 by the Double Negation Rule; 7 is from 6 by a Necessity Rule.

EXERCISES

EXERCISE 2.3.1.   Prove $\Diamond(P \supset \Box P)$ in the **T** system.

EXERCISE 2.3.2.   Prove $(\Box X \land \Box Y) \supset \Box(\Box X \land \Box Y)$ in the **K4** system.

EXERCISE 2.3.3.   Prove $(\Box\Diamond X \land \Box\Diamond Y) \supset \Box\Diamond(\Box\Diamond X \land \Box\Diamond Y)$ in the **S4** system. Can it be proved using the **K4** rules?

EXERCISE 2.3.4.   Prove $\Box P \lor \Box(\Box P \supset Q)$ in each of the **S5** systems.

EXERCISE 2.3.5.   Give an **S5** proof of $\Box(\Diamond P \supset P) \supset \Box(\Diamond\neg P \supset \neg P)$.

EXERCISE 2.3.6.   Prove $\Diamond(P \land \Box Q) \equiv (\Diamond P \land \Box Q)$ in each of the **S5** systems.

2.4. LOGICAL CONSEQUENCE AND TABLEAUS

In Section 1.9 we discussed logical consequence in a modal setting, $S \models_L U \rightarrow X$, which we saw was a more complex thing than its classical counterpart because it involves both global assumptions, S, and local assumptions, U. Recall, the idea was that X should be true at each world of a model at which the members of U are all true, provided the members of S are true at every world of that model. Fortunately, tableau rules for this notion of consequence are both simple and intuitive.

DEFINITION 2.4.1. [Assumption Rules] Let **L** be one of the modal logics for which tableau rules have been given. Also let S and U be sets of formulas. A tableau uses S as *global* assumptions and U as *local* assumptions if the following two additional tableau rules are admitted.

LOCAL ASSUMPTION RULE If Y is any member of U then 1 Y can be added
   to the end of any open branch.

GLOBAL ASSUMPTION RULE If Y is any member of S then $\sigma\, Y$ can be
   added to the end of any open branch on which σ appears as a prefix.

Thus local assumptions can be used with prefixes of 1, and 1 is the prefix associated with the formula we are trying to prove—local assumptions can only be used locally, so to speak. Likewise any prefix can be used with a global assumption, corresponding to the idea that global assumptions are available at every world.

If assumptions, local or global, are involved in a tableau construction for $1\neg X$, we refer to the tableau as a *derivation* of X rather than a proof of X.

EXAMPLE 2.4.2. In Example 1.9.3 we showed directly that $\{\Box P \supset P\} \models_{\mathbf{K}}$ $\emptyset \rightarrow \Box\Box P \supset \Box P$ holds. Now we give a **K** derivation of $\Box\Box P \supset \Box P$ using $\{\Box P \supset P\}$ as the set of global assumptions,

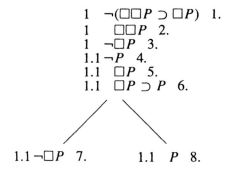

Items 2 and 3 are from 1 by a Conjunction Rule; 4 is from 3 by a Possibility Rule; 5 is from 2 by a Necessity Rule; 6 uses the Global Assumption Rule; 7 and 8 are from 6 by a Disjunctive Rule.

EXERCISES

EXERCISE 2.4.1. Give a **K** derivation of $\Box\Box P \supset P$ using $\Box P \supset P$ as a global assumption.

2.5. TABLEAUS WORK

Each of the modal logics we have been looking at has been specified in two quite different ways: using models, and using tableaus. We need to know these two ways agree. We need to show, for instance, that a formula X is K-valid if and only if X has a tableau proof using the K rules. And similarly for the other modal logics as well. This is a result proved *about* logical machinery, rather than *using* it, and so it is sometimes described as a *metalogical* result.

Generally, one shows a formula is L-valid by giving a proof, and one shows it not L-valid by giving a model. The two approaches complement each other nicely. Finding a flaw in a proof does not show the result is in error, since a different proof might actually establish the result. Once the match between validity and provability is established, giving a counter-model shows no proof will do, and giving a proof shows no counter-model can be found.[12]

Our argument divides itself into two parts, called *soundness* and *completeness*.

Soundness

The word "sound," when used of a proof procedure, means the procedure cannot prove any formula it should not. More specifically, to say the tableau rules for **L** are sound means: if a formula X has a proof using the tableau rules for **L**, then X is **L**-valid. To keep things concrete, we will show this for **L** being **K**—other choices are treated similarly.

DEFINITION 2.5.1. [Satisfiable] Suppose S is a set of prefixed formulas. We say S is *satisfiable* in the model $\langle \mathcal{G}, \mathcal{R}, \Vdash \rangle$ if there is a way, θ, of assigning to each prefix σ that occurs in S some possible world $\theta(\sigma)$ in $\mathcal{G}$ such that:

1. If σ and $\sigma.n$ both occur as prefixes in S, then $\theta(\sigma.n)$ is a world accessible from $\theta(\sigma)$, that is, $\theta(\sigma)\mathcal{R}\theta(\sigma.n)$.
2. If σX is in S, then X is true at the world $\theta(\sigma)$, that is, $\theta(\sigma) \Vdash X$.

We say a tableau branch is satisfiable if the set of prefixed formulas on it is satisfiable in some model. And we say a tableau is satisfiable if some branch of it is satisfiable.

---

[12] In 1858, Abraham Lincoln closed his part of the Charleston Lincoln-Douglas debate with the following observation. "If you have ever studied geometry, you remember that by a course of reasoning Euclid proves that all the angles in a triangle are equal to two right angles. Euclid has shown you how to work it out. Now, if you undertake to disprove that proposition, and to show that it is erroneous, would you prove it to be false by calling Euclid a liar?" (Angle, 1991)

We need two fundamental results about satisfiability, then we can show soundness of the **K** tableau rules. The first is rather simple.

PROPOSITION 2.5.2. A closed tableau is not satisfiable.

Proof Suppose we had a tableau that was both closed and satisfiable. Since it is satisfiable, some branch of it is. Let S be the set of prefixed formulas on that branch. Say S is satisfiable in the model $\langle \mathcal{G}, \mathcal{R}, \Vdash \rangle$, using the mapping θ of worlds to prefixes. Since the tableau is also closed, for some formula X and some prefix σ, both σX and $\sigma \neg X$ are in S. But then both X and $\neg X$ must be true at the world $\theta(\sigma)$ of $\mathcal{G}$, and this is not possible. ∎

The second result needs a little more work to prove.

PROPOSITION 2.5.3. If a tableau branch extension rule is applied to a satisfiable tableau, the result is another satisfiable tableau.

Proof Say $\mathcal{T}$ is a satisfiable tableau, and a branch extension rule is applied to it; more specifically, say a branch extension rule is applied to a formula on branch $\mathcal{B}$ of tableau $\mathcal{T}$.

The proof now divides into several cases. We begin with the one case that is essentially trivial. Some branch of $\mathcal{T}$ is satisfiable. If a branch $\mathcal{B}'$ that is different from $\mathcal{B}$ is satisfiable, applying a branch extension rule on $\mathcal{B}$ will not affect $\mathcal{B}'$, so after the rule application we still have a satisfiable branch, $\mathcal{B}'$, hence a satisfiable tableau.

For the rest of the proof, we assume $\mathcal{B}$ itself is satisfiable. Say the set of prefixed formulas on it is satisfiable in the model $\langle \mathcal{G}, \mathcal{R}, \Vdash \rangle$, using the mapping θ of worlds to prefixes. Now the division into cases depends on what tableau rule we applied.

CONJUNCTIVE CASE Suppose $\sigma X \wedge Y$ occurs on $\mathcal{B}$, and we add σX and σY to the end of $\mathcal{B}$. Since $\mathcal{B}$ is satisfiable using the model $\langle \mathcal{G}, \mathcal{R}, \Vdash \rangle$ and the mapping θ, and $\sigma X \wedge Y$ is on $\mathcal{B}$, we have $\theta(\sigma) \Vdash X \wedge Y$. But then by the definition of truth in a model, $\theta(\sigma) \Vdash X$ and $\theta(\sigma) \Vdash Y$. It follows that $\mathcal{B}$, with σX and σY added, is satisfiable in the same model, using the same mapping. (All the other conjunctive cases, as well as the double negation case, are treated the same way.)

DISJUNCTIVE CASE Suppose $\sigma X \vee Y$ occurs on $\mathcal{B}$, and we split the end of the branch, adding σX to the left fork and σY to the right. Because $\mathcal{B}$ was satisfiable in $\langle \mathcal{G}, \mathcal{R}, \Vdash \rangle$ using the mapping θ, we have $\theta(\sigma) \Vdash X \vee Y$, hence by the definition of truth in a model, either $\theta(\sigma) \Vdash X$ or $\theta(\sigma) \Vdash Y$. In the first case, the left extension of $\mathcal{B}$ is satisfiable in

$\langle \mathcal{G}, \mathcal{R}, \Vdash \rangle$ using the mapping θ. In the second case, the right extension is satisfiable. Either way, one of the extensions of $\mathcal{B}$ is satisfiable, so some branch of the extended tableau is satisfiable, so the tableau itself is satisfiable. (The other disjunctive cases are similar.)

POSSIBILITY CASE Suppose $\sigma \, \Diamond X$ occurs on $\mathcal{B}$ and we add $\sigma.n \, X$ to the end of $\mathcal{B}$, where the prefix $\sigma.n$ did not occur previously on the branch (and so $\theta(\sigma.n)$ is not defined). By our assumptions about the satisfiability of $\mathcal{B}$, we know that $\theta(\sigma) \Vdash \Diamond X$. But then, for some possible world Δ in $\mathcal{G}$, $\theta(\sigma) \mathcal{R} \Delta$ and $\Delta \Vdash X$. Define a new mapping, θ' as follows. For all the prefixes occurring in $\mathcal{B}$, set θ' to be the same as θ. Since $\sigma.n$ did not occur on $\mathcal{B}$, we are free to define θ' on it as we please—set $\theta'(\sigma.n)$ to be Δ.

Since θ and θ' agree on the prefixes of $\mathcal{B}$, all the prefixed formulas of $\mathcal{B}$ have the same behavior in $\langle \mathcal{G}, \mathcal{R}, \Vdash \rangle$ no matter whether we use θ or θ'. Further, $\theta'(\sigma) \mathcal{R} \theta'(\sigma.n)$ since $\theta'(\sigma) = \theta(\sigma)$, $\theta'(\sigma.n) = \Delta$, and $\theta(\sigma) \mathcal{R} \Delta$. So finally, $\theta'(\sigma.n) \Vdash X$. It follows that all prefixed formulas on the branch $\mathcal{B}$ extended with $\sigma.n \, X$ are satisfiable in $\langle \mathcal{G}, \mathcal{R}, \Vdash \rangle$, using the mapping θ'. (The other possibility case is similar.)

NECESSITY CASE We leave this to you.

■

Now soundness of the tableau rules follows easily.

THEOREM 2.5.4. [Tableau Soundness] If X has a tableau proof using the **K** rules, X is **K**-valid.

Proof Suppose X has a **K** tableau proof, but is not **K**-valid; we derive a contradiction.

Since X has a tableau proof, there is a closed tableau $\mathcal{T}$ beginning with the prefixed formula $1 \neg X$. The construction of $\mathcal{T}$ begins with the trivial tableau $\mathcal{T}_0$ consisting of a single node, labeled $1 \neg X$. $\mathcal{T}$ results from $\mathcal{T}_0$ by the application of various branch extension rules.

Since X is not valid, there is some world Γ in some model $\langle \mathcal{G}, \mathcal{R}, \Vdash \rangle$ at which X is not true. Define a mapping θ by setting $\theta(1) = \Gamma$. Using this model and mapping, the set $\{1 \neg X\}$ is clearly satisfiable. But then the tableau $\mathcal{T}_0$ is satisfiable, since the set of formulas on its only branch is satisfiable.

Since $\mathcal{T}_0$ is satisfiable, by Proposition 2.5.3 so is any tableau we get by starting with $\mathcal{T}_0$ and applying branch extension rules. It follows that $\mathcal{T}$ is satisfiable. But $\mathcal{T}$ is also closed, and this is impossible, by Proposition 2.5.2. We have our desired contradiction. ■

Above we only considered tableau rules for the basic system, **K**, and we allowed any model, so our proof showed the soundness of the **K** rules relative to the **K** semantics. To show soundness for one of the other logics, say **L**, we must modify Proposition 2.5.3. Specifically it is enough to show, if we have a tableau that has a branch that is satisfiable in some **L** model, and we apply any of the **L** tableau rules, we get a tableau with a branch that is satisfiable in a **L** model. We leave this to you.

Completeness

If a tableau system is sound, it doesn't prove anything it shouldn't. Of course a tableau system that proves nothing at all will be sound. We also need that it proves everything it should. We want to show that if X is **L**-valid then X has a tableau proof using the rules for **L**. We will sketch a proof of this for **K**, and leave the other logics to you. Our proof, like most completeness arguments, proceeds in the contrapositive direction: if X has no tableau proof, there is a model in which X fails, so X is not valid. We show this by showing how to extract a counter-model from a failed attempt at a tableau proof. A description of how to do this occupies the rest of the section.

DEFINITION 2.5.5. [Saturated] Let us say a **K** tableau is *saturated* if all appropriate tableau rule applications have been made. More precisely, a tableau is saturated provided, for every branch that is not closed:

1. If a prefixed formula other than a possibility or necessity formula occurs on the branch, the applicable rule has been applied to it on the branch.
2. If a possibility formula occurs on the branch, the possibility rule has been applied to it on the branch once.
3. If a necessity formula occurs on the branch, with prefix σ, the necessity rule has been applied to it on the branch once for each prefix $\sigma.n$ that occurs on the branch.

If $\mathcal{T}$ is a saturated **K** tableau, and $\mathcal{B}$ is a branch of it that is not closed, we can say several useful things about it. If $\sigma\ X \wedge Y$ occurs on $\mathcal{B}$, so do both $\sigma\ X$ and $\sigma\ Y$, by item 1. Similarly, if $\sigma\ X \vee Y$ occurs on it, one of $\sigma\ X$ or $\sigma\ Y$, again by 1. If $\sigma\ \Diamond X$ occurs, so will $\sigma.n\ X$ for some n. And if $\sigma\ \Box X$ occurs, so will $\sigma.n\ X$ for every prefix $\sigma.n$ that occurs on $\mathcal{B}$. Similar remarks apply to other formulas like $\neg(X \wedge Y)$ and $X \supset Y$.

What is perhaps less clear is that if we start constructing a **K** tableau for $1\ \neg X$, we can always produce a saturated one. To do this, we simply follow a *systematic* construction procedure. The following one will do. At each stage, pick a branch that has not closed, pick a prefixed formula on it that is not a necessity formula, not atomic, not the negation of an atom, and that has had no rule applied to it on the branch, and do the following. If it is not a possibility formula, simply apply the appropriate rule to it. If it is a possibility formula, say $\sigma \lozenge X$, pick the smallest integer n such that $\sigma.n$ does not occur on the branch, and add $\sigma.n\, X$ to the end of the branch; and then for each necessity formula on the branch having σ as prefix, also add the instance having $\sigma.n$ as prefix (so if $\sigma \,\square Y$ is present, add $\sigma.n\, Y$).

There is a certain amount of flexibility in the procedure just described, since it was left open which unused formula to choose at each stage. But it can be shown that, no matter how the choice is made, the process must terminate. A proof of this can be found in (Fitting, 1983) for instance—we omit it here. Instead of a proof, we give an example of a tableau that has been constructed following this procedure, and is saturated.

EXAMPLE 2.5.6. The following is an attempted proof of $\square(X \wedge \square Y) \supset \square(\square X \wedge Y)$ using the **K** rules.

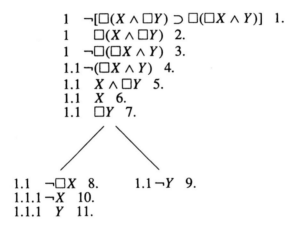

At the start, formula 1 is the only unused one, so we apply a Conjunctive Rule to it, getting 2 and 3. Since 2 is a necessity type formula, we do not use it at this point; we apply a Possibility Rule to 3, getting 4, after which we apply a Necessity Rule to 2, getting 5. We now have a choice between formulas 4 and 5, neither of which has had a rule applied to it. We decide to use 5 first, getting 6 and 7. Then we use 4, getting 8 and 9.

On the right branch every prefixed formula has already been used, or is atomic, or is the negation of an atomic formula, or is a necessity formula. There is nothing more to be done here.

On the left branch, we apply a Possibility Rule to 8, getting 10, then a Necessity Rule to 7 getting 11. This finishes work on the left branch.

Suppose we have a saturated tableau $\mathcal{T}$, and there is a branch $\mathcal{B}$ of it that is not closed. We show how to construct a model in which the branch is satisfiable. The method is, in fact, quite simple.

Let $\mathcal{G}$ be the *collection of prefixes* that occur on the branch $\mathcal{B}$. If σ and $\sigma.n$ are both in $\mathcal{G}$, set $\sigma \mathcal{R} \sigma.n$. Finally, if P is a propositional letter, and σP occurs on $\mathcal{B}$, take P to be true at σ, that is, $\sigma \Vdash P$. Otherwise take P to be false at σ. This completely determines a model $\langle \mathcal{G}, \mathcal{R}, \Vdash \rangle$.

Now, the **Key Fact** we need is: For each formula Z,

$$\text{if } \sigma\, Z \text{ occurs on } \mathcal{B} \text{ then } \sigma \Vdash Z$$
$$\text{if } \sigma\, \neg Z \text{ occurs on } \mathcal{B} \text{ then } \sigma \nVdash Z$$

The proof of this is by induction on the complexity of the formula Z.

Suppose first that Z is a propositional variable, say P. If σP occurs on $\mathcal{B}$, $\sigma \Vdash P$ by definition. And if $\sigma \neg P$ occurs on $\mathcal{B}$, σP does not, since $\mathcal{B}$ is not a closed branch, so $\sigma \nVdash P$, again by definition.

Next, the induction step. Suppose the Key Fact is known for formulas that are simpler than Z; we show it for Z itself. There are several cases, depending on the form of Z. We consider only a representative few.

Suppose Z is $X \wedge Y$. If $\sigma Z = \sigma X \wedge Y$ occurs on $\mathcal{B}$, since the tableau is saturated, both σX and σY also occur on $\mathcal{B}$. Since these are simpler formulas than Z, the induction hypothesis applies, and we have that $\sigma \Vdash X$ and $\sigma \Vdash Y$. But then, $\sigma \Vdash X \wedge Y$, that is, $\sigma \Vdash Z$. On the other hand, if $\sigma \neg Z = \sigma \neg(X \wedge Y)$ occurs on $\mathcal{B}$, again by saturation, one of $\sigma \neg X$ or $\sigma \neg Y$ must also occur. Then by the induction hypothesis again, either $\sigma \nVdash X$ or $\sigma \nVdash Y$. Either way, $\sigma \nVdash (X \wedge Y)$, that is, $\sigma \nVdash Z$.

Suppose Z is $\Box X$. If $\sigma \Box X$ is on $\mathcal{B}$, so is $\sigma.n\, X$ for every prefix $\sigma.n$ that occurs on $\mathcal{B}$. By the induction hypothesis, $\sigma.n \Vdash X$ for every prefix $\sigma.n$ that occurs on $\mathcal{B}$. But this is equivalent to saying that $\sigma' \Vdash X$ for every $\sigma' \in \mathcal{G}$ such that $\sigma \mathcal{R} \sigma'$. It follows that $\sigma \Vdash \Box X$. Similarly, if $\sigma \neg \Box X$ occurs on $\mathcal{B}$, by saturation, $\sigma.n \neg X$ also occurs, for some prefix $\sigma.n$. Thus there is some $\sigma' \in \mathcal{G}$ (namely $\sigma.n$) such that $\sigma \mathcal{R} \sigma'$ and $\sigma' \nVdash X$ (using the induction hypothesis). So $\sigma \nVdash \Box X$.

We leave the remaining cases to you. But now, the completeness argument is essentially done.

THEOREM 2.5.7. [Tableau Completeness] If X is **K**-valid, X has a tableau proof using the **K** rules.

Proof Again we give a contrapositive argument. Suppose X does not have a **K** proof; we show it is not **K**-valid. Since X does not have any proof, if we start a tableau with $1 \neg X$, and carry out the construction to saturation, it will not close. Let $\mathcal{B}$ be an open branch of the resulting saturated tableau. Using the method above, we can create a model $\langle \mathcal{G}, \mathcal{R}, \Vdash \rangle$ for which the Key Fact is true. In particular, if $\sigma \ \neg Z$ is on $\mathcal{B}$, there is a world, σ, in $\mathcal{G}$, at which Z is false. But $1 \ \neg X$ is on $\mathcal{B}$, since it is the prefixed formula we began the tableau construction with, so it is on every branch. So there is a world, namely 1, in our model, at which X is false, so X is not **K**-valid. ∎

EXAMPLE 2.5.8. We continue Example 2.5.6, using the model construction procedure just outlined. Both tableau branches are open—say we work with the left one. We construct a model $\langle \mathcal{G}, \mathcal{R}, \Vdash \rangle$ as follows.

Let $\mathcal{G} = \{1, 1.1, 1.1.1\}$, the set of prefixes on the left branch. Let $1\mathcal{R}1.1$ and $1.1\mathcal{R}1.1.1$. Finally, set $1.1 \Vdash X$, $1.1.1 \Vdash Y$, and in no other cases are propositional letters true at worlds.

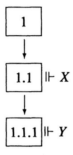

The prefixed formula $1.1 \ \neg \Box X$ is on the left branch. And in fact, $1.1 \ \nVdash \Box X$, since $1.1\mathcal{R}1.1.1$ and $1.1.1 \ \nVdash X$. Similarly $1.1 \ \Box Y$ is present, and $1.1 \Vdash \Box Y$ since the only possible world of the model that is accessible from 1.1 is $1.1.1$, and we have $1.1.1 \Vdash Y$. In this way we work our way up the branch, finally verifying that $\Box(X \wedge \Box Y) \supset \Box(\Box X \wedge Y)$ is not true at world 1, and hence is not valid.

We have now shown completeness of the tableau system for **K**. The other modal tableau systems have similar completeness proofs, but with added complications. Some of these are simple to deal with. For instance, if we are dealing with a logic rules include T, the accessibility relation on prefixes must also take each prefix to be accessible from itself—this gives us reflexivity. Analogous remarks apply to transitivity and symmetry. But there are harder issues to deal with as well. Consider the rules for **K4**; if we apply the tableau construction procedure given above to produce saturated tableaus, we discover that it may never terminate!

EXAMPLE 2.5.9. Suppose we try to prove $\Box X \vee \Diamond \Box X$ in **K4**, and so start a tableau with its negation. We can now proceed as follows.

$$1 \quad \neg(\Box X \vee \Diamond \Box X) \quad 1.$$
$$1 \quad \neg \Box X \quad 2.$$
$$1 \quad \neg \Diamond \Box X \quad 3.$$
$$1.1 \neg X \quad 4.$$
$$1.1 \neg \Box X \quad 5.$$
$$1.1 \neg \Diamond \Box X \quad 6.$$

Formulas 2 and 3, of course, are from 1 by a Conjunctive Rule. Formula 4 is from 2 by a Possibility Rule, and then 5 is from 3 by a Necessity Rule. But now, in **K4**, Rule 4 allows us to add formula 6 using formula 3, and a desire to apply all applicable rules means we must do so. Now, formulas 5 and 6 duplicate formulas 2 and 3, but with lengthened prefixes. We can work with 5 and 6 the same way we did with 2 and 3, getting fresh occurrences of the same formulas, with 1.1.1 prefixes. And so on. In short, we find ourselves in an infinite loop, generating an infinite branch.

The possibility of infinite loops is no problem for completeness proofs. It simply means we construct infinite, not finite, models. But we must be careful with our tableau constructions to ensure that every potentially applicable rule is actually applied. This takes some care, and we omit a proper discussion here. See (Fitting, 1983) or (Goré, 1998) for details.

Logical Consequence

We have shown, in section 2.4, how to modify tableau systems to use global and local assumptions. Extending soundness and completeness arguments to take these into account is quite straightforward. Rather than giving the full details we give one illustrative example, and leave the rest to you.

We attempt a **K** derivation of $Q \supset \Box Q$ from $P \supset \Box P$ as a global assumption. As we will see, this attempt fails, and from the failed attempt we will be able to construct a model showing it is not the case that $\{P \supset \Box P\} \models_K \emptyset \rightarrow Q \supset \Box Q$, or using our abbreviated notation, it is not the case that $\{P \supset \Box P\} \models_K Q \supset \Box Q$.

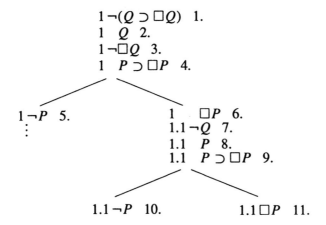

Items 2 and 3 are from 1 by a Conjunction Rule; 4 is a Global Assumption; 5 and 6 are from 4 by a Disjunctive Rule; 7 is from 3 by a Possibility Rule; 8 is from 6 by a Necessity Rule; 9 is a Global Assumption; 10 and 11 are from 9 by a Disjunctive Rule. The branch ending with item 10 is closed. The branch through 5 is not finished yet, but the branch ending with 11 is not closed, and every applicable rule has been applied on it. We use it to construct our counter-model. Doing as the branch instructs, we get the following.

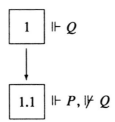

We leave it to you to check that every formula on the branch ending with 11 is satisfied in this model and the model is, in fact, the counter-model we wanted.

EXERCISES

EXERCISE 2.5.1.   Give detailed soundness proofs for the tableau system for **S4** and for both tableau systems for **S5**.

EXERCISE 2.5.2.   Construct a model corresponding to the right branch of the tableau in Example 2.5.6, and show in detail that $\Box(X \wedge \Box Y) \supset \Box(\Box X \wedge Y)$ fails at a world of it.

EXERCISE 2.5.3.   Continue with the pattern of construction begun in Example 2.5.9, and use the infinite tableau that results to produce a **K4** model in which $\Box X \vee \Diamond \Box X$ is not valid.

EXERCISE 2.5.4.   The formula $\Box(\Diamond P \supset P) \supset \Box(\Diamond \neg P \supset \neg P)$ is not provable in **S4**, though it is in **S5** (see Exercise 2.3.5). Extract an **S4** model showing it is not valid from a failed attempt to prove it using a systematic tableau construction.

EXERCISE 2.5.5.   Attempt a **K** derivation of $\Box P \supset \Box \Box P$ from $\Box P \supset P$ as a global assumption. From the failed attempt, construct a model showing it is not the case that $\{\Box P \supset P\} \models_{\mathbf{K}} \emptyset \rightarrow \Box P \supset \Box \Box P$. As it happens, the tableau construction does not terminate, but fortunately after a certain point there is enough regularity to see a pattern.

EXERCISE 2.5.6.   Define a suitable notion of soundness for tableaus that use the Local and Global Assumption Rules, and prove the **K** tableau system with these rules is, in fact, sound.

CHAPTER THREE

AXIOM SYSTEMS

3.1. WHAT IS AN AXIOMATIC PROOF

The axiomatic approach to a logic is quite different from that of tableaus. Certain formulas are simply announced to be theorems (they are called axioms), and various rules are adopted for adding to the stock of theorems by deducing additional ones from those already known. An axiom system is, perhaps, the most traditional way of specifying a logic, though proof discovery can often be something of a fine art. Historically, almost all of the best-known modal logics had axiomatic characterizations long before either tableau systems or semantical approaches were available. While early modal axiom systems were somewhat circuitous by today's standards, Gödel (1933) introduced the modern axiomatic approach and this is now used by almost everybody, including us.

Our axiomatic treatment in this chapter is relatively full. When quantifiers are introduced we give an abbreviated axiomatic version, and after that axiom systems disappear altogether. We feel that as systems get more and more complex, axiomatics provides less and less intuition, while axiomatic proofs get harder and harder to discover. We have decided to concentrate on the tableau approach here. Nonetheless, at the propositional level an axiomatic treatment provides considerable insight into how different modal logics differ, hence the full treatment now.

It is generally convenient, when using axiom systems, to keep the number of connectives down, so in this chapter we will take $\neg$ and $\supset$ as primitive, and the other connectives as defined in the usual way. Further, we will take $\Box$ as primitive, and think of $\Diamond X$ as an abbreviation for $\neg\Box\neg X$. Also we will make use of axiom *schemas*—we will take all formulas having some common form as axioms.

Since we are primarily interested in the modal aspects, we take propositional classical logic for granted. A *tautology* is a formula (without modal operators) that evaluates to T on every line of a truth table. We assume you know all about tautologies. In fact, we broaden the term somewhat. $(P \wedge Q) \supset P$ is a typical tautology. We can substitute complex formulas, involving modal operators, for the propositional variables of a tautology. For instance, we might substitute $\Box(X \vee Y)$ for P and $\Diamond X$ for Q in the tautology we just gave, getting $(\Box(X \vee Y) \wedge \Diamond X) \supset \Box(X \vee Y)$. *We will call the result of such a substitution a tautology too.*

Now we formulate a basic modal axiom system, for the logic **K**.

DEFINITION 3.1.1. [Axioms] The axioms for **K** are in two categories.

CLASSICAL BASIS All tautologies

SCHEMA **K** All formulas of the form $\Box(X \supset Y) \supset (\Box X \supset \Box Y)$

DEFINITION 3.1.2. [Rules of Inference] **K** has two rules of inference.

$$\text{Modus Ponens} \quad \frac{X \quad X \supset Y}{Y} \qquad\qquad \text{Necessitation} \quad \frac{X}{\Box X}$$

Notice what the rule of necessitation does and does not say. It does not say $X \supset \Box X$—if something is true, it is necessary. This is obviously undesirable. What it says is, if something is *provable*, it is necessary. This is quite different. Our axiom system captures a certain notion of logical truth, so if X is provable within the system, X is not simply true, but must be a logical truth, and so we can conclude $\Box X$. It is the Necessitation Rule, introduced by Gödel, that makes possible the simple, elegant modal axiom systems in use today.

DEFINITION 3.1.3. [Proof] An axiomatic *proof* is a finite sequence of formulas, each of which is either an axiom or else follows from earlier items by one of the rules of inference. An axiomatic *theorem* is the last line of a proof.

Before we give examples of proofs, it is convenient to give a *derived rule of inference*. By this we mean a rule of inference that can be safely added to our axiom system because applications of it can be "translated away." That is, we have a standard way of replacing applications of the rule by uses only of the machinery that is an official part of the axiom system.

DEFINITION 3.1.4. [Derived Rule of Regularity]

$$\frac{X \supset Y}{\Box X \supset \Box Y}$$

This is a derived rule because we can always replace an application of it in a proof by the following sequence of steps.

$X \supset Y$	line occurring in a proof
$\Box(X \supset Y)$	Necessitation Rule on previous line
$\Box(X \supset Y) \supset (\Box X \supset \Box Y)$	Axiom **K**
$\Box X \supset \Box Y$	Modus Ponens on previous two lines

Now, here are two examples of proofs in the present axiom system. We have added line numbers and abbreviated explanations, though these are not officially parts of proofs.

EXAMPLE 3.1.5. Axiom system proof of $\Box(X \wedge Y) \supset (\Box X \wedge \Box Y)$.

1	$(X \wedge Y) \supset X$	tautology
2	$\Box(X \wedge Y) \supset \Box X$	Regularity on 1
3	$(X \wedge Y) \supset Y$	tautology
4	$\Box(X \wedge Y) \supset \Box Y$	Regularity on 3
5	$[\Box(X \wedge Y) \supset \Box X]$ $\supset \{[\Box(X \wedge Y) \supset \Box Y]$ $\supset [\Box(X \wedge Y) \supset (\Box X \wedge \Box Y)]\}$	tautology
6	$[\Box(X \wedge Y) \supset \Box Y]$ $\supset [\Box(X \wedge Y) \supset (\Box X \wedge \Box Y)]$	Modus Ponens, 2, 5
7	$\Box(X \wedge Y) \supset (\Box X \wedge \Box Y)$	Modus Ponens, 4, 6

In the example above, line 7 comes from lines 2 and 4 using a common classical logic inference, embodied in the tautology $(A \supset B) \supset ((A \supset C) \supset (A \supset (B \wedge C)))$. In the future we will simply say so in words, and abbreviate the formal proof. We do this in the following; one of the classical tautologies involved, for instance, is $(A \supset B) \supset ((B \supset C) \supset (A \supset C))$.

EXAMPLE 3.1.6. Axiom system proof of $(\Box X \wedge \Box Y) \supset \Box(X \wedge Y)$.

1	$X \supset (Y \supset (X \wedge Y))$	tautology
2	$\Box X \supset \Box(Y \supset (X \wedge Y))$	Regularity on 1
3	$\Box(Y \supset (X \wedge Y)) \supset (\Box Y \supset \Box(X \wedge Y))$	axiom **K**
4	$\Box X \supset (\Box Y \supset \Box(X \wedge Y))$	from 2, 3 classically
5	$(\Box X \wedge \Box Y) \supset \Box(X \wedge Y)$	from 4, classically

We now have enough to conclude that $\Box(X \wedge Y) \equiv (\Box X \wedge \Box Y)$ also has an axiom system proof.

We will eventually establish that the formulas with proofs in the axiom system we have given are exactly the **K**-valid formulas, and hence are exactly the formulas having proofs using the **K** tableau rules.

We are less interested in giving lots of formal axiomatic proofs than we are in using features of the axiomatic treatment, often substantially abbreviated, to manipulate modal formulas in ways that preserve validity. Here is a particularly useful tool for this purpose.

THEOREM 3.1.7. [Replacement] Suppose X, X', Y, and Y' are formulas, X occurs as a subformula of Y, and Y' is like Y except that the occurrence of X has been replaced with an occurrence of X'. If $X \equiv X'$ has an axiomatic proof, so does $Y \equiv Y'$.

This is a well-known theorem classically, and the proof of the modal version is a simple extension of the classical proof. It amounts to showing how an axiomatic proof of $X \equiv X'$ can be transformed into one of $Y \equiv Y'$. We omit the argument here—details can be found in (Chellas, 1980) or (Fitting, 1983). Here is a typical example of an application.

EXAMPLE 3.1.8.   We show $\Diamond\Diamond\neg X \equiv \neg\Box\Box X$ is a theorem. First, recall that here $\Diamond$ is an abbreviated symbol—it stands for $\neg\Box\neg$—so when written out, what we are claiming is theoremhood for $\neg\Box\neg\neg\Box\neg\neg X \equiv \neg\Box\Box X$. Now, $\neg\neg X \equiv X$ is a tautology, hence a theorem, so by the Replacement Theorem, $\neg\Box\neg\neg\Box\neg\neg X \equiv \neg\Box\neg\neg\Box X$ is also a theorem. Similarly, $\neg\neg\Box X \equiv \Box X$ is also a tautology, so by the Replacement Theorem again, $\neg\Box\neg\neg\Box X \equiv \neg\Box\Box X$ is a theorem. Now the result follows immediately, by classical logic.

There are certain replacements that are often useful to make: $\Box\neg X \equiv \neg\Diamond X$ and $\Diamond\neg X \equiv \neg\Box X$ are two good examples. (It is easy to see that each of these is a theorem.) Repeated use of one of these quickly gives the result in the previous example, for instance.

EXERCISES

EXERCISE 3.1.1. Show the following is a derived rule: conclude $\Diamond X \supset \Diamond Y$ from $X \supset Y$.

EXERCISE 3.1.2.   In the last two examples above, abbreviated proofs are given. Give them in full.

EXERCISE 3.1.3.   Give an axiomatic proof of each of the following. (Recall, $\Diamond$ abbreviates $\neg\Box\neg$.)

1. $\Diamond(X \vee Y) \equiv (\Diamond X \vee \Diamond Y)$
2. $\Box(X \supset Y) \supset (\Diamond X \supset \Diamond Y)$
3. $(\Box X \vee \Box Y) \supset \Box(X \vee Y)$
4. $(\Box X \wedge \Diamond Y) \supset \Diamond(X \wedge Y)$
5. $(\Diamond X \supset \Box Y) \supset \Box(X \supset Y)$

EXERCISE 3.1.4. Show how $\wedge$ and $\vee$ can be characterized using $\neg$ and $\supset$.

3.2. MORE AXIOM SYSTEMS

In the previous section we gave an axiom system for **K**. Now we give versions for the other modal logics we have been considering. It is a simple matter to do so because each logic has an axiomatization that is the same as that for **K**, but with the addition of a few extra axioms. To make the presentation easier, we begin by listing the extra axioms we will need, all at once, and giving them names. Then we say which ones give which logics.

Name	Scheme
D	$\Box P \supset \Diamond P$
T	$\Box P \supset P$
4	$\Box P \supset \Box\Box P$
B	$P \supset \Box\Diamond P$
5	$\Diamond P \supset \Box\Diamond P$

Now, axiom systems for each of our logics consist of the system for **K**, from the previous section, plus the additions from the following chart.

Logic	Added Axioms
D	D
T	T
K4	4
B	T, B
S4	$T, 4$
S5	$T, 5$ or $T, 4, B$

This specifies a family of axiomatically characterized logics. We will show in Section 3.4 that they correspond exactly to the semantic characterizations, and hence to the tableau versions as well.

The axioms above can also be given in a variety of alternate forms. For instance, in the axiomatic version of **T**, $X \supset \Diamond X$ is a theorem, by the following argument. $\Box P \supset P$ is an axiom scheme[13] of **T**, so $\Box \neg X \supset \neg X$ is an axiom. Using classical logic (contraposition), $\neg\neg X \supset \neg\Box\neg X$ is a theorem, and using double negation replacement and the definition of $\Diamond$, we have theoremhood for $X \supset \Diamond X$. By a similar argument, we can show that taking $P \supset \Diamond P$ as an axiom scheme allows us to prove all instances of $\Box X \supset X$, so either

---

[13] "nature must obey necessity," *Julius Caesar*, William Shakespeare

$\Box P \supset P$ or $P \supset \Diamond P$ would do to axiomatize **T**. Similarly, $\Diamond \Diamond P \supset \Diamond P$ is a theorem of any system including axiom scheme 4, $\Diamond \Box P \supset \Box P$ is a theorem of any system including axiom scheme 5, and these would do as substitutes for schemes 4 and 5 respectively.

Again without proof, we note that the Replacement Theorem 3.1.7 holds for the axiomatic versions of all these logics too. We make considerable use of this fact.

In Section 1.8 we noted that **D** was a sublogic of **T**. This can be shown for the axiomatic versions as well: it is enough to show that every theorem of axiomatic **D** is also a theorem of **T**, and to show this it is enough to show that every formula of the form $\Box X \supset \Diamond X$ is a theorem of **T**. But, in **T**, $\Box X \supset X$ is, itself, an axiom, and we saw above that $X \supset \Diamond X$ is a theorem of **T**, and $\Box X \supset \Diamond X$ follows immediately from these two, by classical logic. Similarly, axiomatic **T** is a sublogic of **S4** and **S5**, and so on.

We gave two systems for **S5**, so it is necessary to show their equivalence. We give the argument in one direction, and leave the other to you.

Assume we are using the axiom system for **K**, with the addition of axioms T and 5. We show every instance of B and 4 is provable. We begin with B.

$$\begin{array}{ll} P \supset \Diamond P & \text{theorem of } \mathbf{T} \\ \Diamond P \supset \Box \Diamond P & \text{by 5} \\ P \supset \Box \Diamond P & \text{by classical logic} \end{array}$$

Next we show each instance of scheme 4 is also provable. Since we have shown that instances of B are provable, we can make use of them in abbreviated proofs.

$$\begin{array}{ll} \Diamond \Box P \supset \Box P & \text{by 5} \\ \Box \Diamond \Box P \supset \Box \Box P & \text{by (derived) Regularity Rule} \\ \Box P \supset \Box \Diamond \Box P & \text{by } B \\ \Box P \supset \Box \Box P & \text{by classical logic} \end{array}$$

EXERCISES

EXERCISE 3.2.1.   Work in the axiom system **K** plus axiom schemes T, 4 and B, and show each instance of axiom scheme 5 is a theorem.

EXERCISE 3.2.2.   Prove $(\Box X \wedge \Box Y) \supset \Box(\Box X \wedge \Box Y)$ in the **K4** axiom system.

EXERCISE 3.2.3. Prove $\Diamond(P \supset \Box P)$ in the **T** axiom system.

3.3. LOGICAL CONSEQUENCE, AXIOMATICALLY

We have seen that logical consequence is more complex for modal logic than it is for classical logic (Sections 1.9 and 2.4). Local assumptions and global assumptions play quite a different role. This difference is somewhat clarified by the axiomatic rules for using the two kinds of assumptions.

Suppose a formula X is true at every world of some model $\langle \mathcal{G}, \mathcal{R}, \Vdash \rangle$. If Γ is any world of this model, X will be true at every world accessible from Γ (since X is true at every world), so $\Box X$ will be true at Γ. Since Γ was arbitrary, $\Box X$ must be true at every world of the model. Now global assumptions, semantically, are true at every world of a model, so we have just argued that if X is a global assumption, the Rule of Necessitation can be applied to X—we can assume $\Box X$ is also the case.

On the other hand, if X is known to be true only at world Γ of some model, we have no reason to suppose $\Box X$ is also true there. It is easy to construct examples for which this is not so. Thus if X is a local assumption, the Rule of Necessitation cannot be assumed to apply to it.

Formally, the distinction between global and local assumptions in axiomatic derivations comes down to the applicability or not of the Rule of Necessitation.

DEFINITION 3.3.1. [Axiomatic Derivation] By an axiomatic *derivation* in the logic **L** using members of S as global assumptions and members of U as local assumptions we mean a sequence of formulas meeting the following conditions.

1. The sequence is divided into two separate parts, a *global* part and a *local* part, with the global part coming first.
2. In the global part, each formula is either an axiom of **L**, a member of S, or follows from earlier lines by Modus Ponens or Necessitation.
3. In the local part, each formula is either an axiom of **L**, a member of U, or follows from earlier lines by Modus Ponens (but not Necessitation).

If X is the last formula in the sequence, we say it is a derivation of X.

Thus the effect is that the Necessitation Rule can be used with global, but not with local assumptions. Here is an example of a derivation.

EXAMPLE 3.3.2. We show that $\Box P$ has a derivation in the logic **K** from $\{\Box P \supset P\}$ as global and $\{\Box \Box P\}$ as local assumptions. In the following, lines

1 and 2 are in the global part, lines 5 and 6 are in the local part, and lines 3 and 4 can be counted either way.

1	$\Box P \supset P$	global assumption
2	$\Box(\Box P \supset P)$	Necessitation, on 1
3	$\Box(\Box P \supset P) \supset (\Box\Box P \supset \Box P)$	**K** axiom
4	$\Box\Box P \supset \Box P$	Modus Ponens on 2, 3
5	$\Box\Box P$	local assumption
6	$\Box P$	Modus Ponens on 4, 5

Many books on modal logic take an axiomatic approach. When derivation is defined, some books allow Necessitation to be applied throughout, some do it differently. Essentially the difference is whether deduction assumptions are being thought of locally or globally. Few books allow both, and so one must exercise some care in reading more than one book, since things can differ profoundly while the same terminology may be employed.

EXERCISES

EXERCISE 3.3.1. First, give an axiomatic derivation in **K** of $\Box\Box P \supset Q$ using $\{\Box P \supset P\}$ as global and $\{P \supset Q\}$ as local assumptions. Then give a model to show we do not have $\{\Box P \supset P\} \models_K \{P \supset Q\} \rightarrow \Box\Box P \supset \Box Q$.

3.4. AXIOM SYSTEMS WORK TOO

In Section 2.5 we showed tableau systems were sound and complete—now we do the same for axiom systems. Once this is done, we will know that if **L** is any of **K, D, T, K4, B, S4,** or **S5,** a formula X is **L**-valid if and only if it has a tableau proof using the **L** rules if and only if it has an axiomatic proof using the **L** axioms. The work can be extended to take logical consequence into account, but we do not do so here.

Soundness

Soundness for an axiomatic system is easy to prove. We simply show each of the axioms is valid, and the rules of inference produce valid formulas from valid formulas. Then it follows that every line of a proof must be a valid formula, hence the last line—the formula being proved—is also valid.

We begin with the rules of inference. And of these, Modus Ponens is easy. Suppose $P \supset Q$ is true at a world. Then either P is not true there, or Q is. So if P is also true at that world, Q must be. It follows that if P and $P \supset Q$ are both valid in a model (that is, if both are true at every world of the model), Q must also be valid in that model.

The Rule of Necessitation is almost as easy, but we cannot work one world at a time, as we did with Modus Ponens. Suppose P is valid in a model. Let Γ be an arbitrary world of that model, and let Δ be any world that is accessible from Γ. Since P is valid in the model, P is true at Δ. Thus P is true at every world that is accessible from Γ, so $\Box P$ must be true at Γ. Since Γ was an arbitrary world of the model, $\Box P$ is true at *every* world of the model, so $\Box P$ is valid in the model.

We have shown that both rules of inference turn formulas that are valid in a model into other formulas that are also valid in that model. Now we look at the axioms, and we begin with those for **K**.

At each world of a modal model the propositional connectives have their usual classical behavior. It follows immediately that every tautology is true at every possible world of every model—tautologies are valid in every model. Likewise, in Exercise 1.7.4 you were asked to show the validity, in every model, of formulas of the form $\Box(P \supset Q) \supset (\Box P \supset \Box Q)$. Thus all the **K** axioms are valid in all modal models.

We have now shown enough to establish the soundness of the **K** axiom system with respect to the **K** semantics. Extending this to the remaining modal logics is a simple matter. Consider **K4**, for instance. In Example 1.7.7 we showed formulas of the form $\Box P \supset \Box\Box P$ are valid in all transitive models. It follows that all axioms of **K4** are valid in all transitive models, and hence all theorems of **K4** are valid in all transitive models. And for the remaining modal logics, Exercise 1.8.1 contains what is needed for **D**, and Example 1.7.8 (and Exercise 1.7.5) takes care of all the rest of the logics.

THEOREM 3.4.1. [Axiom System Soundness] If X has a proof using the **L** axiom system then X is **L**-valid, where **L** is any of **K, D, T, K4, B, S4,** or **S5**.

Completeness

Completeness is more work, as usual. Perhaps you are familiar with a completeness proof for *classical* axiom systems using maximal consistent sets of formulas. Completeness proofs for modal axiom systems are really extensions of that argument. We sketch the basic ideas, and refer you to (Chellas, 1980), (Fitting, 1983) or (Hughes and Cresswell, 1996) for fuller presentations. We begin with results that hold for any of the modal logics, then we narrow things down to **K**. Finally we discuss briefly how to modify things for the other systems. For what immediately follows, **L** is one of **K, D, T, K4, B, S4**, or **S5**.

In classical logic, $\wedge$ is commutative and associative; that is, $(X \wedge Y) \equiv (Y \wedge X)$ is a tautology, as is $((X \wedge Y) \wedge Z) \equiv (X \wedge (Y \wedge Z))$. It follows that any two ways of forming the conjunction of the formulas $X_1, X_2, \ldots, X_n$ will be equivalent. For instance, $((X_1 \wedge X_2) \wedge (X_3 \wedge X_4))$ and $(X_2 \wedge (X_3 \wedge (X_4 \wedge X_1)))$ are equivalent. Consequently we can leave parentheses out in a conjunction like $(X_1 \wedge X_2 \wedge \ldots \wedge X_n)$, since their exact placement won't matter. We make use of this in the following definition.

DEFINITION 3.4.2. [Consistent] Let $\bot$ be an abbreviation for $(P \wedge \neg P)$. A finite set $\{X_1, X_2, \ldots, X_n\}$ of formulas is **L**-*consistent* if $(X_1 \wedge X_2 \wedge \ldots \wedge X_n) \supset \bot$ is not provable using the **L** axiom system. An infinite set is **L**-consistent if every finite subset is **L**-consistent.

DEFINITION 3.4.3. [Maximal Consistent] A set S of formulas is *maximally* **L**-consistent if S is **L**-consistent, and no proper extension of it is. That is, if $S \subseteq S'$ and S' is also **L**-consistent, then S' not a *proper* extension, so $S = S'$.

THEOREM 3.4.4. [Lindenbaum] If S is **L**-consistent, it can be extended to a maximally **L**-consistent set.

Proof Assume S is **L**-consistent. The entire set of formulas is countable, that is, all formulas can be arranged into a single infinite list: $X_1, X_2, X_3, \ldots$. (This is a standard result of elementary set theory which we do not prove here.) Now, assume this has been done and define an infinite sequence of sets of formulas as follows.

$$S_0 = S$$
$$S_{n+1} = \begin{cases} S_n \cup \{X_{n+1}\} & \text{if } S_n \cup \{X_{n+1}\} \text{ is } \mathbf{L}\text{-consistent} \\ S_n & \text{otherwise} \end{cases}$$

Note that, by construction, each S_n is consistent, and moreover, $S_0 \subseteq S_1 \subseteq S_2 \subseteq \ldots$. Finally, let S^* be the limit, that is, $S^* = S_0 \cup S_1 \cup S_2 \cup \ldots$. We claim S^* is a maximal consistent extension of S.

First, S^* extends S, since it contains all the members of $S_0 = S$.

Second, S^* is consistent. Suppose not; we derive a contradiction. If S^* were not **L**-consistent, it would have a finite subset $\{Z_1, \ldots, Z_k\}$ that is not consistent. Since this is a finite subset of S^*, $Z_1 \in S_{i_1}$ for some index i_1. Likewise $Z_2 \in S_{i_2}$ for some i_2, and so on. But then, if we let m be the largest of $i_1, i_2, \ldots, i_k$, it follows that all of $Z_1, Z_2, \ldots, Z_k$ are in S_m, so S_m has a subset that is not consistent. But this means S_m itself is not consistent, and this is impossible.

Finally, S^* is maximal. Again, we suppose otherwise and derive a contradiction. If S^* were not maximal, there would be a *proper* extension S' of S^* that is L-consistent. Let Z be any formula that is in S' but not in S^*. Then $S^* \cup \{Z\}$ itself must be L-consistent since it is a subset of S' (Exercise 3.4.1). In defining S^* we made use of a list consisting of *all* formulas, so for some n, $Z = X_n$. Now, $S_{n-1} \subseteq S^* \subseteq S'$, and also $X_n = Z \in S'$, so $S_{n-1} \cup \{X_n\} \subseteq S'$. Since any subset of an L-consistent set is L-consistent (Exercise 3.4.1 again), $S_{n-1} \cup \{X_n\} \subseteq S'$ is L-consistent. But then, by definition, $S_n = S_{n-1} \cup \{X_n\}$. Since $X_n \in S_n \subseteq S^*$, $X_n \in S^*$. But $X_n = Z$ and $Z \notin S^*$, and this is our contradiction. ∎

We need one more preliminary result, then we can give the completeness proof.

PROPOSITION 3.4.5. If the set $\{\neg \Box B, \Box A_1, \Box A_2, \ldots\}$ is L-consistent, so is $\{\neg B, A_1, A_2, \ldots\}$.

Proof Suppose $\{\neg B, A_1, A_2, A_3, \ldots\}$ is not L-consistent. Then it has a finite subset that is not L-consistent. If a finite set is not L-consistent, neither is any extension of it, so we can assume we have a finite subset that is inconsistent and includes $\neg B$, and all of $A_1, A_2, \ldots, A_n$, for some n. Now we proceed as follows.

$(\neg B \wedge A_1 \wedge \ldots \wedge A_n) \supset \bot$	assumption
$(A_1 \wedge \ldots \wedge A_n) \supset (\neg B \supset \bot)$	exportation
$(A_1 \wedge \ldots \wedge A_n) \supset B$	since $(\neg X \supset \bot) \equiv X$
$\Box(A_1 \wedge \ldots \wedge A_n) \supset \Box B$	Regularity Rule
$(\Box A_1 \wedge \ldots \wedge \Box A_n) \supset \Box B$	Example 3.1.6
$(\Box A_1 \wedge \ldots \wedge \Box A_n) \supset (\neg \Box B \supset \bot)$	$(\neg X \supset \bot) \equiv X$ again
$(\neg \Box B \wedge \Box A_1 \wedge \ldots \wedge \Box A_n) \supset \bot$	importation

And it follows that $\{\neg \Box B, \Box A_1, \Box A_2, \Box A_3, \ldots\}$ has a finite subset that is not L-consistent, so it is not L-consistent itself. ∎

We now have much of the preliminary work out of the way. Next we say how to construct the *canonical model* for the logic L.

DEFINITION 3.4.6. [Canonical Model for L] We create a model as follows. $\mathcal{G}$ is the set of *all* maximally L-consistent sets of formulas. If Γ and Δ are in $\mathcal{G}$, set $\Gamma \mathcal{R} \Delta$ provided, for each formula in Γ of the form $\square Z$, the corresponding formula Z is in Δ. And finally, for each propositional letter P and each $\Gamma \in \mathcal{G}$, set $\Gamma \Vdash P$ just in case $P \in \Gamma$ (recall, members of $\mathcal{G}$ are sets of formulas, so this condition makes sense). We have now completely defined a model $\mathcal{M} = \langle \mathcal{G}, \mathcal{R}, \Vdash \rangle$. It is called the *canonical* model for **L**.

Here is the main result concerning the canonical model.

PROPOSITION 3.4.7. [Truth Lemma] Let $\langle \mathcal{G}, \mathcal{R}, \Vdash \rangle$ be the canonical model for **L**. For every formula Z and for every $\Gamma \in \mathcal{G}$,

$$\Gamma \Vdash Z \text{ if and only if } Z \in \Gamma$$

Proof The argument is by induction on the complexity of the formula Z. That is, we assume the Truth Lemma is correct for formulas that are simpler than Z (of lower degree), and using this we show it is correct for Z itself. There are several cases.

ATOMIC If P is a propositional variable, $P \in \Gamma$ if and only if $\Gamma \Vdash P$, by definition of the model.

NEGATION Suppose $Z = \neg W$, and the Truth Lemma is known for W. Then $Z \in \Gamma$ if and only if $\neg W \in \Gamma$ if and only if $W \notin \Gamma$ (this is by Exercise 3.4.2, which we leave to you) if and only if $\Gamma \not\Vdash W$ (by the induction hypothesis) if and only if $\Gamma \Vdash \neg W$ if and only if $\Gamma \Vdash Z$.

CONJUNCTION This case is immediate, by Exercise 3.4.2. (Other propositional connective cases are similar.)

NECESSITATION Suppose $Z = \square W$. This requires a two-part argument. Assume first that $Z \in \Gamma$, that is, $\square W \in \Gamma$. Let Δ be an arbitrary member of $\mathcal{G}$ such that $\Gamma \mathcal{R} \Delta$. By our definition of $\mathcal{R}$, since $\square W$ is in Γ, W is in Δ. By the induction hypothesis, $\Delta \Vdash W$. Since Δ was arbitrary, W is true at *every* member of $\mathcal{G}$ accessible from Γ, hence $\Gamma \Vdash \square W$, that is, $\Gamma \Vdash Z$.

Now assume that $Z \notin \Gamma$, that is, $\square W \notin \Gamma$. By Exercise 3.4.2, $\neg \square W \in \Gamma$. Let $\{\square Y_1, \square Y_2, \square Y_3, \dots\}$ be the set of all members of Γ that begin with $\square$. Then $\{\neg \square W, \square Y_1, \square Y_2, \square Y_3, \dots\}$ is a subset of Γ, and so is **L**-consistent. By Proposition 3.4.5, $\{\neg W, Y_1, Y_2, Y_3, \dots\}$ is **L**-consistent. Using Theorem 3.4.4, extend it to a maximal **L**-consistent set Δ. Then $\Delta \in \mathcal{G}$. Also, if $\square Y$ is in Γ, Y is in Δ, by construction, so $\Gamma \mathcal{R} \Delta$. Finally, $\neg W \in \Delta$, so $W \notin \Delta$. Then by the induction hypothesis, $\Delta \not\Vdash W$. But then, $\Gamma \not\Vdash \square W$, so $\Gamma \not\Vdash Z$.

This completes the proof of the Truth Lemma. ∎

We have all the preliminary work out of the way. Here is the main event. We give the argument for **K**, and say how to adapt it to other logics.

THEOREM 3.4.8. [Axiom System Completeness] If X is **K**-valid then X has a proof using the **K** axiom system.

Proof We show the contrapositive, and in doing so we can use all the results above, since **L** could have been **K**. Suppose X is a formula that has no axiomatic **K** proof—we show it is not **K**-valid.

Since X has no proof, the set $\{\neg X\}$ is **K**-consistent (because if it were not, $\neg X \supset \bot$ would be provable, and this is equivalent to X). Using Proposition 3.4.4, extend $\{\neg X\}$ to a maximally **K**-consistent set Γ_0.

Let $\langle \mathcal{G}, \mathcal{R}, \Vdash \rangle$ be the canonical model for **K**. We know $\Gamma_0 \in \mathcal{G}$. But also, $\neg X \in \Gamma_0$, since Γ_0 extends $\{\neg X\}$, so $X \notin \Gamma_0$. But then, by the Truth Lemma, $\Gamma_0 \nVdash X$, so X is not **K**-valid, which is what we wanted to show. ∎

Now that completeness has been established for **K**, what about the other modal logics? We discuss **T** as representative, and leave the rest to you. We construct the canonical model for **T**, and use it just as we used the canonical model for **K** in the proof of Theorem 3.4.8. Everything works as before, but in addition, *the canonical model for* **T** *is, in fact, a* **T** *model*. This fact is enough to get us the completeness of the axiom system for **T**.

Showing that the canonical model for **T** is reflexive is rather easy. Let $\langle \mathcal{G}, \mathcal{R}, \Vdash \rangle$ be the canonical **T** model, and let $\Gamma \in \mathcal{G}$. Suppose $\Box Z \in \Gamma$. Now, $\Box Z \supset Z$ is an axiom of **T**, hence a theorem, and so $(\Box Z \supset Z) \in \Gamma$ by Exercise 3.4.2. Then, again by Exercise 3.4.2, $Z \in \Gamma$. We have argued that whenever $\Box Z \in \Gamma$ we also have $Z \in \Gamma$. It follows from the definition of $\mathcal{R}$ for a canonical model that $\Gamma \mathcal{R} \Gamma$, and so the canonical model for **T** is reflexive, that is, it is a **T** model.

Completeness for the other modal logics follows the same lines, and we leave the verification to you as an exercise.

EXERCISES

EXERCISE 3.4.1. Show every subset of a **L**-consistent set is **L**-consistent.

EXERCISE 3.4.2. Assume that S is maximally **L**-consistent.
1. Show that if $S \cup \{Z\}$ is **L**-consistent, then $Z \in S$.
2. Show that if X is a theorem of the **L** axiom system, then $X \in S$.

3. Show that if $X \in S$ and $(X \supset Y) \in S$ then $Y \in S$.
4. Show that exactly one of Z or $\neg Z$ is in S.
5. Show that $(Z \wedge W) \in S$ if and only if both $Z \in S$ and $W \in S$.

EXERCISE 3.4.3.   For **L** being any of **K4**, **B**, **S4**, or **S5**, show the canonical model for **L** is, in fact, a **L** model.

CHAPTER FOUR

QUANTIFIED MODAL LOGIC

This is a book on first-order modal logic. Adding quantifier machinery to classical propositional logic yields first-order classical logic, fully formed and ready to go. For modal logic, however, adding quantifiers is far from the end of the story, as we will soon see. But certainly adding quantifiers is the place to start. As it happens, even this step presents complications that do not arise classically. We say what some of these are after we get language syntax matters out of the way.

4.1. FIRST-ORDER FORMULAS

We have the same *propositional connectives* that we had in Chapter 1, and the same *modal operators*. We add to these two *quantifiers*, $\forall$ (*for all*, the universal quantifier) and $\exists$ (*there is*, the existential quantifier). Just as with the modal operators, these turn out to be interdefinable, so sometimes it will be convenient to take one as basic and the other as defined.

We assume we have available an infinite list of *one place relation symbols*, $P_1^1, P_2^1, P_3^1, \ldots$, an infinite list of *two place relation symbols*, P_1^2, P_2^2, P_3^2, $\ldots$, an infinite list of *three place relation symbols*, $P_1^3, P_2^3, P_3^3, \ldots$, and so on. An n-place relation symbol is said to be *of arity n*. We will generally be informal in our notation and use P, R, or something similar for a relation symbol, with its arity determined from context. One place relation symbols are sometimes referred to as *predicate symbols*.

We also assume we have available an infinite list of *variables*, v_1, v_2, v_3, $\ldots$. Here too we will generally be informal in our notation, and write x, y, z, and the like, for variables.

DEFINITION 4.1.1. [Atomic Formula] An *atomic formula* is any expression of the form $R(x_1, x_2, \ldots , x_n)$, where R is an n-place relation symbol and x_1, $x_2, \ldots , x_n$ are variables.

The propositional letters of Chapter 1 were stand-ins for propositions, which were not further analyzed. In a first-order logic, atomic formulas have more structure to them. Think of a relation symbol as standing for a relation, such as *is-the-brother-of* (two-place) or *is-a-number-between* (three-place). An atomic formula like $R(x, y)$ is supposed to mean: the relation that R

stands for holds of the things that are the values of x and y. This will be given a precise meaning when semantics is discussed.

Next we define the notion of *first-order formula*, and simultaneously we define the notion of *free variable occurrence*. Anticipating notation that we are about to define, we will see that $((\forall x)P(x) \supset Q(x))$ is a formula. In it, the occurrence of x in $P(x)$ is within the scope of the quantifier $(\forall x)$, and is not counted as free. What x is supposed to stand for makes no difference in the interpretation of $(\forall x)P(x)$. If we say x stands for the king of spades, or we say x stands for the king of rock and roll, it doesn't affect the interpretation of $(\forall x)P(x)$—indeed, we could have written $(\forall z)P(z)$ just as well. The formula simply says everything has property P, and variables allow us to say this clearly. On the other hand, the occurrence of x in $Q(x)$ is not within the scope of any quantifier, so what this x stands for can make a difference. Thus in the formula $((\forall x)P(x) \supset Q(x))$ only one occurrence of x is to be taken as free, and so is affected by the value assigned to x. Now for the details.

DEFINITION 4.1.2. [First-Order Modal Formulas] The sets of *first-order formula* and *free variable occurrence* are determined by the following rules.

1. Every atomic formula is a formula; every occurrence of a variable in an atomic formula is a free occurrence.
2. If X is a formula, so is $\neg X$; the free variable occurrences of $\neg X$ are those of X.
3. If X and Y are formulas and $\circ$ is a binary connective, $(X \circ Y)$ is a formula; the free variable occurrences of $(X \circ Y)$ are those of X together with those of Y.
4. If X is a formula, so are $\Box X$ and $\Diamond X$; the free variable occurrences of $\Box X$ and of $\Diamond X$ are those of X.
5. If X is a formula and v is a variable, both $(\forall v)X$ and $(\exists v)X$ are formulas; the free variable occurrences of $(\forall v)X$ and of $(\exists v)X$ are those of X, *except* for occurrences of v.

Any variable occurrences in a formula that are not free are said to be *bound*.

EXAMPLE 4.1.3. Consider $(\exists y)((\forall x)P(x, y) \supset \Box Q(x, y))$, where P and Q are both two-place relation symbols. We sketch why this is a formula, and determine which variable occurrences are free. To help make things clear, we display free variable occurrences in formulas using bold-face.

Both $P(\mathbf{x}, \mathbf{y})$ and $Q(\mathbf{x}, \mathbf{y})$ are atomic formulas, hence both are formulas, and all variable occurrences are free.

Since $Q(\mathbf{x}, \mathbf{y})$ is a formula, so is $\Box Q(\mathbf{x}, \mathbf{y})$, and it has the same free variable occurrences.

Since $P(x, y)$ is a formula, so is $(\forall x) P(x, y)$, and the free variable occurrences are those of $P(x, y)$, except for occurrences of x, hence only the occurrence of **y** is free.

Now $((\forall x) P(x, y) \supset \Box Q(x, y))$ is a formula, with free variable occurrences as indicated.

Finally, $(\exists y)((\forall x) P(x, y) \supset \Box Q(x, y))$ is a formula, and its free variable occurrences are those of $((\forall x) P(x, \mathbf{y}) \supset \Box Q(\mathbf{x}, y))$, except for occurrences of y. Thus finally, only one occurrence of **x** is free.

When models are defined later on, we will see that a formula without free variables is simply true at a world of a model, or not. But for a formula with free variables, additional information must be supplied before truth can be determined, namely values for the free variables. This suggests that formulas without free variables will play a special role.

DEFINITION 4.1.4. [Sentence] We call a formula with no free variable occurrences a *sentence* or a *closed formula*.

Finally, we will continue our earlier practice of informally omitting outer parentheses from formulas, and using differently shaped parentheses.

EXERCISES

EXERCISE 4.1.1.   Verify that the following are formulas, and determine which are the free variable occurrences. Also, determine which of the following are sentences.

1. $((\forall x) \Diamond P(x, y) \supset (\exists y) \Box Q(x, y))$
2. $(\exists x)(\Box P(x) \supset (\forall x) \Box P(x))$
3. $(\forall x)((\exists y) R(x, y) \supset R(y, x))$

4.2. AN INFORMAL INTRODUCTION

Let's look at the kinds of things we can say with first-order formulas before quantifiers are taken into account.

Contrast the formula

(4.1) $F(x)$

with the formula

(4.2) $\Box F(x)$.

(4.1) says that an object x has the property F—in the actual world, of course. (4.2) says that an object x has the property F necessarily, i.e., in every possible world.[14] The x that occurs in both (4.1) and (4.2) is a free variable, not a name or a description or a function name of any sort. Our formal machinery will include a mechanism for assigning values—objects—to free variables. Informally, we will just refer to the object x. It is important to keep in mind that it is the object itself—the value of x—that we are speaking about, and not the variable.

The contrast between (4.1) and (4.2) is just the contrast, for example, between saying that *God exists* and saying that *God necessarily exists*. If F stands for "exists" and x has God as its value, then (4.1) says that God exists in this world. It does not, of course, say that he exists in every possible world. And if he did not exist in every possible world, then his existence would only be contingent. This is a matter of considerable importance in classical theology. For, it has been claimed that God differs from other things in that it is part of God's essence that he exists. This is to say that it is an essential property of him that he exists. And, indeed, (4.2) can be read as *F is an essential property of x.*

On the crudest version of the *Ontological Argument*, the claim that it is part of God's essence that he exists is interpreted as (4.2), and so the fact of his existence, (4.1), follows immediately *via* $\Box F(x) \supset F(x)$ (which is valid in a frame at least as strong as **T**).

But there are more subtle ways of understanding the claim that it is part of God's essence that he exists. Consider

(4.3) $F(x) \supset \Box F(x)$,

which says that x has F necessarily if *it has F at all*. This is not generally true: just because x is F in the actual world, it does not follow that it is F in every possible world. But something special happens for the case of God's existence. (4.2) categorically asserts that x is F in every possible world; (4.3) asserts this only hypothetically, i.e., on the condition that x is F in this world. So, for the case of God's existence, the contrast is between saying that *God necessarily exists* and saying that *God necessarily exists if he exists*. The distinction is important. It has been a matter of some concern among Descartes' commentators which of these two is the proper conclusion of the version of the *Ontological Argument* he presents in Meditation V. For if it is (4.3), then, of course, God's existence has not been established.

We can make yet finer discriminations here if we contrast (4.3) with

(4.4) $\Box(F(x) \supset \Box F(x))$.

---

[14] Understood, of course, as "in every *accessible* possible world." We will generally suppress "accessible" in our informal discussion.

(4.4) contains one $\square$ nested inside another. This is where informally interpreting modal formulas becomes difficult; it is also where possible world semantics is most helpful. (4.4) says that (4.3) is necessary, i.e., that it holds in every possible world. Now, (4.3) says that the following holds in the actual world:

> Either x isn't F or is F in every possible world.

So, (4.4) says that this holds in *any world*.

Returning to our example, the contrast is between saying that *God necessarily exists if he exists* and saying that *Necessarily, God necessarily exists if he exists*. Once again, one might balk at the idea that it is only a contingent property of him that he necessarily exists if he exists.

Our first-order language enables us to make very subtle distinctions that greatly enrich our understanding of traditional modal claims In this section, we have focussed on open sentences, i.e., sentences that contain free variables. In the next section, we will look at quantified sentences.

EXERCISES

EXERCISE 4.2.1. Consider the claim

It is necessary that an omniscient being is essentially omniscient

Does this require that an omniscient being exist in more than one possible world (if it exists at any)? Suppose it is possible that there is an omniscient being. Does it follow that one exists?

4.3. NECESSITY DE RE AND DE DICTO

The sentence

(4.5)   Everything is necessarily F.

contains an ambiguous construction that has caused much confusion in the history of modal logic. We can express these two readings of (4.5) a bit more clearly as

(4.6)   It is a necessary truth that everything is F.

and

(4.7)   Each thing is such that it has F necessarily.

The medieval logicians were aware of these two interpretations. (4.6) expresses that a proposition [*dictum*] is necessary, and so it is an example of what they called *necessity de dicto*. (4.7) expresses that a thing [*res*] has a property necessarily. It is an example of *necessity de re*.

We find the same *de re/de dicto* ambiguity in

(4.8)   Something necessarily exists

On the one hand, (4.8) could mean

(4.9)   It is necessarily true that something exists.

This is the *de dicto* reading, and it is true in our semantics simply because, following standard procedure, quantifier domains are taken to be non-empty. On the other hand, the *de re* reading of (4.8),

(4.10) Something has the property of existence essentially,

is quite controversial. (4.10) commits us to a necessary existent—perhaps God, but, in any event, an object that exists in every possible world.

Here is an example familiar to those who have followed the recent philosophical literature on modal logic,

(4.11) The number of planets is necessarily odd.

On the *de dicto* reading, (4.11) says that a certain proposition is necessary, namely, the proposition *that the number of planets is odd*. And on this *de dicto* interpretation, (4.11) is clearly false, for surely it is only a contingent fact that the number of planets is odd. Had history been different, there could have been 4, or 10, or 2 planets in our solar system. But the *de re* reading of (4.11) says, of the number of planets, that that number is necessarily odd. And as a matter of fact, the number of planets is 9, and it is an essential property of the number 9 that it is odd.

It turns out that the *de re/de dicto* distinction is readily expressible in our first-order modal language as a *scope* distinction. We take the *de dicto* (4.6) to be

(4.12) $\Box(\forall x)F(x)$

which asserts, of the statement $(\forall x)F(x)$, that it is necessary. And we take the *de re* (4.7) to be

(4.13) $(\forall x)\Box F(x)$

which asserts, of each thing, that it has F necessarily. (4.12) says, *In every possible world, everything is F*. (4.13) says, *Everything is, in every possible world, F*. We take the *de dicto* and *de re* readings of

> Something is necessarily F

to be, respectively,

(4.14) $\Box(\exists x)F(x)$.

and

(4.15) $(\exists x)\Box F(x)$.

(4.14) says, *In every possible world, something is F*. (4.15) says, *Something is, in every possible world, F*.

If we regard a modal operator as a kind of quantifier over possible worlds, then the *de re/de dicto* distinction corresponds to a permutation of two types of quantifiers. This way of viewing the matter makes the difference readily apparent. For example, (4.15) requires that there be some one thing (at least)—let's call it a—which is, in every possible world, F. So, a must be F in world Γ, a must be F in world Δ, and so on. (4.14) requires only that in each world there be something that is F, but it does not require that it be the same thing in each world.

The sort of ambiguity exhibited in (4.11) rarely arises in classical logic, but it can—most famously when the item picked out by a definite description fails to exist. Whitehead and Russell (1925) introduced notation as part of their theory of definite descriptions to mark just such a distinction of scope, similar to the scope indicated by quantifier placement. In Chapter 9 we will formally introduce a related notation we call *predicate abstraction* that will enable us to mark this distinction explicitly in our modal language. Anticipating a bit, where t is a singular term (say, "the number of planets") we will distinguish the *de dicto*

(4.16) $\Box\langle\lambda x.x \text{ is odd}\rangle(t)$

which says, "It is necessary *that* t is odd," from the *de re*

(4.17) $\langle\lambda x.\Box(x \text{ is odd})\rangle(t)$

which says, "t is such that it is necessary *of it* that it is odd."

The *de dicto* reading (4.16) is false. As we noted, it is a contingent, not a necessary, fact about the universe that there are exactly 9 planets. The expression "the number of planets" picks out a number via one of its contingent properties. If $\Box$ is to be sensitive to the quality of the truth of a proposition in its scope, then it must be sensitive as well to differences in the quality of terms designating objects—that is, it will be sensitive as to whether the object is picked out by an essential property or by a contingent one. Since "the number of planets" picks out a number by means of one of its contingent properties, it can pick out different numbers in different possible worlds. In the actual world, that number is 9, but there might have been 10 planets, so in another possible world that number could be 10. The proposition *that the number of planets is odd* therefore can come out true in some worlds and false in other. It is not necessarily true, and (4.16) is false.

But the *de re* (4.17) says, "The number of planets is, in every possible world, odd." This is true. There are exactly 9 planets, and so "the number of planets" designates the number 9. Unlike the *de dicto* case, the designation of the term has been fixed in the actual world as 9. And that number, in every possible world, is odd.

Predicate abstraction machinery will get a full treatment in Chapter 9. The remarks above must remain informal until then.

Note that there is no *de re/de dicto* distinction for an open sentence like

(4.18) $\Box(x$ is odd$)$.

(4.18) is true of a given object x iff in every possible world that object is odd. (4.18) is true of the number 9; it is false of the number 10. The mode of specification of the object that is the value of x is irrelevant to the truth value of the open sentence; the open sentence is true of *it*, or not, as the case may be.

Unlike free variables, singular terms—proper names, definite descriptions, function names—require considerable care in modal contexts. To avoid difficulties and confusions, we have chosen to introduce our first-order modal language first without constant or function names, and reserve for later chapters the special problems introduced by them. So, in this chapter we will discuss *de re/de dicto* distinctions of the sort exemplified by (4.12) and (4.13). Distinctions such as those of (4.16) and (4.17) must wait until Chapter 9.

In the next section we will show how the failure to observe the *de re/de dicto* distinction has led to serious errors about the very possibility of coherently doing quantified modal logic.

EXERCISES

EXERCISE 4.3.1.   Interpreting $\Box$ as *At all future times*, show informally:

1. $\Box(\forall x)F(x)$ fails to imply $(\forall x)\Box F(x)$
2. $(\forall x)\Box F(x)$ fails to imply $\Box(\forall x)F(x)$

EXERCISE 4.3.2.   Does the sentence "Some things are necessarily F" mask a *de re*/*de dicto* distinction? If so, what are the two readings of the sentence?

4.4.   Is Quantified Modal Logic Possible?

For much of the latter half of the twentieth century, there has been considerable antipathy toward the development of modal logic in certain quarters. Many of the philosophical objectors find their inspiration in the work of W. V. O. Quine, who as early as (Quine, 1943) expressed doubts about the coherence of the project. We will find that one of the main sources of these doubts rests on the failure to carefully observe the distinction between *de re* and *de dicto* readings of necessity.

Quantified modal logic makes intelligible the idea that objects themselves, irrespective of how we speak about them, have properties necessarily or contingently.[15] Our semantics for modal logic will not require that any particular object have any particular substantial property either accidentally or essentially, but only that it makes sense to speak this way. We leave it to metaphysicians to fill in their details.

Quine does not believe that quantified modal logic can be done coherently because it takes to be a feature of reality what is actually a feature of language. He says, for example, "Being necessarily or possibly thus and so is in general not a trait of the object concerned, but depends on the manner of referring to the object." (Quine, 1961a, p. 148) It is instructive to go over Quine's reasons for holding this view.

Quine (1953) sets up the problem by identifying three interpretations of $\Box$ on which the modality is progressively more deeply involved in our world outlook. On Grade 1, the least problematic level of involvement, $\Box$ is taken to be a metalinguistic predicate that attaches to a name of a sentence, in the same way as the Tarski reading of "is true." To say that a sentence is necessarily true is no more than to say it is a theorem (of a formal system reasonably close to logic—perhaps including set theory), and the distinction between theorems and nontheorems is clear. But on this reading, there can be no iteration of modal operators, and as a result there is no need for a specifically modal

---

[15] Quine (1953) defines "Aristotelian Essentialism" as "the doctrine that some of the attributes of a thing (quite independently of the language in which the thing is referred to, if at all) may be essential to the thing, and others accidental." (p. 173-4) This is just quantified modal logic. There is nothing essentially Aristotelian about any of it! However, one begins to look a bit Aristotelian if one holds that there *are* essential properties, and even more so if one holds that there are *non-trivial* essential properties.

logic. There is no interpretation for the many propositional modal logics we studied earlier in this book.

On Grade 2, which is the interpretation we relied on for our discussion of propositional modal logic, □ is an operator like ¬ that attaches to *closed* formulas, that is, to sentences. But there is an important difference between the logical operators □ and ¬. When P and Q have the same truth value, ¬P and ¬Q also have the same truth value, but □P and □Q need not have. For example, although the two sentences "9 > 7" and "The number of planets > 7" are both true, only the former is necessarily true. This should not surprise us. Modal logic differs from classical logic in its sensitivity to the quality of a statement's truth, so one would hardly expect □ to be indifferent to sentences merely happening to have the same truth value.

At Grade 3, □ is allowed to attach to *open* formulas, as in □$(x > 7)$. This is the level needed to combine modality with quantifiers, for we need to say such things as "Something is such that *it* is necessarily greater than 7." And it is in the passage from Grade 2 to Grade 3 involvement that Quine finds his problems.[16]

Quine finds the contrast between the two sentences,

$$9 > 7$$

and

The number of planets > 7

puzzling. The first is necessary, but the second is not. Why is this so? We are speaking about the same thing each time, since the number of the planets is 9. The only difference is in the way in which the thing is picked out. It appears that whether or not the claim is necessary depends, not on the thing talked about, but on the way in which it is talked about. And if so, Quine argues, there can be no clear understanding of whether an open sentence like

$$x > 7$$

is necessarily true or not, for the terms "9" and "the number of planets" on which our intuitions about necessity relied are no longer available.

Here is another example of the phenomenon that has puzzled Quine, this time involving the notion of *belief*. Although Dr. Jekyll and Mr. Hyde are one and the same person (so the story says), we can very well understand how Robert might believe that Dr. Jekyll is a good citizen and yet not believe that Mr. Hyde is a good citizen. The sentence

Dr. Jekyll is believed by Robert to be a good citizen

---

[16] Of course, Quine has other problems with the connection between necessity and analyticity.

is true, but the sentence

> Mr. Hyde is believed by Robert to be a good citizen

is false. But then it is a mystery what the open sentence

> *x* is believed by Robert to be a good citizen

is true *of*. Which individual is this? Dr. Jekyll? Mr. Hyde? Which individual this is appears to depend upon how he is specified, not on the individual himself. The quantified statement "*Someone* is such that *he* is believed by Robert to be a good citizen" seems totally uninterpretable.

Grade 2 involvement appears to imply that the behavior of □ depends on the way things are picked out; Grade 3 involvement requires that □ be independent of the way things are picked out. Quine argues that we cannot have it both ways. This difference between Grade 2 and Grade 3 involvement looks very much like the distinction between necessity *de dicto* and necessity *de re*. But it isn't, and it is important we see that it isn't. If one assumes there is only *de dicto* necessity at Grade 2, which is surrepticiously what Quine has done, then one will have severe problems with *de re* necessity, as has happened. If one starts out by assuming that necessity has to do with language and not things, then one will certainly run into problems interpreting modal logic as having do with things and not language.

As we have pointed out on a number of occasions, the complications for names, descriptions and quantifiers indicates no special problem in interpreting open sentences. In particular, it does not show that necessity is more closely connected with how we specify objects than with the objects themselves. One final example using a temporal interpretation *always* for □ should make this clear. Contrast the two sentences:

(4.19) The U.S. President will always be a Democrat

(4.20) Bill Clinton will always be a Democrat

Suppose, for simplicity, that Bill Clinton never changes his party affiliation. Understood *de re*, then, (4.19) and (4.20) are both true. For that man, Bill Clinton or the President of the United States—it does not matter how we specify him—will forever be a Democrat. Understood *de dicto*, (4.20) is true, but (4.19) is not—for it is highly unlikely that the Democratic party will have a lock on the Presidency forever. Does the difference in truth value show that *temporality* has more to do with how an object is specified than with the object itself? Hardly. It depends on the fact that the Presidency will be changing hands, and Bill Clinton only temporarily holds that office. In the present world, the two coincide; but in later worlds, they won't. [17]

EXERCISES

EXERCISE 4.4.1.   Suppose necessity were intrinsically related to the way in which we pick things out. Then we would have to locate the *de re/de dicto* distinction in differences in the way designators designate. Discuss how this distinction might be drawn.

4.5. WHAT THE QUANTIFIERS QUANTIFY OVER

In first-order modal logic, we are concerned with the logical interaction of the modal operators $\Box$ and $\Diamond$ with the first-order quantifiers $\forall$ and $\exists$. From the perspective of possible world semantics, this is the interaction of two types of quantifiers: quantifiers that range over possible worlds, and quantifiers that range over the objects in those worlds. This interaction leads to complications. In classical logic, Universal Instantiation,

$$(4.21) \quad (\forall x)\varphi(x) \supset \varphi(x)$$

is valid. But the validity of (4.21) in a modal context depends on which particular possible world semantics we choose, i.e., on what we take our quantifiers to quantify over.

In the language of possible world semantics, the formula

$$(4.22) \quad \Box\varphi(x) \supset \varphi(x)$$

says that *If x is φ in every possible world, then x is φ*. What is this x? Whatever it is, we will suppose it to be something that occurs in at least one of the possible worlds in our model. And whatever it is, we will suppose that it makes sense to speak of it as existing in more than one possible world, although it need not so exist.

At a very general level, possible world semantics does not *require* that objects exist in more than one possible world. There is a fairly well-known

---

[17] There is an alternative reading of a singular term like "the President of the United States" which takes it to refer to an *intentional entity*. The entity would consist partly of George Washington, partly of Thomas Jefferson, ... , that is, of each of the individuals who occupied the office when they did. See (Hughes and Cresswell, 1996) for a discussion of this idea.

alternative semantics for first-order modal logic, most vividly put forward by Lewis (1968), which denies that objects can exist in more than one possible world. Lewis takes objects to be "worldbound." An object in one world, however, can have a *counterpart* in another possible world. This will be the object in the other world that is (roughly) most similar to the object in this one. (An object will be its own counterpart in any given world.) On the counterpart interpretation, (4.22) says *If in every accessible possible world x's counterpart is φ, then x is φ.* The main technical problem with counterpart theory is that the *being-a-counterpart-of* relation is, in general, neither symmetric nor transitive, and so no natural logic of equality is forthcoming.

What does it mean for an object to exist in more than one possible world? Here is an example. An assassination attempt was made upon Ronald Reagan's life. He was hit by the bullet, but only injured. Eventually he recovered from his wounds and continued on as President of the United States for many years. But he was almost killed by that bullet; and he could have been killed. This invites us to consider a counterfactual situation in which that man, Ronald Reagan, was indeed killed by that assassin. We are not considering a situation in which a person just like him was killed by that assassin. We are considering a situation in which that very person, Ronald Reagan, was killed by the assassin. We are imagining the very same person in a counterfactual situation. Since what we mean by a possible world is often just such a counterfactual situation, we are imagining an object—Ronald Reagan—to exist in more than one possible world.

But we have only argued for the coherence of allowing objects to exist in more than one possible world. We have not argued that *all* objects do, let alone that *any* of them do. This is actually a choice to be made in setting up modal models: Since a modal model contains many possible worlds, we need to decide whether the domain of discourse should be fixed for the whole of a modal model, or be allowed to vary from world to world, within the model? Taking the domain to be fixed for the whole model provides the simplest formal semantics. We refer to this as *constant domain* semantics. Allowing modal models to have world-dependent domains gives the greatest flexibility. We refer to this as *varying domain* semantics. In constant domain semantics, the domain of each possible world is the same as every other; in varying domain semantics, the domains need not coincide, or even overlap. In either case, what we call the *domain of the model* is the union of the domains of the worlds in the model. (For an interesting application of constant domain modal models in mathematics see (Smullyan and Fitting, 1996), where they provide appropriate machinery for establishing various independence results in set theory.)

We are assuming that free variables have as values objects in the domain of the model, not necessarily in the domain of the world we are in. If we suppose we are dealing with constant domain semantics, then (4.21) holds; more generally,

(4.23) $(\forall x)\varphi(x) \supset \varphi(y)$

is logically valid. Whenever we speak about an object, that object exists in at least one possible world. In constant domain semantics, what exists at one world exists at all. So any value assigned to y will be in the range of the quantifier $\forall x$. This is analogous to classical logic.

In varying domain semantics, however, the situation is a bit different. No longer does (4.23) hold. Just because everything in this world is φ it doesn't follow that y is φ, because y might exist only at worlds other than this one, and so not be in the range of the quantifier $(\forall x)$, which we take as ranging over what exists "here." This is quite unclassical.

There are, accordingly, two very different ways it could happen that x is φ at a world Γ:

1. x is φ at Γ and x is in Γ
2. x is φ at Γ but x need not be in Γ

In the classical situation, we speak of something only if it is in the domain of the model, and therefore in the domain of the quantifier. In the modal situation, however, we want to speak about things that do not exist but could, like Pegasus or the golden mountain or a tenth planet in the solar system. This is where we have a choice. One solution is to open up the domain of the actual world to all possible objects and keep the classical quantifier rules like Universal Instantiation intact: this is what has sometimes been called *possibilist* quantification. The other solution is to allow the domain of the actual world to differ from the domain of other possible worlds and abandon classical quantifier rules like Universal Instantiation: this is *actualist* quantification. The possibilist quantifier is evaluated for every element of the model $\mathcal{M}$; the actualist quantifier is evaluated at a world Γ only for elements of the domain of the world Γ. Possibilist quantification and actualist quantification correspond to constant domain and varying domain models, respectively. The connection is clear. Constant domain semantics models our intuitions about modality most naturally if we take the domain to consist of possible existents, not just actual ones, for otherwise we would be required to treat every existent as a necessary existent.

EXERCISE 4.5.1. Discuss informally how constant domain semantics will differ from varying domain semantics when the modality is given a temporal interpretation. Which is more natural in the temporal reading?

EXERCISE 4.5.2. "In constant domain semantics, possible objects exist, so there is no distinction between what is possible and what is actual." Discuss this claim.

EXERCISE 4.5.3. "In varying domain semantics, an object can have properties in a world even though it does not exist at that world." Write an essay either defending or criticizing this claim.

4.6. CONSTANT DOMAIN MODELS

We begin our formal treatment of semantics for quantifiers with *constant domain* models—*possibilist quantification*. In these models, the domain of quantification is the same from world to world. Technically this is somewhat simpler than allowing the domain to vary, and pedagogically it is easier to explain as a first approach. What we present is essentially from (Kripke, 1963b), with some modification of notation. We begin by enhancing the notion of a *frame*, from Definition 1.6.1.

DEFINITION 4.6.1. [Augmented Frame] A structure $\langle \mathcal{G}, \mathcal{R}, \mathcal{D} \rangle$ is a *constant domain augmented frame* if $\langle \mathcal{G}, \mathcal{R} \rangle$ is a frame and $\mathcal{D}$ is a non-empty set, called the *domain* of the frame.

The domain of an augmented frame is the set of things over which quantifiers can range, no matter at what world.

To turn a propositional frame into a model, all we had to say was which propositional letters were true at which worlds. The analog of propositional letters now is atomic formulas, and they have a structure that must be taken into account. More specifically, they involve *relation symbols*, which should stand for relations. But since more than one possible world can be involved, we should say which relation each relation symbol represents, at *each* of the worlds.

DEFINITION 4.6.2. [Interpretation] $\mathcal{I}$ is an *interpretation* in a constant domain augmented frame $\langle \mathcal{G}, \mathcal{R}, \mathcal{D} \rangle$ if $\mathcal{I}$ assigns to each n-place relation symbol R and to each possible world $\Gamma \in \mathcal{G}$, some n-place relation on the domain $\mathcal{D}$ of the frame. Thus, $\mathit{l}(R, \Gamma)$ is an n-place relation on $\mathcal{D}$, and so each n-tuple $\langle d_1, d_2, \ldots, d_n \rangle$ of members of $\mathcal{D}$ either is in the relation $\mathit{l}(R, \Gamma)$ or is not. If $\langle d_1, d_2, \ldots, d_n \rangle$ is in the relation $\mathit{l}(R, \Gamma)$ we will write $\langle d_1, d_2, \ldots, d_n \rangle \in \mathit{l}(R, \Gamma)$, following the standard mathematical practice of thinking of an n-place relation as a set of n-tuples.

DEFINITION 4.6.3. [Model] A constant domain *first-order model* is a structure $\mathcal{M} = \langle \mathcal{G}, \mathcal{R}, \mathcal{D}, \mathit{l} \rangle$ where $\langle \mathcal{G}, \mathcal{R}, \mathcal{D} \rangle$ is a constant domain augmented frame and l is an interpretation in it. By the *domain of the model* $\mathcal{M}$ we mean the domain of its augmented frame, $\mathcal{D}$. We say $\mathcal{M}$ is a constant domain first-order model for a modal logic **L** if the frame $\langle \mathcal{G}, \mathcal{R} \rangle$ is an **L** frame in the propositional sense.

EXAMPLE 4.6.4. Here is our first example of a constant domain first-order model. Let $\mathcal{G}$ consist of three possible worlds, Γ, Δ, and Ω, with $\Gamma \mathcal{R} \Delta$, $\Gamma \mathcal{R} \Omega$, and $\mathcal{R}$ holding in no other cases. Let $\mathcal{D} = \{a, b\}$. Let P be a one-place relation symbol. It is the only relation symbol we are interested in for now, so we will specify an interpretation only for it, and leave things unspecified for other relation symbols. Now, let $\mathit{l}(P, \Gamma)$ be the empty set (that is, nothing is in this relation); let $\mathit{l}(P, \Delta)$ consist of just a; let $\mathit{l}(P, \Omega)$ consist of just b. This specifies a constant domain first-order model $\mathcal{M} = \langle \mathcal{G}, \mathcal{R}, \mathcal{D}, \mathit{l} \rangle$. We represent it schematically as follows.

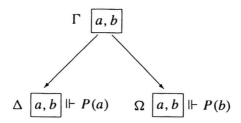

We have shown the domain, $\{a, b\}$, explicitly at each possible world, emphasizing that this is a constant domain model. Notice that we wrote $P(a)$ at the Δ part of the schematic. *This is not a formula!* It is not, since a is a member of the domain associated with Δ, but is not a variable of our formal language. Nonetheless this abuse of notation is handy for specifying models, provided its limitations are understood. Similar remarks apply to $P(b)$ at Ω, of course.

Next we must specify truth in constant domain first-order models. Not surprisingly, this is more complicated than it was in the propositional case, and we hinted at the source of the complications in the example above. We would like to have $(\forall x)P(x)$ be true at a possible world, say Γ, just in case $P(x)$ is true at Γ for all members of the domain $\mathcal{D}$. But, we cannot express this by saying that we want $P(c)$ to be true at Γ for all $c \in \mathcal{D}$, because $P(c)$ will not generally be a formula of our language. A way around this was introduced many years ago by Tarski for classical logic, and is easily adapted to modal models. For $(\forall x)P(x)$ to be true at Γ, we will require that $P(x)$ be true at Γ no matter what member of $\mathcal{D}$ we might have assigned to x as its value. But this means we cannot confine ourselves to the truth of *sentences* at worlds, but instead we must deal with the broader notion of the truth of formulas containing free variables, when values have been assigned to those free variables. This is done using a *valuation* function—akin to what computer scientists call an *environment*.

DEFINITION 4.6.5. [Valuation] Let $\mathcal{M} = \langle \mathcal{G}, \mathcal{R}, \mathcal{D}, \mathcal{I} \rangle$ be a constant domain first-order model. A *valuation* in the model $\mathcal{M}$ is a mapping v that assigns to each free variable x some member $v(x)$ of the domain $\mathcal{D}$ of the model.

What we are about to define is denoted thus:

$$\mathcal{M}, \Gamma \Vdash_v \Phi$$

where $\mathcal{M}$ is a constant domain first-order model, Γ is a possible world of the model, Φ is a formula, possibly with free variables, and v is a valuation. Read it as: formula Φ is true at world Γ of model $\mathcal{M}$ with respect to valuation v, where v tells us what values have been assigned to free variables. (Other books may use different notation.) We need one additional piece of technical terminology, then we can go ahead.

DEFINITION 4.6.6. [Variant] Let v and w be two valuations. We say w is an x-*variant* of v if v and w agree on all variables except possibly the variable x.

Now, here is the fundamental definition.

DEFINITION 4.6.7. [Truth in a Model] Let $\mathcal{M} = \langle \mathcal{G}, \mathcal{R}, \mathcal{D}, \mathcal{I} \rangle$ be a constant domain first-order modal model. For each $\Gamma \in \mathcal{G}$ and each valuation v in $\mathcal{M}$:

1. If R is an n-place relation symbol, $\mathcal{M}, \Gamma \Vdash_v R(x_1, \ldots, x_n)$ provided $\langle v(x_1), \ldots, v(x_n) \rangle \in \mathcal{I}(R, \Gamma)$.
2. $\mathcal{M}, \Gamma \Vdash_v \neg X \iff \mathcal{M}, \Gamma \not\Vdash_v X$.
3. $\mathcal{M}, \Gamma \Vdash_v (X \wedge Y) \iff \mathcal{M}, \Gamma \Vdash_v X$ and $\mathcal{M}, \Gamma \Vdash_v Y$.

4. $\mathcal{M}, \Gamma \Vdash_v \Box X \Longleftrightarrow$ for every $\Delta \in \mathcal{G}$, if $\Gamma \mathcal{R} \Delta$ then $\mathcal{M}, \Delta \Vdash_v X$.

5. $\mathcal{M}, \Gamma \Vdash_v \Diamond X \Longleftrightarrow$ for some $\Delta \in \mathcal{G}$, $\Gamma \mathcal{R} \Delta$ and $\mathcal{M}, \Delta \Vdash_v X$.

6. $\mathcal{M}, \Gamma \Vdash_v (\forall x)\Phi \Longleftrightarrow$ for every x-variant w of v in $\mathcal{M}$, $\mathcal{M}, \Gamma \Vdash_w \Phi$.

7. $\mathcal{M}, \Gamma \Vdash_v (\exists x)\Phi \Longleftrightarrow$ for some x-variant w of v in $\mathcal{M}$, $\mathcal{M}, \Gamma \Vdash_w \Phi$.

This definition should be compared with Definition 1.6.3. Propositional connectives other than $\wedge$ and $\neg$ have their usual defined behavior.

The last two items are the key new ones, of course. Item 6 says we should take $(\forall x)\Phi$ to be true at Γ, relative to a valuation v, provided Φ is true at Γ no matter what member of the domain $\mathcal{D}$ we assign to x (keeping the values assigned to other variables unchanged, of course). Likewise 7 says a similar thing about $(\exists x)\Phi$.

PROPOSITION 4.6.8. Suppose $\mathcal{M} = \langle \mathcal{G}, \mathcal{R}, \mathcal{D}, \mathcal{I} \rangle$ is a constant domain model, $\Gamma \in \mathcal{G}$, v_1 and v_2 are two valuations in $\mathcal{M}$, and Φ is a formula. If v_1 and v_2 agree on all the free variables of Φ, then

$$\mathcal{M}, \Gamma \Vdash_{v_1} \Phi \Longleftrightarrow \mathcal{M}, \Gamma \Vdash_{v_2} \Phi$$

Essentially, this Proposition says that if two valuations agree on the free variables actually present in a formula, the behavior of that formula with respect to the two valuations is the same. The proof is an induction argument on the complexity of Φ. It is routine, but complicated, and we omit it.

The next Proposition says, roughly, that one variable is as good as another provided we adjust valuations appropriately. In fact, the Proposition below is actually a generalization of the one we just gave. It too has a routine but complicated proof, and it too is omitted.

PROPOSITION 4.6.9. Suppose $\mathcal{M} = \langle \mathcal{G}, \mathcal{R}, \mathcal{D}, \mathcal{I} \rangle$ is a constant domain model, $\Gamma \in \mathcal{G}$, and v_1 and v_2 are valuations in $\mathcal{M}$. Further suppose that $\Phi(x)$ is a formula in which x may have some free occurrences, but y has no occurrences at all, and $\Phi(y)$ is the result of replacing all free occurrences of x with occurrences of y. Finally, suppose v_1 and v_2 agree on all the free variables of $\Phi(x)$ except for x, and $v_1(x) = v_2(y)$. Then

$$\mathcal{M}, \Gamma \Vdash_{v_1} \Phi(x) \Longleftrightarrow \mathcal{M}, \Gamma \Vdash_{v_2} \Phi(y)$$

Here is the reason this is a generalization of the previous Proposition. Suppose we take for x a variable that does not actually occur in the formula we are denoting by $\Phi(x)$. Then the hypothesis that v_1 and v_2 agree on all free variables of $\Phi(x)$ except for x simply says that they agree on all the free variables present. Thus if we assume that x has no occurrences in $\Phi(x)$, the hypothesis of Proposition 4.6.9 becomes the hypothesis of Proposition 4.6.8.

But also, if x does not occur in $\Phi(x)$, then $\Phi(x)$ and $\Phi(y)$ are the same formula; let us write it more simply as Φ. So the conclusion of Proposition 4.6.9, in this special case, is also the conclusion of Proposition 4.6.8.

DEFINITION 4.6.10. [True at Γ] Let $\mathcal{M} = \langle \mathcal{G}, \mathcal{R}, \mathcal{D}, \mathcal{I} \rangle$ be a constant domain model with $\Gamma \in \mathcal{G}$. For a *sentence* Φ, if $\mathcal{M}, \Gamma \Vdash_v \Phi$ for some valuation v in $\mathcal{M}$ then $\mathcal{M}, \Gamma \Vdash_v \Phi$ for every valuation v in $\mathcal{M}$ (by Proposition 4.6.8), and conversely. We abbreviate notation in this situation by writing $\mathcal{M}, \Gamma \Vdash \Phi$, and we say Φ *is true at* Γ.

Now the terminology of Definition 1.8.1 is simply carried over to first-order models. Φ is *valid* in a model if it is true at every world of the model, and so on.

EXAMPLE 4.6.11. We continue Example 4.6.4, and show $(\forall x) \Diamond P(x) \supset \Diamond (\forall x) P(x)$ is not valid in the model we gave. All references to $\mathcal{M}$ are to the model of Example 4.6.4.

Since $(\forall x) \Diamond P(x) \supset \Diamond (\forall x) P(x)$ is a sentence, to show

$$\mathcal{M}, \Gamma \not\Vdash (\forall x) \Diamond P(x) \supset \Diamond (\forall x) P(x)$$

we must show

$$\mathcal{M}, \Gamma \not\Vdash_v (\forall x) \Diamond P(x) \supset \Diamond (\forall x) P(x)$$

where v is any valuation in $\mathcal{M}$. We do this as follows.

Let w be like v on all variables except that $w(x) = a$. Now, $\mathcal{I}(P, \Delta)$ is the one-place relation that holds of just a, and $w(x) = a$, so $\langle w(x) \rangle \in \mathcal{I}(P, \Delta)$ and by definition we have

$$\mathcal{M}, \Delta \Vdash_w P(x)$$

and consequently

$$\mathcal{M}, \Gamma \Vdash_w \Diamond P(x).$$

In the same way we can show that, if w' is like v on all variables, except that $w'(x) = b$, then

$$\mathcal{M}, \Gamma \Vdash_{w'} \Diamond P(x).$$

Now w and w' are all the x-variants of v that there are, since the domain of $\mathcal{M}$ is $\{a, b\}$, so by item 6 of Definition 4.6.7,

$$\mathcal{M}, \Gamma \Vdash_v (\forall x) \Diamond P(x).$$

On the other hand, suppose we had

$$\mathcal{M}, \Gamma \Vdash_v \Diamond(\forall x)P(x).$$

Then we would have one of

$$\mathcal{M}, \Delta \Vdash_v (\forall x)P(x) \quad \text{or} \quad \mathcal{M}, \Omega \Vdash_v (\forall x)P(x).$$

Say we had the first—the argument for the second is similar. Let w' be the same x-variant of v as above—$w'(x) = b$. By item 6 of Definition 4.6.7 again, we must have

$$\mathcal{M}, \Delta \Vdash_{w'} P(x)$$

but we do not, since b is not in $\mathcal{I}(P, \Delta)$.

We have shown that

$$\mathcal{M}, \Gamma \Vdash_v (\forall x)\Diamond P(x) \quad \text{and} \quad \mathcal{M}, \Gamma \nVdash_v \Diamond(\forall x)P(x)$$

and it follows that

$$\mathcal{M}, \Gamma \nVdash_v (\forall x)\Diamond P(x) \supset \Diamond(\forall x)P(x).$$

Hence the sentence is not valid in the model $\mathcal{M}$.

EXAMPLE 4.6.12.   This time we show the converse implication $\Diamond(\forall x)P(x) \supset (\forall x)\Diamond P(x)$ *is* valid in all constant domain first-order modal models. Let $\mathcal{M} = \langle \mathcal{G}, \mathcal{R}, \mathcal{D}, \mathcal{I} \rangle$ be a model, let Γ be an arbitrary member of $\mathcal{G}$, and let v be a valuation in $\mathcal{M}$. To show

$$\mathcal{M}, \Gamma \Vdash_v \Diamond(\forall x)P(x) \supset (\forall x)\Diamond P(x)$$

it is enough to show that

$$\mathcal{M}, \Gamma \Vdash_v \Diamond(\forall x)P(x) \implies \mathcal{M}, \Gamma \Vdash_v (\forall x)\Diamond P(x).$$

Suppose

$$\mathcal{M}, \Gamma \Vdash_v \Diamond(\forall x)P(x).$$

Then for some Δ with $\Gamma \mathcal{R} \Delta$,

$$\mathcal{M}, \Delta \Vdash_v (\forall x)P(x),$$

and so if w is any x-variant of v in $\mathcal{M}$,

$$\mathcal{M}, \Delta \Vdash_w P(x),$$

but $\Gamma \mathcal{R} \Delta$, and hence

$$\mathcal{M}, \Gamma \Vdash_w \Diamond P(x).$$

Since w is *any* x-variant of v, we have

$$\mathcal{M}, \Gamma \Vdash_v (\forall x)\Diamond P(x).$$

EXERCISES

EXERCISE 4.6.1. Assume all models are **K** models, that is, no special assumptions are made about the accessibility relation. Which of the following sentences are valid in all constant domain first-order modal **K** models and which are not?

1. $[(\exists x)\Diamond P(x) \wedge \Box(\forall x)(P(x) \supset Q(x))] \supset (\exists x)\Diamond Q(x).$
2. $(\forall x)\Box P(x) \supset \Box(\forall x)P(x).$
3. $\Box(\forall x)P(x) \supset (\forall x)\Box P(x).$
4. $(\exists x)\Box P(x) \supset \Box(\exists x)P(x).$
5. $\Box(\exists x)P(x) \supset (\exists x)\Box P(x).$
6. $(\exists x)\Diamond[\Box P(x) \supset (\forall x)\Box P(x)].$
7. $(\exists x)\Diamond[P(x) \supset (\forall x)\Box P(x)].$
8. $(\exists x)(\forall y)\Box R(x, y) \supset (\forall y)\Box(\exists x)R(x, y).$

EXERCISE 4.6.2. Give a proof of Proposition 4.6.8.

4.7. VARYING DOMAIN MODELS

A more general notion of model than that of constant domains allows quantifier domains to vary from world to world. This gives us *actualist quantification*. Think of the domain associated with a world as what actually exists there, so quantifiers at each world range over the actually existent. Allowing domains to vary complicates the machinery somewhat, but not terribly much. Essentially we just replace the domain *set* $\mathcal{D}$ with a domain *function* that can assign a different domain to each world.

DEFINITION 4.7.1. [Augmented Frame] A structure $\langle \mathcal{G}, \mathcal{R}, \mathcal{D} \rangle$ is a *varying domain augmented frame* if $\langle \mathcal{G}, \mathcal{R} \rangle$ is a frame and $\mathcal{D}$ is a function mapping members of $\mathcal{G}$ to non-empty sets. The function $\mathcal{D}$ is called a *domain function*.

As we said above, think of a domain function as associating with each possible world of the frame the set of things that exist at that world, that is, the set over which quantifiers quantify *at that world*. We often refer to $\mathcal{D}(\Gamma)$ as the domain of the world Γ.

We can think of a constant domain model as a special kind of varying domain model: we simply have a domain function $\mathcal{D}$ that assigns the same set to each possible world. Consequently, anything established about varying domain models will apply automatically to constant domain ones. From now on, when appropriate, we will speak indifferently about a constant domain *set* $\mathcal{D}$, or a domain *function* that is constant, with $\mathcal{D}$ as its value at each world. The difference won't matter.

We do face a significant complication before we can properly define varying domain models though. Suppose a formula $\Box(P(x) \vee \neg P(x))$ is true at some possible world Γ, under a valuation v that assigns c to x, where c is something in the domain of world Γ. If $\Box$ is to have its standard interpretation, then for any world Δ that is accessible from Γ, we should have that $P(x) \vee \neg P(x)$ is true at Δ under the valuation v. But, *how do we know that $v(x)$, that is, c, exists at Δ?* More precisely, how do we know $c \in \mathcal{D}(\Delta)$? The answer, of course, is that we don't. And so we find ourselves required to consider the truth of a formula at a world when free variables of the formula have values that don't exist at the world in question!

There are two ways out of this problem. One is to allow *partial* models: take $P(x) \vee \neg P(x)$ to be neither true nor false at Δ under valuation v, when $v(x)$ is not in the domain associated with Δ. The other approach, which is the one we follow, is to say that even though $v(x)$ might not exist in the domain associated with Δ, it does exist under alternative circumstances we are willing to consider, and consequently talk about $v(x)$ is meaningful. Then at Δ, either the property P is true of $v(x)$ or is false of it, and in any event, $P(x) \vee \neg P(x)$ holds. As we said, this is the approach we follow.

DEFINITION 4.7.2. [Frame Domain] Let $\mathcal{F} = \langle \mathcal{G}, \mathcal{R}, \mathcal{D} \rangle$ be a varying domain augmented frame. The *domain of the frame* is the set $\cup \{\mathcal{D}(\Gamma) \mid \Gamma \in \mathcal{G}\}$. That is, we put together the domains associated with all the possible worlds of the frame. We write $\mathcal{D}(\mathcal{F})$ for the domain of the frame $\mathcal{F}$.

In a varying domain frame $\mathcal{F} = \langle \mathcal{G}, \mathcal{R}, \mathcal{D} \rangle$, if $\Gamma \in \mathcal{G}$, think of $\mathcal{D}(\Gamma)$ as the set of things that actually exist at world Γ, or in situation or state Γ. And think of $\mathcal{D}(\mathcal{F})$ as the things it makes sense to talk about at Γ, though these things may or may not exist at Γ.

Note that if our varying domain frame happens to be constant domain, if $\mathcal{D}$ assigns the same domain to each possible world, what we are now calling the domain of the frame turns out to be the domain of the frame as we used the terminology in Section 4.6.

Now, the definition of interpretation is essentially what it was before (Definition 4.6.2), with obvious minor modifications.

DEFINITION 4.7.3. [Interpretation] I is an *interpretation* in varying domain augmented frame $\mathcal{F} = \langle \mathcal{G}, \mathcal{R}, \mathcal{D} \rangle$ if I assigns, to each n-place relation symbol R, and to each possible world $\Gamma \in \mathcal{G}$, some n-place relation on the domain $\mathcal{D}(\mathcal{F})$ of the frame.

Finally, the definition of model is word for word as it was earlier (Definition 4.6.3), except that we use varying domain augmented frames instead of constant domain ones.

DEFINITION 4.7.4. [Model] A varying domain *first-order model* is a structure $\mathcal{M} = \langle \mathcal{G}, \mathcal{R}, \mathcal{D}, \mathit{I} \rangle$ where $\langle \mathcal{G}, \mathcal{R}, \mathcal{D} \rangle$ is a varying domain augmented frame and I is an interpretation in it. We say $\mathcal{M}$ is a varying domain first-order model for a modal logic **L** if the frame $\langle \mathcal{G}, \mathcal{R} \rangle$ is an **L** frame.

Incidentally, just as with constant domain models, we use the terminology *domain of a model* and mean by it the domain of the underlying frame. We write $\mathcal{D}(\mathcal{M})$ for the domain of the model $\mathcal{M}$.

EXAMPLE 4.7.5. Here is an example of a varying domain first-order model. Let $\mathcal{G}$ consist of two possible worlds, Γ and Δ, with $\Gamma \mathcal{R} \Delta$, and $\mathcal{R}$ holding in no other cases. Let $\mathcal{D}(\Gamma) = \{a\}$, and $\mathcal{D}(\Delta) = \{a, b\}$. Let P be a one-place relation symbol. Finally, let $\mathit{I}(P, \Gamma)$ be the empty set (that is, nothing is in this relation), and let $\mathit{I}(P, \Delta)$ consist of just b. This specifies a varying domain first-order model $\mathcal{M} = \langle \mathcal{G}, \mathcal{R}, \mathcal{D}, \mathit{I} \rangle$. We represent it schematically as follows.

We return to this example shortly.

To define truth in varying domain models we modify earlier machinery slightly, to ensure that quantifiers really do quantify over things that exist.

DEFINITION 4.7.6. [Valuation] As before, a *valuation* in a varying domain model $\mathcal{M}$ is a mapping v that assigns to each free variable x some member $v(x)$ of the domain of the model $\mathcal{D}(\mathcal{M})$.

DEFINITION 4.7.7. [Variant] The notion of x-variant is exactly as in Definition 4.6.6. But in addition, we say a valuation w is an x-variant of v *at* Γ if $w(x)$ is a member of $\mathcal{D}(\Gamma)$.

DEFINITION 4.7.8. [Truth in a Model] Let $\mathcal{M} = \langle \mathcal{G}, \mathcal{R}, \mathcal{D}, \mathcal{I} \rangle$ be a varying domain first-order modal model. For each $\Gamma \in \mathcal{G}$ and each valuation v in $\mathcal{D}(\mathcal{M})$:

1–5 Exactly as in Definition 4.6.7.

6. $\mathcal{M}, \Gamma \Vdash_v (\forall x)\Phi \Longleftrightarrow$ for every x-variant w of v at Γ, $\mathcal{M}, \Gamma \Vdash_w \Phi$.

7. $\mathcal{M}, \Gamma \Vdash_v (\exists x)\Phi \Longleftrightarrow$ for some x-variant w of v at Γ, $\mathcal{M}, \Gamma \Vdash_w \Phi$.

The essential difference between the definition above and the earlier Definition 4.6.7 is that, above, we require x-variants to assign to x something that is in the domain associated with the possible world we are considering. This, of course, was unnecessary for constant domain models, since all worlds had the same domains.

As before, the behavior of a formula at a world with respect to a valuation is not affected by values the valuation may assign to variables that are not free in the formula. More precisely, Propositions 4.6.8 and 4.6.9 carry over to the varying domain case. Consequently whether a *sentence* is true or false at a world is completely independent of which valuation we may choose, and so we can suppress mention of valuations when dealing with sentences.

EXAMPLE 4.7.9. We continue with Example 4.7.5, and show the sentence $\Diamond(\exists x)P(x) \supset (\exists x)\Diamond P(x)$ is not valid in the model we gave.

Let v be any valuation in $\mathcal{M}$. Let w be like v on all variables except that $w(x) = b$. Now, $\mathcal{I}(P, \Delta)$ is the one-place relation that holds of just b, and $w(x) = b$, so by definition, we have

$$\mathcal{M}, \Delta \Vdash_w P(x).$$

Since $b \in \mathcal{D}(\Delta)$, w is an x-variant of v at Δ and it follows that we have

$$\mathcal{M}, \Delta \Vdash_v (\exists x)P(x)$$

and consequently we have

$$\mathcal{M}, \Gamma \Vdash_v \Diamond(\exists x)P(x).$$

On the other hand, suppose we had

$$\mathcal{M}, \Gamma \Vdash_v (\exists x)\lozenge P(x).$$

Then for some x-variant w of of v at Γ we would have

$$\mathcal{M}, \Gamma \Vdash_w \lozenge P(x).$$

But $w(x)$ must be a, since a is the only member of the domain associated with Γ. It follows that

$$\mathcal{M}, \Delta \Vdash_w P(x)$$

which can only happen if $w(x)$ is in the relation $\mathcal{I}(P, \Delta)$, but $w(x) = a$, and a is not in this relation. This contradiction shows our supposition was wrong, and so

$$\mathcal{M}, \Gamma \not\Vdash_v (\exists x)\lozenge P(x).$$

Now it follows that

$$\mathcal{M}, \Gamma \not\Vdash_v \lozenge(\exists x)P(x) \supset (\exists x)\lozenge P(x)$$

and so the sentence $\lozenge(\exists x)P(x) \supset (\exists x)\lozenge P(x)$ is not true at Γ, and hence is not valid in the model $\mathcal{M}$.

EXERCISES

EXERCISE 4.7.1.   Which of the sentences of Exercise 4.6.1 are valid in all *varying domain* models and which are not?

4.8. DIFFERENT MEDIA, SAME MESSAGE

We have now seen two radically different semantics: constant domain, and varying domain. Which is, or should be, primary? It turns out, in a very precise sense, that it doesn't matter. We can formalize the same philosophical ideas either way, with a certain amount of care.

In varying domain semantics we think of the domain of a world as what "actually exists" at that world, in that state, under those circumstances. A quantifier ranges over what actually exists. Thus $(\exists x)\ldots$ means, *there is something, x, that actually exists such that* .... The domain of the *model* consists of what exists at all worlds, collectively, and so represents what *might* exist. While quantifiers do not range over the domain of the model, free variables take their values there. Thus a formula $\Phi(x)$, with x free, generally has a different behavior from its universal closure, $(\forall x)\Phi(x)$, in marked contrast to classical logic.

Constant domain semantics, on the other hand, has the same domain for every world. This common domain is analogous to the domain of the model, when varying domain semantics are used. Then, in the constant domain approach, quantifiers range over what does and what might exist. Now $(\exists x)\ldots$ can be read, "there is something, x, that could exist, such that $\ldots$."

Suppose, in constant domain semantics, we had an "existence predicate" that tells us which of the things that could exist actually exist and which do not. Then we could relativize constant domain quantifiers to this predicate, and achieve an effect rather like being in a varying domain model. Based on the development in (Hughes and Cresswell, 1996), this can be made precise as follows.

DEFINITION 4.8.1. [Existence Relativization] Let $\mathcal{E}$ be a one-place relation symbol. (We intend this to be an existence primitive, and will not use it for any other purpose.) The *existence relativization* of a formula Φ, denoted $\Phi^{\mathcal{E}}$, is defined by the following conditions.

1. If A is atomic, $A^{\mathcal{E}} = A$.
2. $(\neg X)^{\mathcal{E}} = \neg(X^{\mathcal{E}})$.
3. For a binary connective $\circ$, $(X \circ Y)^{\mathcal{E}} = (X^{\mathcal{E}} \circ Y^{\mathcal{E}})$.
4. $(\Box X)^{\mathcal{E}} = \Box X^{\mathcal{E}}$.
5. $(\Diamond X)^{\mathcal{E}} = \Diamond X^{\mathcal{E}}$.
6. $((\forall x)\Phi)^{\mathcal{E}} = (\forall x)(\mathcal{E}(x) \supset \Phi^{\mathcal{E}})$;
7. $((\exists x)\Phi)^{\mathcal{E}} = (\exists x)(\mathcal{E}(x) \wedge \Phi^{\mathcal{E}})$.

The intuition is that the relativized quantifiers are restricted by a predicate intended to mean: actually exists. Now, the following says to what extent this stratagem succeeds.

PROPOSITION 4.8.2. Let Φ be a sentence not containing the symbol $\mathcal{E}$. Then: Φ is valid in every varying domain model if and only if $\Phi^{\mathcal{E}}$ is valid in every constant domain model.

Proof The proof of this Proposition amounts to applying formally the motivating ideas, and doing so in a straightforward way.

Part I. Suppose Φ is not valid in some varying domain model $\mathcal{M} = \langle \mathcal{G}, \mathcal{R}, \mathcal{D}, \mathcal{I} \rangle$. We construct a constant domain model in which $\Phi^{\mathcal{E}}$ is not valid. This will establish half the Proposition.

Let $\mathcal{M}'$ be the constant domain model $\langle \mathcal{G}', \mathcal{R}', \mathcal{D}', \mathcal{I}' \rangle$ constructed as follows. $\mathcal{G}'$ and $\mathcal{R}'$ are the same as $\mathcal{G}$ and $\mathcal{R}$ respectively. $\mathcal{D}'$ is the domain of the original model $\mathcal{M}$. $\mathcal{I}'$ is like $\mathcal{I}$ on all relation symbols except $\mathcal{E}$, and for that, $\mathcal{I}'(\mathcal{E}, \Gamma) = \mathcal{D}(\Gamma)$. Note that since the varying domain model $\mathcal{M}$ and the constant domain model $\mathcal{M}'$ have the same model domains, any valuation in one model is a valuation in both.

Now to complete Part I it is enough to show the following. For any formula X not containing $\mathcal{E}$ (possibly with free variables, though), for any world $\Gamma \in \mathcal{G}$, and for any valuation v,

$$\mathcal{M}, \Gamma \Vdash_v X \Longleftrightarrow \mathcal{M}', \Gamma \Vdash_v X^{\mathcal{E}}.$$

This is proved by induction on the complexity of X. The base case, where X is atomic, is directly by the construction of $\mathcal{M}'$—in particular the definition of $\mathcal{I}'$. The various induction cases are straightforward. We give the existential quantifier case only. Thus, suppose X is $(\exists x)Y$, and the result is known for Y.

Suppose first that $\mathcal{M}, \Gamma \Vdash_v (\exists x)Y$. Then for some x-variant w of v at Γ, $\mathcal{M}, \Gamma \Vdash_w Y$, so by the induction hypothesis we have $\mathcal{M}', \Gamma \Vdash_w Y^{\mathcal{E}}$. Since w is an x-variant at Γ, $w(x) \in \mathcal{D}(\Gamma)$. Then by definition, $w(x) \in \mathcal{I}'(\mathcal{E}, \Gamma)$, so $\mathcal{M}', \Gamma \Vdash_w \mathcal{E}(x)$. Thus we have $\mathcal{M}', \Gamma \Vdash_w \mathcal{E}(x) \wedge Y^{\mathcal{E}}$, so $\mathcal{M}', \Gamma \Vdash_v (\exists x)(\mathcal{E}(x) \wedge Y^{\mathcal{E}})$.

If we assume $\mathcal{M}', \Gamma \Vdash_v (\exists x)(\mathcal{E}(x) \wedge Y^{\mathcal{E}})$, it follows by a similar argument that $\mathcal{M}, \Gamma \Vdash_v (\exists x)Y$, and this completes the existential quantifier case for Part I.

Part II. Suppose $\Phi^{\mathcal{E}}$ is not valid in some constant domain model. A varying domain model in which Φ is not valid must be constructed. We leave this to you as an exercise. ∎

Essentially, this Proposition says that instead of working with varying domain models we could work with constant domain models, provided we suitably relativize quantifiers. Relativization provides us an embedding of varying domain semantics into constant domain semantics. We will see below and in Chapter 8 that there is a way of going in the other direction as well, though it is more complicated.

Since each of varying domain and constant domain semantics can simulate the other, semantical machinery does not dictate a solution to us for the problem of what quantifiers must quantify over. Whether we take quantifiers as ranging over the actually existent or over the possibly existent is, in a precise sense, just a manner of speaking. Which way to speak is, finally, a choice to be made based on what seems most natural for what one wants to say.

In this work we treat both varying and constant domain systems—both actualist and possibilist quantification. For some of what we do, it won't matter which version we choose. When the formal details differ we will say so, and specify which version is appropriate at that point.

EXERCISES

EXERCISE 4.8.1.   Let Φ be the formula

$$(\exists x)(\Box P(x) \supset \Box(\forall x)P(x))$$

where $P(x)$ is atomic. What is Φ^g?

EXERCISE 4.8.2.   Give the argument for Part II of Proposition 4.8.2.

4.9. BARCAN AND CONVERSE BARCAN FORMULAS

We have seen there is a natural embedding of varying domain modal logic into the constant domain version. Going the other way is more complex, but the route goes through interesting territory. Further, it is territory whose exploration began early in the development of quantified modal logic, and whose connections with the issues of present concern to us were not realized until some time later.

The modal operators $\Box$ and $\Diamond$ are like disguised quantifiers. To say X is necessary at a world is to say X is true at *all* accessible worlds; to say X is possible is to say X is true at *some* accessible world. The move to first-order modal logic adds the quantifiers $\forall$ and $\exists$. We know that in classical first-order logic some quantifier permutations are allowed, but others are not: $(\forall x)(\forall y)\Phi$ and $(\forall y)(\forall x)\Phi$ are equivalent; $(\exists y)(\forall x)\Phi$ implies $(\forall x)(\exists y)\Phi$; $(\forall x)(\exists y)\Phi$ does not imply $(\exists y)(\forall x)\Phi$. These are important facts. A natural question, then, is: what permutations hold between the conventional first-order quantifiers and the modal operators?

Marcus (1946) wrote a study of quantified modal logic—one of the first. Since first-order modal semantics had not yet been invented, all work at the time was axiomatic. It turned out that a particular assumption concerning quantifier/modality permutability was found to be useful. That assumption, or rather, a modernized version of it, has come to be called the *Barcan formula*. (Barcan was the name Marcus was known by at the time of writing (Marcus, 1946).) Properly speaking, it is not a formula but a scheme.

DEFINITION 4.9.1.   [Barcan Formula] All formulas of the following forms are Barcan formulas:

1. $(\forall x)\Box \Phi \supset \Box(\forall x)\Phi$;
2. $\Diamond(\exists x)\Phi \supset (\exists x)\Diamond\Phi$.

There is some (convenient) redundancy in the definition above.

$$\Diamond(\exists x)F \supset (\exists x)\Diamond F.$$

is a Barcan formula of form 2. It is equivalent to its contrapositive:

$$\neg(\exists x)\Diamond F \supset \neg\Diamond(\exists x)F.$$

But $\neg\Diamond X \equiv \Box\neg X$, and $\neg(\exists x)X \equiv (\forall x)\neg X$, so this formula in turn is equivalent to

$$(\forall x)\Box\neg F \supset \Box(\forall x)\neg F,$$

and this is a Barcan formula of form 1. In fact, if we take *either* form as basic, we get equivalent versions of the other. We simply adopt them both, for convenience.

Over the years it has become common to refer to implications that go the other way as *Converse* Barcan formulas. It was observed that, for certain natural ways of axiomatizing first-order modal logics, Converse Barcan formulas were provable, though this was not the case for Barcan formulas. Consequently Converse Barcan formulas were thought to be of lesser importance than Barcan formulas, since only assuming the truth of Barcan formulas made a difference. Eventually this point of view proved misleading, and both versions are now seen to play a significant role.

DEFINITION 4.9.2. [Converse Barcan Formula] All formulas of the following forms are Converse Barcan formulas:

1. $\Box(\forall x)\Phi \supset (\forall x)\Box\Phi$;
2. $(\exists x)\Diamond\Phi \supset \Diamond(\exists x)\Phi$.

There is the same redundancy in our definition of Converse Barcan formula that there was for Barcan formulas themselves.

The situation concerning Barcan and Converse Barcan formulas is clarified by possible world semantics. It turns out that these formulas really correspond to fundamental semantic properties of frames, and so the fact that they turned up from time to time over the years was no coincidence. We are about to present the semantical connections, but first a convenient piece of terminology. If we say the Barcan formula is valid (or valid under certain circumstances), we mean *all* Barcan formulas are; if we say the Barcan formula is not valid, we mean *at least one* Barcan formula is not. Similarly for the

Converse Barcan formula. It is customary to speak of these formulas as if they were single entities.

The first thing to observe is that neither the Barcan formula nor the Converse Barcan formula is valid generally, that is, in the family of all varying domain **K** models. We have already seen this for the Barcan formula, in Example 4.7.5, continued in Example 4.7.9. For the Converse Barcan formula, Example 4.9.3 below gives a model in which $(\exists x)\Diamond P(x) \supset \Diamond(\exists x)P(x)$ fails. We leave it to you to verify this.

EXAMPLE 4.9.3. A varying domain counterexample to

$$(\exists x)\Diamond P(x) \supset \Diamond(\exists x)P(x).$$

Now we introduce a class of frames intermediate between constant domain and varying domain. They turn up frequently in the literature, are very natural mathematically, but do not seem to have much justification from a philosophical point of view.

DEFINITION 4.9.4. [Monotonic Frame] The varying domain augmented frame $\langle \mathcal{G}, \mathcal{R}, \mathcal{D} \rangle$ is *monotonic* provided, for every $\Gamma, \Delta \in \mathcal{G}$, if $\Gamma\mathcal{R}\Delta$ then $\mathcal{D}(\Gamma) \subseteq \mathcal{D}(\Delta)$. A model is monotonic if its frame is.

We have already seen some examples. Example 4.6.4, continued in Example 4.6.11, is constant domain, hence monotonic. $(\forall x)\Diamond P(x) \supset \Diamond(\forall x)P(x)$ is not valid in it. Example 4.7.5, continued in Example 4.7.9, is varying domain and, as it happens, monotonic. It shows that $\Diamond(\exists x)P(x) \supset (\exists x)\Diamond P(x)$ is not valid in all monotonic models. However, its converse is.

EXAMPLE 4.9.5. We show $(\exists x)\Diamond P(x) \supset \Diamond(\exists x)P(x)$ is valid in all monotonic models. Let $\mathcal{M} = \langle \mathcal{G}, \mathcal{R}, \mathcal{D}, \mathcal{I} \rangle$ be a monotonic model. We show

$$\mathcal{M}, \Gamma \Vdash_v (\exists x)\Diamond P(x) \implies \mathcal{M}, \Gamma \Vdash_v \Diamond(\exists x)P(x)$$

where $\Gamma \in \mathcal{G}$ and v is a valuation in $\mathcal{D}(\mathcal{M})$.

Assume

$$\mathcal{M}, \Gamma \Vdash_v (\exists x)\Diamond P(x),$$

then for some x-variant w of v at Γ,

$$\mathcal{M}, \Gamma \Vdash_w \Diamond P(x)$$

and hence for some Δ such that $\Gamma \mathcal{R} \Delta$,

$$\mathcal{M}, \Delta \Vdash_w P(x).$$

Now, $\Gamma \mathcal{R} \Delta$ and the model is monotonic, so $\mathcal{D}(\Gamma) \subseteq \mathcal{D}(\Delta)$. w is an x-variant of v at Γ, so $w(x) \in \mathcal{D}(\Gamma)$, and hence $w(x) \in \mathcal{D}(\Delta)$. Then w is also an x-variant of v at Δ, so we have

$$\mathcal{M}, \Delta \Vdash_v (\exists x) P(x)$$

and hence

$$\mathcal{M}, \Gamma \Vdash_v \Diamond(\exists x) P(x).$$

It turns out this Example is no coincidence. There is an exact correspondence between the Converse Barcan formula and monotonicity.

PROPOSITION 4.9.6. A varying domain augmented frame is monotonic if and only if every model based on it is one in which the Converse Barcan formula is valid.

Proof One direction has already been done. If the frame is monotonic the Converse Barcan formula must be valid in every frame based on it, as was shown in Example 4.9.5 (in that example, $P(x)$ could have been *any* formula).

The other direction requires a new argument. Suppose $\langle \mathcal{G}, \mathcal{R}, \mathcal{D} \rangle$ is *not* monotonic. We produce a model based on it in which one particular Converse Barcan formula is not valid.

Since $\langle \mathcal{G}, \mathcal{R}, \mathcal{D} \rangle$ is not monotonic, there must be $\Gamma, \Delta \in \mathcal{G}$ with $\Gamma \mathcal{R} \Delta$, but $\mathcal{D}(\Gamma) \not\subseteq \mathcal{D}(\Delta)$. Say $c \in \mathcal{D}(\Gamma)$ but $c \notin \mathcal{D}(\Delta)$. Let P be a one-place relation symbol, and define an interpretation $\mathcal{I}$ as follows. $c \in \mathcal{I}(P, \Delta)$, but for any $\Omega \in \mathcal{G}$ other than Δ, $\mathcal{I}(P, \Omega)$ is empty. We claim that, in the model $\mathcal{M} = \langle \mathcal{G}, \mathcal{R}, \mathcal{D}, \mathcal{I} \rangle$,

$$\mathcal{M}, \Gamma \nVdash (\exists x) \Diamond P(x) \supset \Diamond(\exists x) P(x).$$

Let v be any valuation in $\mathcal{M}$, and let w be the x-variant of it such that $w(x) = c$. Since $c \in \mathcal{I}(P, \Delta)$,

$$\mathcal{M}, \Delta \Vdash_w P(x);$$

since $\Gamma \mathcal{R} \Delta$,

$$\mathcal{M}, \Gamma \Vdash_w \Diamond P(x);$$

and since $c \in \mathcal{D}(\Gamma)$, w is an x-variant of v at Γ, so

$$\mathcal{M}, \Gamma \Vdash_v (\exists x) \Diamond P(x).$$

On the other hand, $(\exists x) P(x)$ is not true at *any* world. More precisely, if Ω is any member of $\mathcal{G}$,

$$\mathcal{M}, \Omega \nVdash_v (\exists x) P(x).$$

For otherwise, for some x-variant v' of v at Ω, we would have

$$\mathcal{M}, \Omega \Vdash_{v'} P(x)$$

and so $v'(x) \in \mathcal{I}(P, \Omega)$. This is not the case if $\Omega \neq \Delta$ because then $\mathcal{I}(P, \Omega)$ is empty. And this is not the case if $\Omega = \Delta$ because the only member of $\mathcal{I}(P, \Delta)$ is c, so $v'(x)$ would have to be c, but $c \notin \mathcal{D}(\Delta)$, so v' would not be an x-variant of v at Δ.

Now, since $(\exists x) P(x)$ is not true at any world of the model, it is not true at any world accessible from Γ, and hence

$$\mathcal{M}, \Gamma \nVdash_v \Diamond (\exists x) P(x)$$

and this establishes our claim. ∎

Notice that while monotonicity gets us the Converse Barcan formula, it is still not enough to ensure the Barcan formula itself, as Examples 4.7.5 and 4.7.9 show.

DEFINITION 4.9.7. [Anti-Monotonicity] We call a varying domain augmented frame $\langle \mathcal{G}, \mathcal{R}, \mathcal{D} \rangle$ *anti-monotonic* provided, for $\Gamma, \Delta \in \mathcal{G}$, $\Gamma \mathcal{R} \Delta$ implies $\mathcal{D}(\Delta) \subseteq \mathcal{D}(\Gamma)$. (Note that the order of inclusion has been reversed.)

PROPOSITION 4.9.8. A varying domain augmented frame is anti-monotonic if and only if every model based on it is one in which the Barcan formula is valid.

We leave it to you to prove this, as Exercise 4.9.2.

We still have not brought constant domain frames into the picture. These, too, fit naturally, connected via a somewhat weaker notion.

DEFINITION 4.9.9. [Locally Constant Domain] Call a varying domain augmented frame $\langle \mathcal{G}, \mathcal{R}, \mathcal{D} \rangle$ *locally constant domain* provided, for $\Gamma, \Delta \in \mathcal{G}$, if $\Gamma \mathcal{R} \Delta$ then $\mathcal{D}(\Gamma) = \mathcal{D}(\Delta)$. A model is locally constant domain if its frame is.

In a true constant domain augmented frame, $\mathcal{D}(\Gamma) = \mathcal{D}(\Delta)$ for all Γ, Δ, whether or not $\Gamma \mathcal{R} \Delta$. Certainly every constant domain frame is locally so, but the converse is not true. Now, it follows from the combination of Propositions 4.9.6 and 4.9.8 that having *both* the Barcan formula and the Converse Barcan formula valid corresponds to the *locally* constant domain condition. The following puts the final piece in place.

PROPOSITION 4.9.10. A sentence X is valid in all locally constant domain models if and only if X is valid in all constant domain models.

Proof One direction is trivial. If X is valid in all locally constant domain models, among these are all constant domain models, so it is valid in all of them as well.

In the other direction, suppose X is *not* valid in some locally constant domain model $\mathcal{M} = \langle \mathcal{G}, \mathcal{R}, \mathcal{D}, \mathcal{I} \rangle$; say it is not true at Γ. We will produce a constant domain model in which X is also not valid.

Let us say there is a *path* in the model $\mathcal{M}$ from Δ_1 to Δ_2 if there is a sequence of worlds, starting with Δ_1, finishing with Δ_2, with each world in the sequence accessible from the one before it. And let us say a world Δ_2 is *relevant to* Δ_1 if it is Δ_1 itself, or if there is a path from Δ_1 to Δ_2.

If we are evaluating the truth or falsity of a sentence Z at a possible world Δ, we generally must consider the truth or falsity of various subformulas of Z at certain other worlds, but it is easy to see that *all those worlds will be relevant to* Δ. Whatever happens at worlds not relevant to Δ can have no effect at Δ itself.

Now, recall that formula X was not true at $\Gamma \in \mathcal{G}$. Let $\mathcal{G}'$ consist of all members of $\mathcal{G}$ *that are relevant to* Γ. Let $\mathcal{R}'$ be $\mathcal{R}$ restricted to members of $\mathcal{G}'$, and similarly for $\mathcal{D}'$ and $\mathcal{I}'$. This gives us a new model, $\mathcal{M}'$, which is a *submodel* of $\mathcal{M}$. In it the truth or falsity of a formula at Γ is the same as in the original model, since in either model only worlds that are relevant to Γ need be considered when evaluating truth of a formula at Γ. Consequently X is false at Γ in the new model $\mathcal{M}'$.

Finally, $\mathcal{M}'$ must be constant domain. For, suppose $\Delta \in \mathcal{G}'$. If Δ is not Γ itself, there must be a path from Γ to Δ, and the locally constant domain condition ensures that all worlds along a path will have the same domain, hence all worlds in $\mathcal{M}'$ have the same domain as Γ. ∎

Remarkably, as we have seen, simple conditions on frames correspond exactly to the Barcan and the Converse Barcan formulas. Thus these formulas have an importance that goes considerably beyond the technical issue of quantifier permutation. They really say something about the existence assumptions we are making in our semantics. The Converse Barcan formula says that, as we move to an alternative situation, nothing passes out of existence. The Barcan formula says that, under the same circumstances, nothing comes into existence. The two together say the same things exist no matter what the situation.

Prior (1957) had raised doubts about the Barcan formula. On the temporal reading, it says that if everything that now exists will at all future times be φ, then at all future times everything that then exists will be φ. But this holds generally only if no new things come into existence, and things are always coming into existence. It is important to see that Prior's quantifier is actually relativized to *things that now exist* and *things that then exist*. Relativising the possibilist quantifier turns it into the actualist quantifier. Where there is no relativization, as when we take the quantifier to range over all objects that have been, are, or will be, the Barcan formula will hold.

The very same distinction is operating in the following seeming counterexample to the claim that the Barcan formula holds in every constant domain model. The example is from epistemic logic. Even if

(4.24)   Everything is known to be F.

is true, nonetheless, it does not follow that

(4.25)   It is known that everything is F

is also true. The reason is, we might not know that we have everything. We might be able to prove of each number that it is F without being able to prove that every number is F. In fact the quantifier is tacitly relativized in (4.24) to what is known to exist, and this need not be the same as what exists in a world compatible with everything that is known.

In Section 4.8 we saw how to embed varying domain modal logic into the constant domain version. We now have the possibility of going the other way. We briefly sketch how this can be done. We do not give formal details, since it is not an issue that is fundamental here.

We introduced a somewhat complicated notion of logical consequence for propositional modal logic in Section 1.9. That notion extends naturally to a first-order version. Recall, there were really two kinds of deduction assumption: local and global. Based on what was established above, it is not hard to establish that a sentence X is valid in all constant domain models if and only if X is a consequence of the Barcan and the Converse Barcan formulas as *global* assumptions, in varying domain logic. This allows us to turn questions of validity for constant domain logic into corresponding questions about varying domain logic. Unfortunately, things are not as simple as we would like. Both the Barcan and the Converse Barcan formulas are really *schemes*, with infinitely many instances. For a given sentence X, which instances should we try working with? Fortunately, we will see that once equality has been introduced, both families of formulas collapse to *single* instances (Section 8.8. This fact, finally, will give us useful embedding machinery from constant to varying domain logic.

EXERCISES

EXERCISE 4.9.1.   Which of the sentences of Exercise 4.6.1 are valid in all *monotonic domain* models and which are not? Similarly for *anti-monotonic domain* models.

EXERCISE 4.9.2.   Prove Proposition 4.9.8.

EXERCISE 4.9.3.   Show the Converse Barcan formula need not be valid in a model whose frame is anti-monotonic.

EXERCISE 4.9.4.   Here are four formula schemes:

1. $(\exists x)\Box P(x) \supset \Box(\exists x)P(x)$
2. $\Box(\exists x)P(x) \supset (\exists x)\Box P(x)$
3. $(\forall x)\Diamond P(x) \supset \Diamond(\forall x)P(x)$
4. $\Diamond(\forall x)P(x) \supset (\forall x)\Diamond P(x)$

Just as we gave two versions of the Barcan formula, and observed they came in pairs, the same is true for the schemes above. Determine which pairs of schemes constitute equivalent assumptions.

EXERCISE 4.9.5.   For the formula schemes in Exercise 4.9.4, determine the status of their validity assuming: constant domains; varying domains; monotonic domains; anti-monotonic domains.

FIRST-ORDER TABLEAUS

In Section 2.2 we gave prefixed tableau rules for several propositional modal logics. Now we extend these to deal with quantifiers. But we have considered two versions of quantifier semantics: varying domain and constant domain. Not surprisingly, these correspond to different versions of tableau rules for quantifiers. Also not surprisingly, the rules corresponding to constant domain semantics are somewhat simpler, so we will start with them.

The basic idea behind all quantifier rules is a simple one, and is well-known for classical logic—see (Smullyan, 1968; Fitting, 1996a) for extensive treatments. If, in some classical model, $(\forall x) P(x)$ is true, then $P(x)$ is true no matter what x stands for in the domain of the model, so we should be able to replace x with anything. On the other hand, if $(\exists x) P(x)$ is true, some value of x in the domain of the model makes $P(x)$ true. Following standard mathematical procedure, we can introduce a *name* for an object making x true, say c, and conclude that $P(c)$ is true. But of course, c should be a name that has no prior commitments—we should not have used it for anything previously. We guarantee this "newness" by two devices. First, we create a completely fresh list of free variables, just for the purpose of instantiating existential quantifiers—these new free variables are called *parameters*. Second, whenever we introduce a parameter into a tableau proof, we require that it has not appeared previously in that particular proof (or at least, in that part of the proof that concerns us).

Adapting this treatment of quantifiers to modal logic is rather straight-forward. The only complications come from whether we want a single list of parameters (appropriate for constant domain semantics), or multiple lists (appropriate for varying domain semantics). There are no surprises in the details.

5.1. CONSTANT DOMAIN TABLEAUS

In constant domain models, quantifiers range over the same domain no matter what possible world we consider. For tableaus, this means we want a *single* list of additional free variables.

DEFINITION 5.1.1. [Parameters] In addition to the list (see Section 4.1) of free variables $v_1, v_2, v_3, \ldots$, we now add a second list $w_1, w_2, w_3, \ldots$, of new free variables. These are called *parameters*. They may appear in formulas in the same way as the original list of free variables *but we never quantify them.*

To make proofs easier to read, we use x, y, z, and other letters from the end of the alphabet to denote the free variables we have been using up to this point (which can appear quantified), and p, q, r, and other letters from the middle region of the alphabet to denote parameters (which cannot appear quantified). Only free variables from the original list will be allowed to occur in the formulas we are attempting to prove, while parameters may be introduced in the course of tableau construction. In effect, this is enforced by two conditions: we will only give tableau proofs of *sentences*; and parameters cannot be bound, and so cannot appear in sentences.

In the following quantifier rules we use the notational convention that $\Phi(x)$ is a formula with some (possibly no) occurrences of the free variable x, and $\Phi(y)$ is the result of replacing *all free occurrences* of x with occurrences of y. Now, here are the rules that are to be added to the earlier propositional ones, to produce a tableau system for *constant domain* modal logics. (Which modal logic depends on which propositional rules are used.)

DEFINITION 5.1.2. [Universal Rules—Constant Domain] In the following, p is any parameter whatsoever.

$$\frac{\sigma\ (\forall x)\Phi(x)}{\sigma\ \Phi(p)} \qquad \frac{\sigma\ \neg(\exists x)\Phi(x)}{\sigma\ \neg\Phi(p)}$$

DEFINITION 5.1.3. [Existential Rules—Constant Domain] In the following, p is a parameter that is *new to the tableau branch.*

$$\frac{\sigma\ (\exists x)\Phi(x)}{\sigma\ \Phi(p)} \qquad \frac{\sigma\ \neg(\forall x)\Phi(x)}{\sigma\ \neg\Phi(p)}$$

EXAMPLE 5.1.4. Here is a proof, using the constant domain tableau rules, of a Barcan formula. We use the propositional rules for **K**.

1	$\neg[\Diamond(\exists x)\Phi(x) \supset (\exists x)\Diamond\Phi(x)]$ 1.
1	$\Diamond(\exists x)\Phi(x)$ 2.
1	$\neg(\exists x)\Diamond\Phi(x)$ 3.
1.1	$(\exists x)\Phi(x)$ 4.
1.1	$\Phi(p)$ 5.
1	$\neg\Diamond\Phi(p)$ 6.
1.1	$\neg\Phi(p)$ 7.

In this, 2 and 3 are from 1 by a Conjunctive Rule; 4 is from 2 by a Possibility Rule; 5 is from 4 by an Existential Rule (note that the parameter p is new at this point); 6 is from 3 by a Universal Rule (p is not new now, but for a Universal Rule, anything is allowed); and 7 is from 6 by a Necessity Rule.

EXAMPLE 5.1.5. This time we show a constant domain non-proof, of

$$[\Diamond(\exists x)P(x) \wedge \Diamond(\exists x)Q(x)] \supset (\exists x)\Diamond[P(x) \wedge Q(x)].$$

Once again the propositional rules are those of **K**. The proof is displayed in Figure 5.

Here 2 and 3 are from 1 by a Conjunctive Rule, as are 4 and 5 from 2; 6 is from 4 and 7 is from 5 by a Possibility Rule; 8 is from 6 and 9 is from 7 by the Existential Rule (notice, both parameters are new at the point of introduction); 10 and 11 are from 3 by the Universal Rule; 12 and 13 are from 10, and 14 and 15 are from 11 by the Necessity Rule; 16 and 17 are from 12 by a Disjunctive Rule; as are 18 and 19 from 15; 20 and 21 from 13; 22 and 23 from 14; and 24 and 25 from 14.

The branches ending with 16 and 19 are closed, but the other branches are not. Note that this does not establish the sentence is unprovable, since we could have tried things in a different order, or tried different things altogether.

EXERCISES

EXERCISE 5.1.1. Show the interdefinability of quantifiers can be proven using tableaus. That is, prove the following (use the propositional **K** rules):

1. $(\forall x)\Phi \supset \neg(\exists x)\neg\Phi$;
2. $\neg(\exists x)\neg\Phi \supset (\forall x)\Phi$;
3. $(\exists x)\Phi \supset \neg(\forall x)\neg\Phi$;
4. $\neg(\forall x)\neg\Phi \supset (\exists x)\Phi$.

EXERCISE 5.1.2. In Exercise 4.6.1 you were asked to determine **K** validity of various sentences. Now determine which of them have tableau proofs using the propositional **K** rules and the constant domain quantifier rules.

5.2. VARYING DOMAIN TABLEAUS

For tableaus corresponding to varying domain semantics there is only one change needed: we introduce a whole family of lists of parameters, one for each prefix, rather than a single list. More specifically, we assume that to each

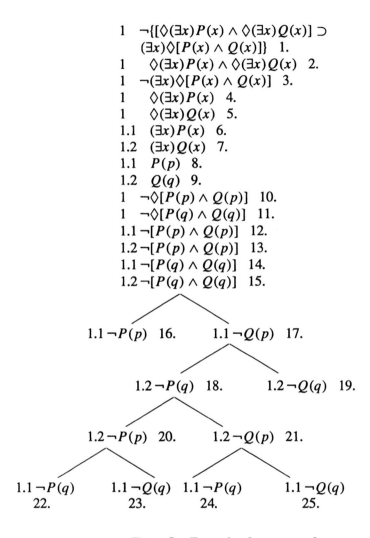

Figure 5.   Example of a non-proof

prefix σ there is associated an infinite list of free variables, called parameters, in such a way that different prefixes never have the same parameter associated with them.

Informally, we write p_σ to indicate that p is a parameter associated with the prefix σ. Now we give the varying domain quantifier rules.

DEFINITION 5.2.1. [Universal Rules—Varying Domain] In the following,

p_σ can be any parameter that is associated with the prefix σ.

$$\frac{\sigma\ (\forall x)\Phi(x)}{\sigma\ \Phi(p_\sigma)} \qquad \frac{\sigma\ \neg(\exists x)\Phi(x)}{\sigma\ \neg\Phi(p_\sigma)}$$

DEFINITION 5.2.2. [Existential Rules—Varying Domain] In the following, p_σ is a parameter associated with the prefix σ, subject to the condition that p_σ is new to the tableau branch.

$$\frac{\sigma\ (\exists x)\Phi(x)}{\sigma\ \Phi(p_\sigma)} \qquad \frac{\sigma\ \neg(\forall x)\Phi(x)}{\sigma\ \neg\Phi(p_\sigma)}$$

EXAMPLE 5.2.3.　We give a varying domain **K** tableau proof of

$$[\lozenge(\exists x)\Box A(x) \land \Box(\forall x)\lozenge B(x)] \supset \lozenge(\exists x)\lozenge[A(x) \land B(x)].$$

1	$\neg\{[\lozenge(\exists x)\Box A(x) \land \Box(\forall x)\lozenge B(x)] \supset$
	$\lozenge(\exists x)\lozenge[A(x) \land B(x)]\}$ 1.
1	$\lozenge(\exists x)\Box A(x) \land \Box(\forall x)\lozenge B(x)$ 2.
1	$\neg\lozenge(\exists x)\lozenge[A(x) \land B(x)]$ 3.
1	$\lozenge(\exists x)\Box A(x)$ 4.
1	$\Box(\forall x)\lozenge B(x)$ 5.
1.1	$(\exists x)\Box A(x)$ 6.
1.1	$(\forall x)\lozenge B(x)$ 7.
1.1	$\neg(\exists x)\lozenge[A(x) \land B(x)]$ 8.
1.1	$\Box A(p_{1.1})$ 9.
1.1	$\lozenge B(p_{1.1})$ 10.
1.1	$\neg\lozenge[A(p_{1.1}) \land B(p_{1.1})]$ 11.
1.1.1	$B(p_{1.1})$ 12.
1.1.1	$A(p_{1.1})$ 13.
1.1.1	$\neg[A(p_{1.1}) \land B(p_{1.1})]$ 14.

$$1.1.1\ \neg A(p_{1.1})\ \ 15. \qquad 1.1.1\ \neg B(p_{1.1})\ \ 16.$$

Here 2 and 3 are from 1 by a Conjunctive Rule; as are 4 and 5 from 2; 6 is from 4 by a Possibility Rule; 7 is from 5 and 8 from 3 by a Necessity Rule; 9 is from 6 by an Existential Rule; 10 is from 7 and 11 is from 8 by a Universal Rule; 12 is from 10 by a Possibility Rule; 13 is from 9 and 14 is from 11 by a Necessity Rule; and finally 15 and 16 are from 14 by a Disjunctive Rule.

And we also give an example of a non-proof—of a Barcan formula.

EXAMPLE 5.2.4. The following is a failed varying domain **K** attempt to prove the Barcan formula: $(\forall x)\Box A(x) \supset \Box(\forall x)A(x)$.

$$
\begin{array}{ll}
1 & \neg[(\forall x)\Box A(x) \supset \Box(\forall x)A(x)] \quad 1. \\
1 & (\forall x)\Box A(x) \quad 2. \\
1 & \neg\Box(\forall x)A(x) \quad 3. \\
1.1 & \neg(\forall x)A(x) \quad 4. \\
1.1 & \neg A(p_{1.1}) \quad 5. \\
1 & \Box A(q_1) \quad 6. \\
1.1 & A(q_1) \quad 7.
\end{array}
$$

In this, 2 and 3 are from 1 by a Conjunctive Rule; 4 is from 3 by a Possibility Rule; 5 is from 4 by an Existential Rule (notice that the parameter has a subscript of 1.1, the same as the prefix); 6 is from 2 by a Universal Rule (notice that we can *not* use 1.1 as a subscript, since the prefix involved is 1); 7 is from 6 by a Necessity Rule. There is no closure.

We will return to this example again.

EXERCISES

EXERCISE 5.2.1.   Attempt to prove a Converse Barcan formula using the varying domain tableau rules. Explain why you cannot.

5.3. TABLEAUS STILL WORK

It is time to show our tableau rules exactly capture the quantifier behavior specified by first-order modal models. That is, we must prove *soundness* and *completeness* of the various tableau systems. This continues what we did in the propositional case, in Section 2.5. We present proofs for the varying domain tableau rules. The constant domain rules have a similar soundness and completeness proof, though the details are slightly simpler.

Soundness

We show soundness of the **K** system using the varying domain quantifier rules. Recall, soundness means: anything provable is valid. Although tableaus only prove *sentences*, formulas that contain parameters turn up in proofs so we must take them into account. We begin with a definition that should be compared with Definition 2.5.1.

DEFINITION 5.3.1. [Satisfiable] Suppose S is a set of prefixed formulas (where members of S may contain parameters). We say S is *satisfiable* in the varying domain model $\mathcal{M} = \langle \mathcal{G}, \mathcal{R}, \mathcal{D}, \mathcal{I} \rangle$, with respect to valuation v, if there is a way, call it θ, of assigning to each prefix σ that occurs in S some possible world $\theta(\sigma)$ in $\mathcal{G}$, such that:

1. If σ and $\sigma.n$ both occur as prefixes in S, $\theta(\sigma.n)$ is a world that is accessible from $\theta(\sigma)$, that is, $\theta(\sigma)\mathcal{R}\theta(\sigma.n)$.
2. If the parameter p_σ occurs in S, $v(p_\sigma) \in \mathcal{D}(\theta(\sigma))$.
3. If $\sigma\, X$ is in S, then $\mathcal{M}, \theta(\sigma) \Vdash_v X$.

The rest of Definition 2.5.1 is carried over directly: a tableau branch is satisfiable if the set of prefixed formulas on it is satisfiable (in some model, with respect to some valuation); a tableau is satisfiable if some branch of it is satisfiable.

Proposition 2.5.2 said a propositional tableau that was closed could not be satisfiable. This carries over to the first-order case directly, with the same proof.

Next we need a version of Proposition 2.5.3 that takes quantifiers into account. The wording is exactly the same, and the proof builds on that of the earlier proposition.

PROPOSITION 5.3.2. If a varying domain tableau branch extension rule is applied to a satisfiable tableau, the result is another satisfiable tableau.

Proof The argument has several cases, depending on what branch extension rule was applied. All the propositional and modal cases are treated exactly as in the proof of Proposition 2.5.3; we only need to add the quantifier cases, and there are just two of these.

Assume the set of prefixed formulas on the branch $\mathcal{B}$ of tableau $\mathcal{T}$ is satisfiable in the varying domain model $\mathcal{M} = \langle \mathcal{G}, \mathcal{R}, \mathcal{D}, \mathcal{I} \rangle$ with respect to the valuation v, using the mapping θ. And now suppose a varying domain quantifier rule is applied on $\mathcal{B}$.

EXISTENTIAL CASE Say $\sigma\, (\exists x)\Phi(x)$ occurs on $\mathcal{B}$, and we add $\sigma\, \Phi(p_\sigma)$ to the end of $\mathcal{B}$, where p_σ is a parameter that is new to the branch. We must show the extended branch, call it $\mathcal{B}'$, is still satisfiable. Note that $\mathcal{B}'$ consists of the prefixed formulas of $\mathcal{B}$, together with $\sigma\, \Phi(p_\sigma)$.

By our assumptions, $\mathcal{M}, \theta(\sigma) \Vdash_v (\exists x)\Phi(x)$, and so for some x-variant w of v at $\theta(\sigma)$, $\mathcal{M}, \theta(\sigma) \Vdash_w \Phi(x)$. Now, define a new valuation v' as follows. For each free variable y (including parameters),

$$v'(y) = \begin{cases} v(y) & \text{if } y \neq p_\sigma \\ w(x) & \text{if } y = p_\sigma \end{cases}$$

That is, v' is the same as v on all variables except p_σ, and $v'(p_\sigma) = w(x)$. Note that since w is an x-variant of v at $\theta(\sigma)$, then $w(x)$, and hence $v'(p_\sigma)$, is in $\mathcal{D}(\theta(\sigma))$; thus v' meets condition 2 of the definition of satisfiability.

According to Proposition 4.6.8 (which carries over to varying domain models), if two valuations agree on all the free variables of X, then at each possible world X behaves the same way with respect to each valuation. Now, v' and v agree on all variables except p_σ, and p_σ is new to the branch $\mathcal{B}$, and so does not occur in any prefixed formula on $\mathcal{B}$. Then, if $\sigma_i \, X_i$ occurs on $\mathcal{B}$,

$$\mathcal{M}, \theta(\sigma_i) \Vdash_v X_i \iff \mathcal{M}, \theta(\sigma_i) \Vdash_{v'} X_i.$$

Since the set of prefixed formulas on $\mathcal{B}$ was satisfiable in $\mathcal{M}$ with respect to the valuation v using the mapping θ, this is also the case if we use the valuation v' instead of v. We will be finished if we show $\mathcal{M}, \theta(\sigma) \Vdash_{v'} \Phi(p_\sigma)$, for then all the prefixed formulas on the extended branch $\mathcal{B}'$ will be satisfied with respect to v', using the mapping θ.

We intend to make use of Proposition 4.6.9 (which also carries over to varying domain models), so we begin setting things up for it. Since w is an x-variant of v, the two agree on all variables except for x. By definition, v' and v agree on all variables except for p_σ. It follows that v' and w agree on all variables except for x and p_σ. Also, p_σ does not occur in $\Phi(x)$, since p_σ is new to the branch $\mathcal{B}$, and $\Phi(x)$ occurred on the branch, as part of $\sigma \, (\exists x)\Phi(x)$. Thus v' and w agree on all the free variables of $\Phi(x)$, except for x. And, by definition, $v'(p_\sigma) = w(x)$. Now, by Proposition 4.6.9,

$$\mathcal{M}, \theta(\sigma) \Vdash_{v'} \Phi(p_\sigma) \iff \mathcal{M}, \theta(\sigma) \Vdash_w \Phi(x)$$

Since we know $\mathcal{M}, \theta(\sigma) \Vdash_w \Phi(x)$, we have $\mathcal{M}, \theta(\sigma) \Vdash_{v'} \Phi(p_\sigma)$, and so we are done with the existential case.

UNIVERSAL CASE This is an easier argument, and we leave it as an exercise.

■

With this key Proposition established, tableau soundness follows immediately, using precisely the same argument we gave in the propositional case—see Theorem 2.5.4.

Completeness

Once again our arguments follow the pattern that began with propositional modal logic. The idea is to extract a countermodel from a failed tableau proof attempt. Before giving the general argument, let us show how the ideas apply in a particular case.

EXAMPLE 5.3.3. In Example 5.2.4 we presented an attempt at a varying domain **K** proof of a Barcan formula, $(\forall x)\Box A(x) \supset \Box(\forall x)A(x)$. For convenience, we repeat the tableau here.

$$
\begin{array}{ll}
1 & \neg[(\forall x)\Box A(x) \supset \Box(\forall x)A(x)] \quad 1. \\
1 & (\forall x)\Box A(x) \quad 2. \\
1 & \neg\Box(\forall x)A(x) \quad 3. \\
1.1 & \neg(\forall x)A(x) \quad 4. \\
1.1 & \neg A(p_{1.1}) \quad 5. \\
1 & \Box A(q_1) \quad 6. \\
1.1 & A(q_1) \quad 7.
\end{array}
$$

Obviously the tableau is not closed. Let us use it to construct a countermodel to the Barcan formula. First the possible worlds are, simply, the prefixes that occur on the tableau branch: $\mathcal{G} = \{1, 1.1\}$. The accessibility relation is the one corresponding to the prefix structure: $1\mathcal{R}1.1$, and $\mathcal{R}$ does not hold otherwise. The domains associated with the possible worlds are just the parameters we have introduced, placed according to subscript: $\mathcal{D}(1) = \{q_1\}$ and $\mathcal{D}(1.1) = \{p_{1.1}\}$. Finally, we interpret the relation symbol A the way the branch "says" we should: $\mathcal{I}(A, 1.1) = \{q_1\}$ (because of 7), and $\mathcal{I}(A, 1) = \emptyset$. We have created a model $\mathcal{M} = \langle \mathcal{G}, \mathcal{R}, \mathcal{D}, \mathcal{I} \rangle$ with two possible worlds. Here is the model displayed schematically.

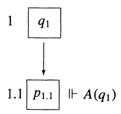

Now let v_0 be a valuation that assigns, to each parameter, itself: $v_0(r) = r$ for each parameter r (what v_0 does on other free variables won't matter to the argument, and can be left unspecified). And let θ be the prefix mapping that maps each prefix on the branch to itself: $\theta(\sigma) = \sigma$ for each prefix σ. We will show the entire branch is satisfied in $\mathcal{M}$ with respect to the valuation v_0, using the mapping θ. Once this has been shown for the entire branch, because of line 1 we conclude that

$$\mathcal{M}, 1 \Vdash_{v_0} \neg[(\forall x)\Box A(x) \supset \Box(\forall x)A(x)]$$

and hence $(\forall x)\Box A(x) \supset \Box(\forall x)A(x)$ is not valid since there is a varying domain model in which it can be falsified.

Now for the details of showing satisfiability for the branch, which we do from bottom up.

7. $\mathcal{M}, 1.1 \Vdash_{v_0} A(q_1)$ because $v_0(q_1) = q_1$ and q_1 is in $\mathcal{I}(A, 1.1)$.

6. Since 1.1 is the only world accessible from world 1, and we have $\mathcal{M}, 1.1 \Vdash_{v_0} A(q_1)$, we also have $\mathcal{M}, 1 \Vdash_{v_0} \Box A(q_1)$.

5. $\mathcal{M}, 1.1 \not\Vdash_{v_0} A(p_{1.1})$ since $v_0(p_{1.1}) = p_{1.1}$ and $p_{1.1}$ is not in $\mathcal{I}(A, 1.1)$. Thus $\mathcal{M}, 1.1 \Vdash_{v_0} \neg A(p_{1.1})$.

4. Suppose we had $\mathcal{M}, 1.1 \Vdash_{v_0} (\forall x)A(x)$. Let w the x-variant of v_0 such that $w(x) = p_{1.1}$. Then we must have $\mathcal{M}, 1.1 \Vdash_w A(x)$. Note that $w(x) = v_0(p_{1.1})$ since v_0 maps each parameter to itself. But then using Proposition 4.6.9 (varying domain version) we should also have $\mathcal{M}, 1.1 \Vdash_{v_0} A(p_{1.1})$ and by 5 we do not. Thus $\mathcal{M}, 1.1 \Vdash_{v_0} \neg(\forall x)A(x)$.

3. If we had $\mathcal{M}, 1 \Vdash_{v_0} \Box(\forall x)A(x)$ we would also have $\mathcal{M}, 1.1 \Vdash_{v_0} (\forall x)A(x)$ and by 4 we do not. Thus we have $\mathcal{M}, 1 \Vdash_{v_0} \neg\Box(\forall x)A(x)$.

2. We want $\mathcal{M}, 1 \Vdash_{v_0} (\forall x)\Box A(x)$. The domain $\mathcal{D}(1)$ has only q_1 as a member, so if we let w be the x-variant of v_0 such that $w(x) = q_1$, it is enough to verify that we have $\mathcal{M}, 1 \Vdash_w \Box A(x)$. Since $v_0(q_1) = q_1 = w(x)$, by Proposition 4.6.9 again, we will have what we need provided we have $\mathcal{M}, 1 \Vdash_{v_0} \Box A(q_1)$, and we have this by 6.

1. $\mathcal{M}, 1 \Vdash_{v_0} \neg[(\forall x)\Box A(x) \supset \Box(\forall x)A(x)]$, by 2 and 3.

Now it is time to prove the completeness of first-order tableaus. We want to show that what we did in the example above can always be done. We sketch the details for the varying domain version of **K**; other modal logics can be dealt with similarly.

Suppose Φ is a sentence that has no varying domain tableau proof. We intend to show it is not valid, and to do this we extract a countermodel from a failed attempt at proving Φ. However, not just any failed attempt will do— we might have done something stupid and missed finding a proof that was there all along. Instead, we must be sure we have tried everything that could be tried. We begin by describing a particular *systematic* tableau construction procedure, then we characterize what sort of a tableau it actually constructs.

To begin, there is an infinite list of parameters associated with each prefix, and there are infinitely many prefixes possible. But we have a *countable* alphabet, and this implies that it is possible to combine all parameters, no matter what prefix is involved, into a single list: ρ_1, ρ_2, ρ_3, ... (The subscripts are not the associated prefixes here; they just mark the position of the parameter in the list.) Thus, for each prefix, and for each parameter associated with that prefix, the parameter occurs in the list somewhere. (A proof of this involves some simple set theory, and would take us too far afield here. Take our word for it—this can be done.)

Systematic Construction Procedure Here is a systematic tableau construction procedure for varying domain **K**. It is readily adapted to other logics. It is not the only one possible, but we only need one. It goes in *stages*. For stage 1, we simply put down $1 \neg \Phi$, getting a one-branch, one-formula tableau. (Recall, Φ is the sentence that we are assuming has no tableau proof.)

Having completed stage n, if the tableau has not closed, here is stage $n+1$. Start with the leftmost open branch, and for each prefixed formula F currently on it (going from top to bottom, say) do the following.

1. If F is $\sigma \neg\neg X$, add σX to the end of each open branch on which this occurrence of $\sigma \neg\neg X$ lies, provided σX does not already occur.
2. If F is $\sigma X \wedge Y$, add σX and σY to the end of each open branch on which this occurrence of $\sigma X \wedge Y$ lies, provided they do not already occur. Similarly for the other conjunctive cases.
3. If F is $\sigma X \vee Y$, for each open branch on which this occurrence of $\sigma X \vee Y$ lies and on which neither σX nor σY occurs, split the end of the branch to σX, σY. Similarly for the other disjunctive cases.
4. If F is $\sigma \Diamond X$, for each open branch on which this occurrence of $\sigma \Diamond X$ lies and on which for no n do we have $\sigma.n\ X$, choose the *smallest* integer k such that $\sigma.k$ does not appear on the branch, and add $\sigma.k\ X$ to the branch end. Similarly for the other possibility case.
5. If F is $\sigma \Box X$, for each open branch on which this occurrence of $\sigma \Box X$ lies, add to the end of the branch every prefixed formula of the form $\sigma.n\ X$ where the prefixed formula itself does not already occur on the branch, but $\sigma.n$ does occur as a prefix somewhere on the branch. Similarly for the other necessity case.
6. If F is $\sigma (\exists x)\Psi(x)$, for each open branch on which this occurrence of $\sigma (\exists x)\Psi(x)$ lies, if $\sigma \Psi(\rho)$ does not occur on the branch for any parameter ρ associated with the prefix σ, then choose the *first* parameter ρ_i associated with σ in the list ρ_1, ρ_2, ... , and add $\sigma \Psi(\rho_i)$ to the branch end. Similarly for the other existential case.

7. If F is $\sigma\,(\forall x)\Psi(x)$, for each open branch on which this occurrence of $\sigma\,(\forall x)\Psi(x)$ lies, add to the end of the branch *every* prefixed formula of the form $\sigma\,\Psi(\rho_i)$ where: ρ_i is a parameter associated with prefix σ; $\sigma\,\Psi(\rho_i)$ does not already occur on the branch; and ρ_i occurs among the first n items of our parameter list, that is, ρ_i is one of $\rho_1, \rho_2, \ldots, \rho_n$. Similarly for the other universal case.

After we have done all this with the leftmost open branch, do the same for the second from the left, then the next, and so on to the rightmost open branch. This completes stage $n + 1$.

We have now completed the description of our systematic tableau construction procedure. It is complicated, and we do not expect people to really construct tableaus this way since there is no place for intuition or insight. But it does guarantee that everything that could be tried will sooner or later be tried. That is all we care about now. (Computer scientists call an algorithm of this sort *fair*.)

If this systematic construction procedure produced a closed tableau, it would be a proof of Φ, and we are assuming there is no such proof. But the procedure could still terminate because it ran out of things to do, since the rules only add prefixed formulas to branches if they are not already present. If the construction does terminate, we can choose an open branch $\mathcal{B}$. But what if the construction does not terminate and we find ourselves constructing an infinite tableau? In this case, how do we know there must be an infinite open branch? Let's see.

In a tableau, a prefixed formula F that is not at a branch end either has a single prefixed formula immediately below it, or the tableau has a split at F, and there are two prefixed formulas immediately below it. It is customary to call the one or two prefixed formulas immediately below F the *children* of F. Children of a prefixed formula, children of children, children of children of children, and so on, are called *descendants*. Now, call an occurrence of a prefixed formula F in the tableau we are constructing *good* if the systematic procedure produces infinitely many descendants for it.

Our tableau construction procedure is an attempt to prove the sentence Φ. Suppose it never terminates. Then the prefixed formula $1\,\neg\Phi$ we begin with must be good, because every other prefixed formula in the tableau is a descendant of it, and there are infinitely many of them since the construction process does not terminate.

Suppose some prefixed formula occurrence F is good. Then the construction procedure yields infinitely many descendants for F. But F itself can have at most two children. It follows that at least one of the children of F must itself have infinitely many descendants. Consequently, if F is good, F must have a good child.

Here, then, is our infinite branch construction assuming the systematic procedure does not terminate. The initial prefixed formula, $1 \neg \Phi$, is good. The construction procedure must produce a good child for $1 \neg \Phi$, and this in turn must in turn have a good child, and so on. In this way an infinite branch (of good prefixed formula occurrences) is created.

There is an important issue that should be pointed out here. The systematic tableau construction is entirely constructive—at each stage it is fully determined what to do next. But the selection of an infinite branch is not constructive at all. A good prefixed formula occurrence must have a good child, but if there are two children, we have no way of knowing which must be good! This is because the definition of goodness is in terms of the completed behavior of the construction procedure: infinitely many descendants will be produced. At a finite stage we don't generally know which branches will terminate and which will continue, so we don't generally know which parts of the tableau will continue to grow and which will not. This point plays no role in the completeness argument, but it does mean we should not expect to get a way of *deciding* on provability or non-provability of sentences from our systematic tableau construction procedure. In fact, it can be proved that no such way of deciding is possible, by this or any other means.

Now, back to the main development. Whether the tableau construction terminates or not, since Φ is not provable there must be an open branch $\mathcal{B}$ generated, possibly finite, possibly infinite. What can we say about this branch?

Branch Conditions For an open branch $\mathcal{B}$ of a systematically constructed tableau:

1. Not both σX and $\sigma \neg X$ are on $\mathcal{B}$, because the branch is open (indeed, we will only need this for atomic X).
2. If $\sigma \neg\neg X$ is on $\mathcal{B}$, so is σX.
3. If $\sigma X \wedge Y$ is on $\mathcal{B}$, so are σX and σY, and similarly for the other conjunctive cases.
4. If $\sigma X \vee Y$ is on $\mathcal{B}$, so is one of σX or σY, and similarly for the other disjunctive cases.
5. If $\sigma \Diamond X$ is on $\mathcal{B}$, so is $\sigma.n\, X$ for some n, and similarly for the other possibility case.
6. If $\sigma \Box X$ is on $\mathcal{B}$, so is $\sigma.n\, X$ for every n such that $\sigma.n$ occurs as a prefix somewhere on $\mathcal{B}$, and similarly for the other necessity case.
7. If $\sigma (\exists x)\Psi(x)$ is on $\mathcal{B}$, so is $\sigma \Psi(p_\sigma)$ for some parameter p_σ associated with σ, and similarly for the other existential case.
8. If $\sigma (\forall x)\Psi(x)$ is on $\mathcal{B}$, so is $\sigma \Psi(p_\sigma)$ for every parameter p_σ associated with σ, and similarly for the other universal case.

We use these properties of $\mathcal{B}$ to construct a countermodel, exactly as in Example 5.3.3.

Model Construction Let $\mathcal{G}$ be the set of prefixes that occur on $\mathcal{B}$. If σ and $\sigma.n$ are both in $\mathcal{G}$, set $\sigma \mathcal{R} \sigma.n$. For each $\sigma \in \mathcal{G}$, take $\mathcal{D}(\sigma)$ to be the set of all parameters p_σ, associated with σ, that occur on $\mathcal{B}$. Finally, for each k-place relation symbol R, define $\mathcal{I}(R, \sigma)$ to be the set

$$\{\langle t_1, \ldots, t_k \rangle \mid \sigma R(t_1, \ldots, t_k) \text{ occurs on } \mathcal{B}\}.$$

We have defined a model $\mathcal{M} = \langle \mathcal{G}, \mathcal{R}, \mathcal{D}, \mathcal{I} \rangle$.

Again, as in Example 5.3.3, define a valuation v_0 by setting $v_0(r) = r$ for each parameter r and letting v_0 be arbitrary on other free variables. Also define a prefix mapping θ by $\theta(\sigma) = \sigma$ for each $\sigma \in \mathcal{G}$. We claim that, just as in the example, each prefixed formula on $\mathcal{B}$ is satisfied in $\mathcal{M}$ with respect to the valuation v_0, using the mapping θ. More precisely, we show the following.

Key Fact For v_0 and θ as above:

$$\sigma X \text{ on } \mathcal{B} \implies \mathcal{M}, \sigma \Vdash_{v_0} X$$

$$\sigma \neg X \text{ on } \mathcal{B} \implies \mathcal{M}, \sigma \nVdash_{v_0} X$$

This is shown by induction on the complexity of the formula. There are lots of cases; we only consider a few.

The base case is that of atomic formulas, and their negations. Suppose $\sigma R(t_1, \ldots, t_k)$ is on $\mathcal{B}$, where R is a k-place relation symbol. By definition, $\mathcal{I}(R, \sigma)$ contains $\langle t_1, \ldots, t_k \rangle$. Since each t_i must be a parameter, $v_0(t_i) = t_i$, it follows immediately that $\mathcal{M}, \sigma \Vdash_{v_0} R(t_1, \ldots, t_k)$.

For negations of atomic formulas, if $\sigma \neg R(t_1, \ldots, t_k)$ is on $\mathcal{B}$ then by Branch Condition 1, $\sigma R(t_1, \ldots, t_k)$ is not, so $\langle t_1, \ldots, t_k \rangle$ is not in $\mathcal{I}(R, \sigma)$, and it follows that $\mathcal{M}, \sigma \nVdash_{v_0} R(t_1, \ldots, t_k)$.

Now for the induction cases. Suppose X is not atomic, and the Key Fact is known for simpler formulas—we show it for X itself. There are several cases, corresponding to the various propositional connectives, modal operators, and quantifiers. We only consider the quantifier cases, since the others are treated exactly as they were in Section 2.5, when we proved the completeness of propositional tableaus as Theorem 2.5.7.

Suppose $\sigma (\forall x) \Psi(x)$ is on $\mathcal{B}$, and the Key Fact is known for simpler formulas. By item 8, $\sigma \Psi(p_\sigma)$ is on $\mathcal{B}$ for every parameter p_σ that is associated with the prefix σ. Since $\Psi(p_\sigma)$ is less complex than $(\forall x)\Psi(x)$ the induction hypothesis applies, and we have that $\mathcal{M}, \sigma \Vdash_{v_0} \Psi(p_\sigma)$ holds for every parameter of the form p_σ. Now, let w be any x-variant of v_0 at σ. By definition of $\mathcal{D}(\sigma)$, $w(x)$ must be some parameter p_σ. Since $\mathcal{M}, \sigma \Vdash_{v_0} \Psi(p_\sigma)$ and $w(x) = p_\sigma = v_0(p_\sigma)$, it follows from Proposition 4.6.9 that $\mathcal{M}, \sigma \Vdash_w \Psi(x)$. Finally, since w was an arbitrary x-variant of v_0 at σ, by definition, $\mathcal{M}, \sigma \Vdash_{v_0} (\forall x)\Psi(x)$.

The existential quantifier case is similar, and we leave it to you as an exercise.

THEOREM 5.3.4. [First-Order Tableau Completeness] If the sentence Φ is valid in all varying domain first-order **K** models, Φ has a tableau proof using the varying domain **K** rules.

Proof Exactly as in the propositional case. Suppose Φ has no tableau proof. Apply the systematic tableau construction procedure given above. It will not produce a closed tableau, so an open branch $\mathcal{B}$ is generated. Using the method described, we produce a model $\mathcal{M}$ in which all the formulas on the branch are satisfied (and in which the possible worlds are just prefixes). Since the branch $\mathcal{B}$ begins with $1 \neg\Phi$, at world 1 of $\mathcal{M}$ the sentence Φ will be false, hence Φ is not **K**-valid. ∎

We have now proved completeness for the varying domain version of the **K** quantifier rules. We note that we have actually proved something stronger than was stated, in two respects. First, we have shown that if a sentence is **K** valid, it must have a *systematic* tableau proof. And second, we have shown that if a sentence does not have an **K** tableau proof, there is a countermodel *in which domains of different worlds are disjoint*.

Completeness for the constant domain version and for other choices of propositional rules is similar. We leave this to you—see Exercises 5.3.3 and 5.3.4.

EXERCISES

EXERCISE 5.3.1.   Complete the proof of Proposition 5.3.2 by doing the Universal Case.

EXERCISE 5.3.2.   In verifying the Key Fact above, we only did the universal quantifier case in detail. Do the case corresponding to $(\exists x)\Phi(x)$.

EXERCISE 5.3.3.   In Section 5.1 we gave an example of a failed tableau using the constant domain rules, Example 5.1.5, in which we attempted to prove $[\Diamond(\exists x)P(x) \wedge \Diamond(\exists x)Q(x)] \supset (\exists x)\Diamond[P(x) \wedge Q(x)]$. Actually, we did not quite complete that example—we stopped short of applying disjunctive rules. Apply them, then take an unclosed branch of the resulting tableau, and use it to construct a constant domain countermodel showing the invalidity of the sentence. Use Example 5.3.3 as a guide, but remember, the result is to be *constant domain*.

EXERCISE 5.3.4. State and prove a completeness theorem for the constant domain tableau rules.

FIRST-ORDER AXIOM SYSTEMS

In order to keep this book to a reasonable size we have decided not to give an extensive treatment of subjects that can be found elsewhere and that are somewhat peripheral to our primary interests. That policy begins to have an effect now. Proving completeness of first-order axiomatizations can be quite complex—see (Garson, 1984) for a full discussion of the issues involved. Indeed, a common completeness proof that can cover constant domains, varying domains, and models meeting other conditions, does not seem available, so a thorough treatment would have to cover things separately for each version. Instead we simply present some appropriate axiomatizations, make a few pertinent remarks, and provide references. For machinery presented in later chapters we omit an axiomatic treatment altogether.

6.1. A CLASSICAL FIRST-ORDER AXIOM SYSTEM

Before bringing modal complexities into the picture, we give an axiomatization of first-order *classical* logic, so that basic quantifier machinery can be understood more easily. Even this is not as straightforward as might be expected. Before we say why, though, let us get some elementary material out of the way.

Syntactic machinery is straightforward. Instead of propositional formulas, we now use first-order ones as defined in Section 4.1. Even if we are interested in proving *sentences*, proofs of sentences will involve formulas with free variables. There is no benefit for axiom systems in using a special list of parameters, as we did with tableaus, so we will just use the same list for free and bound variables.

When developing propositional modal axiom systems, it was convenient to take $\neg$, $\supset$, and $\Box$ as primitive, and the other connectives and modal operator as defined. With quantifiers it is also convenient to take $\forall$ as primitive, and treat $\exists$ as defined: $(\exists x)\Phi$ abbreviates $\neg(\forall x)\neg\Phi$.

A little more notation and terminology, and then we can get to serious issues. If we write $\Phi(x)$, we intend that this be a formula in which the variable x may have free occurrences. If, subsequently, we write $\Phi(y)$, we mean the formula that results from $\Phi(x)$ when all free occurrences of x have been replaced by occurrences of y. Now, a first impression is that one free variable is as good as another, and so we could use $\Phi(y)$ just as well as $\Phi(x)$ provided we suitably modify our discussion of the formulas. But this is not quite so. If $\Phi(x)$ is the formula $(\forall y)R(x, y)$ then $\varphi(y)$ is the formula $(\forall y)R(y, y)$, and this is very different indeed. Our choice of $\varphi(x)$ contained one free occurrence of x, but the result of substitution, $\varphi(y)$, contains no free variables at all. We avoided such problems when working with tableaus because we introduced a new set of free variables, *parameters*, and these could not be bound. Here we must guard against problems like this more directly.

DEFINITION 6.1.1. [Substitutability] We say a free variable y is *substitutable* for x in $\varphi(x)$ provided no free occurrence of x in $\varphi(x)$ is in a subformula beginning with $(\forall y)$. (This is often described as: no free occurrence of x in $\varphi(x)$ is within the scope of $(\forall y)$.)

If y is substitutable for x in $\varphi(x)$, then indeed $\varphi(x)$ and $\varphi(y)$ say essentially the same things except that in discussing $\varphi(y)$ we must say "y" where in discussing $\varphi(x)$ we would say "x."

Now, let us turn to the serious issues. In most axiomatizations of classical logic, formulas of the form $(\forall x)\varphi(x) \supset \varphi(y)$ are taken as axioms, provided y is substitutable for x in $\varphi(x)$. *Such formulas are not valid in the varying domain modal semantics of Chapter 4.* The reasons are quite straightforward. Valuations in modal models assign to free variables members of the domain of the *model*, while quantifiers quantify over the domain of a particular *world*. Thus, in a varying domain model $\mathcal{M}$ we might have $\mathcal{M}, \Gamma \Vdash_v (\forall x)\varphi(x)$, because $\varphi(x)$ is true at Γ when x is given any value in $\mathcal{D}(\Gamma)$, the domain of world Γ. But at the same time we might also have $\mathcal{M}, \Gamma \not\Vdash_v \varphi(y)$ because v assigns to y something in the domain of the model $\mathcal{M}$ that is not in $\mathcal{D}(\Gamma)$, and for which $\varphi(y)$ fails at Γ.

The problem is easy to identify. In $(\forall x)\varphi(x) \supset \varphi(y)$, the quantifier ranges over a set that can be smaller than the set of allowed values for y. But, the values assignable to y can be restricted by also quantifying over that variable: $(\forall y)[(\forall x)\varphi(x) \supset \varphi(y)]$. This formula *is* valid in our modal setting. Classically the two versions are almost interchangeable. In our context they are not, so care must be exercised.

What follows is an axiomatization of classical logic that is a slight variant of one given in (Hughes and Cresswell, 1996), which in turn derives from (Kripke, 1963b). It is designed to avoid the difficulties pointed out above.

DEFINITION 6.1.2. [Axioms] All formulas of the following forms are taken as axioms.

TAUTOLOGIES Any instance of a classical tautology.

VACUOUS QUANTIFICATION $(\forall x)\varphi \equiv \varphi$ where x does not occur free in φ.

UNIVERSAL DISTRIBUTIVITY $(\forall x)[\varphi \supset \psi] \supset [(\forall x)\varphi \supset (\forall x)\psi]$.

PERMUTATION $(\forall x)(\forall y)\varphi \equiv (\forall y)(\forall x)\varphi$.

UNIVERSAL INSTANTIATION $(\forall y)[(\forall x)\varphi(x) \supset \varphi(y)]$, where y is substitutable for x in $\varphi(x)$.

DEFINITION 6.1.3. [Rules of Inference] We have the following two rules.

MODUS PONENS $\dfrac{\varphi \quad \varphi \supset \psi}{\psi}$.

UNIVERSAL GENERALIZATION $\dfrac{\varphi}{(\forall x)\varphi}$.

We take a formula such as $(\forall x)R(x, y) \vee \neg(\forall x)R(x, y)$ to be a substitution instance of a classical tautology, hence it is an axiom. Note that free variables can occur in axioms, so $(\forall x)R(y, z) \equiv R(y, z)$ is an axiom of the Vacuous Quantification type. In most axiomatizations of classical logic Permutation is a provable theorem. But in most axiomatizations of classical logic, Universal Instantiation is stated differently, without the occurrence of $(\forall y)$. With it stated as we did, Permutation can no longer be proved.

As usual, a *proof* is a sequence of formulas, each an axiom or following from earlier lines by a rule of inference. A *theorem* is the last line of a proof.

Here are a few examples of proofs in this classical first-order axiom system. More precisely, we give abbreviated versions of proofs, enough to make a convincing argument that real proofs exist.

EXAMPLE 6.1.4. The formula $[(\forall x)\varphi \wedge (\forall x)\psi] \supset (\forall x)[\varphi \wedge \psi]$ is a theorem. Here is an abbreviated proof. We have left out steps that are easily justified using propositional logic.

1. $\varphi \supset [\psi \supset (\varphi \wedge \psi)]$ Tautology
2. $(\forall x)\{\varphi \supset [\psi \supset (\varphi \wedge \psi)]\}$ Univ. Gen. on 1
3. $(\forall x)\varphi \supset (\forall x)[\psi \supset (\varphi \wedge \psi)]$ Univ. Distrib. and 2
4. $(\forall x)[\psi \supset (\varphi \wedge \psi)] \supset [(\forall x)\psi \supset (\forall x)(\varphi \wedge \psi)]$ Univ. Distrib.
5. $(\forall x)\varphi \supset [(\forall x)\psi \supset (\forall x)[\varphi \wedge \psi]]$ from 3, 4
6. $[(\forall x)\varphi \wedge (\forall x)\psi] \supset (\forall x)[\varphi \wedge \psi]$ from 5

EXAMPLE 6.1.5. Suppose $\Phi(x)$ is a formula, y does not occur free in it, but is substitutable for x. Then $(\forall x)\Phi(x) \supset (\forall y)\Phi(y)$ is a theorem.

1. $(\forall y)[(\forall x)\Phi(x) \supset \Phi(y)]$ Univ. Ins.
2. $(\forall y)(\forall x)\Phi(x) \supset (\forall y)\Phi(y)$ Univ. Dist. on 1
3. $(\forall x)\Phi(x) \supset (\forall y)\Phi(y)$ Vac. Quant. on 2

Now, $(\forall x)\varphi(x) \supset \varphi(y)$ is not a theorem of the system above, though it is *classically* valid. So in what sense is this an axiomatization of classical logic? The system above is certainly classically sound, so any theorem of it is classically valid. In the other direction it is possible (and not difficult) to show that every *sentence* that is classically valid is provable. Thus the system is sound and complete with respect to the set of classically valid *closed* formulas.

EXERCISES

EXERCISE 6.1.1. Give axiomatic proofs of the following.

1. $(\forall x)[\Phi(x) \wedge \Psi(x)] \supset [(\forall x)\Phi(x) \wedge (\forall x)\Psi(x)]$.
2. $(\forall x)[\Phi(x) \wedge \Psi(x)] \equiv [(\forall x)\Phi(x) \wedge (\forall x)\Psi(x)]$
3. If x does not occur free in Φ, $[\Phi \vee (\forall x)\Psi(x)] \supset (\forall x)[\Phi \vee \Psi(x)]$.
4. If x does not occur free in Φ, $(\forall x)[\Phi \vee \Psi(x)] \supset [\Phi \vee (\forall x)\Psi(x)]$. Hint: $[A \supset (B \vee C)] \equiv [(A \wedge \neg B) \supset C]$.
5. If x does not occur free in Φ, $[\Phi \vee (\forall x)\Psi(x)] \equiv (\forall x)[\Phi \vee \Psi(x)]$.
6. If x does not occur free in Φ, $[\Phi \wedge (\exists x)\Psi(x)] \equiv (\exists x)[\Phi \wedge \Psi(x)]$. Recall that $(\exists x)$ abbreviates $\neg(\forall x)\neg$.
7. $(\exists x)[\Phi(x) \supset (\forall x)\Phi(x)]$. Hint: $\neg(P \supset Q) \equiv (P \wedge \neg Q)$.
8. $[(\forall x)\Phi(x) \wedge (\exists x)\Psi(x)] \supset (\exists x)[\Phi(x) \wedge \Psi(x)]$. Hint: you will primarily need Universal Distributivity.

6.2. VARYING DOMAIN MODAL AXIOM SYSTEMS

In Chapter 3 we gave axiom systems for several propositional modal logics. Now we extend them to take quantification into account by simply combining them with the machinery of the previous section. Thus formulas of the form $\Box X \supset X$ were axioms of the propositional modal logic **T**, so we now take all *first-order* formulas of this form as axioms of first-order **T**. Similarly for the other modal logics considered axiomatically in Chapter 3. Doing so produces systems corresponding to the varying domain semantics.

DEFINITION 6.2.1. [Varying Domain Axiomatic Proof] By a *varying domain axiomatic proof* for the modal logic **L** we mean a sequence of formulas each of which is a first-order instance of an **L** axiom (as given in Chapter 3), or is a classical first-order axiom (as given in Section 6.1), or follows from earlier lines by one of the rules of inference: Necessitation, Modus Ponens, Universal Generalization.

EXAMPLE 6.2.2.   Here is a varying domain proof using the **T** rules of the
sentence $(\forall x)\square[\Phi(x) \wedge \Psi(x)] \supset (\forall y)\Phi(y)$. As usual, we actually present
an abbreviated version, letting you fill in the missing steps.

1. $[\Phi(y) \wedge \Psi(y)] \supset \Phi(y)$ tautology
2. $\square[\Phi(y) \wedge \Psi(y)] \supset [\Phi(y) \wedge \Psi(y)]$ **T** axiom
3. $\square[\Phi(y) \wedge \Psi(y)] \supset \Phi(y)$ from 1, 2
4. $(\forall y)\{\square[\Phi(y) \wedge \Psi(y)] \supset \Phi(y)\}$ Univ. Gen. on 3
5. $(\forall y)\square[\Phi(y) \wedge \Psi(y)] \supset (\forall y)\Phi(y)$ 4, Univ. Dist.
6. $(\forall y)\{(\forall x)\square[\Phi(x) \wedge \Psi(x)] \supset \square[\Phi(y) \wedge \Psi(y)]\}$ Univ. Inst.
7. $(\forall y)(\forall x)\square[\Phi(x) \wedge \Psi(x)] \supset (\forall y)\square[\Phi(y) \wedge \Psi(y)]$ 6, Univ. Dist.
8. $(\forall y)(\forall x)\square[\Phi(x) \wedge \Psi(x)] \supset (\forall y)\Phi(y)$ from 5 and 7
9. $(\forall x)\square[\Phi(x) \wedge \Psi(x)] \supset (\forall y)\Phi(y)$ 8, Vac. Quant.

EXERCISES

EXERCISE 6.2.1.   Give a varying domain axiomatic **K** proof of the following.

$$[\lozenge(\forall x)A(x) \wedge \square(\exists x)B(x)] \supset \lozenge(\exists x)[A(x) \wedge B(x)]$$

EXERCISE 6.2.2.   Give a varying domain axiomatic **D** proof of the following.

$$\square(\forall x)[A(x) \vee B(x)] \supset [\lozenge(\forall x)A(x) \vee \lozenge(\exists x)B(x)]$$

6.3. CONSTANT DOMAIN SYSTEMS

The Barcan and Converse Barcan formulas were introduced in Section 4.9,
and it was shown that they correspond to anti-monotonicity and monotonicity
respectively. So it should be no surprise that they are precisely the additional
assumptions we must make to get axiom systems suitable for constant domain
logics.

DEFINITION 6.3.1.   [Constant Domain Axiomatic Proof] A *constant domain* axiom proof for **L** is as in Definition 6.2.1 with the addition: every
Converse Barcan formula and every Barcan formula is also an axiom.

EXAMPLE 6.3.2. The following is a proof in constant domain **T** of

$$[(\forall x)\Box\Box\Phi(x) \wedge \Diamond(\exists x)\Psi(x)] \supset \Diamond(\exists x)[\Phi(x) \wedge \Psi(x)].$$

As usual, it is considerably abbreviated.

1. $\Box\Box\Phi(x) \supset \Box\Phi(x)$ **T** axiom
2. $(\forall x)\Box\Box\Phi(x) \supset (\forall x)\Box\Phi(x)$ 1, Univ. Gen. and Univ. Dist.
3. $(\forall x)\Box\Phi(x) \supset \Box(\forall x)\Phi(x)$ Barcan formula
4. $(\forall x)\Box\Box\Phi(x) \supset \Box(\forall x)\Phi(x)$ from 2,3
5. $[(\forall x)\Box\Box\Phi(x) \wedge \Diamond(\exists x)\Psi(x)] \supset$
    $[\Box(\forall x)\Phi(x) \wedge \Diamond(\exists x)\Psi(x)]$ from 4
6. $[\Box(\forall x)\Phi(x) \wedge \Diamond(\exists x)\Psi(x)] \supset$
    $\Diamond[(\forall x)\Phi(x) \wedge (\exists x)\Psi(x)]$ theorem of propositional **K**
7. $[(\forall x)\Phi(x) \wedge (\exists x)\Psi(x)] \supset$
    $(\exists x)[\Phi(x) \wedge \Psi(x)]$ Exercise 6.1.1
8. $\Diamond[(\forall x)\Phi(x) \wedge (\exists x)\Psi(x)] \supset$
    $\Diamond(\exists x)[\Phi(x) \wedge \Psi(x)]$ from 7 by Exercise 3.1.1
9. $[(\forall x)\Box\Box\Phi(x) \wedge \Diamond(\exists x)\Psi(x)] \supset$
    $\Diamond(\exists x)[\Phi(x) \wedge \Psi(x)]$ from 5, 6, 8

EXERCISES

EXERCISE 6.3.1. Give constant domain axiomatic **K** proofs of the following.

1. $\Diamond(\forall x)A(x) \supset (\forall x)\Diamond A(x)$.
2. $(\exists x)\Box A(x) \supset \Box(\exists x)A(x)$.
3. $[\Diamond(\forall x)A(x) \wedge \Box(\exists x)B(x)] \supset (\exists x)\Diamond[A(x) \wedge B(x)]$.

EXERCISE 6.3.2. Give constant domain axiomatic **S4** proofs of the following.

1. $\Box(\forall x)\Box A(x) \equiv (\forall x)\Box A(x)$.
2. $\Box(\exists x)\Box A(x) \equiv (\exists x)\Box A(x)$.

EXERCISE 6.3.3. Give a constant domain axiomatic **S5** proof of the following.

$$\Box(\forall x)A(x) \equiv \Diamond(\forall x)\Box A(x)$$

6.4. MISCELLANY

In Section 4.9 we noted the correspondence between the Converse Barcan formula and monotonicity, and likewise between the Barcan formula and anti-monotonicity. Now, in a model whose accessibility relation is *symmetric*, monotonicity and anti-monotonicity are equivalent—we either have both properties, or neither. Further, symmetry corresponds to assuming the axiom called B in Section 3.2: $P \supset \Box\Diamond P$. Consequently, if we assume the B axiom schema and the Converse Barcan axiom schema, each instance of the Barcan schema should be provable, and also the other way around. We now present a much abbreviated proof of this in one direction: symmetry together with Converse Barcan allows the proof of Barcan. We use axiom schema B in the form $P \supset \Box\Diamond P$ for line 9, and in the equivalent version $\Diamond\Box P \supset P$ for line 5. The Converse Barcan schema is used for line 3.

1. $(\forall x)\,[(\forall x)\Box\Phi(x) \supset \Box\Phi(x)]$ Univ. Inst.
2. $\Box(\forall x)\,[(\forall x)\Box\Phi(x) \supset \Box\Phi(x)]$ Nec. rule on 1
3. $(\forall x)\Box\,[(\forall x)\Box\Phi(x) \supset \Box\Phi(x)]$ from 2 using Conv. Barcan
4. $(\forall x)\,[\Diamond(\forall x)\Box\Phi(x) \supset \Diamond\Box\Phi(x)]$ from 3 using Ex. 3.1.1
5. $(\forall x)\,[\Diamond(\forall x)\Box\Phi(x) \supset \Phi(x)]$ from 4 using B axiom
6. $(\forall x)\Diamond(\forall x)\Box\Phi(x) \supset (\forall x)\Phi(x)$ from 5 using Univ. Dist.
7. $\Diamond(\forall x)\Box\Phi(x) \supset (\forall x)\Phi(x)$ from 6 using Vac. Quant.
8. $\Box\Diamond(\forall x)\Box\Phi(x) \supset \Box(\forall x)\Phi(x)$ from 7 using Reg. Rule
9. $(\forall x)\Box\Phi(x) \supset \Box(\forall x)\Phi(x)$ from 8 using B axiom

Essentially this argument is due to Prior. Now, there should be a proof going in the other direction as well. That is, assuming B and the Barcan schema, each instance of the Converse Barcan schema should be provable. We have not been able to find such a proof. We do, however, have a proof that uses an equivalent formulation of the Barcan formula, involving equality. We will present this proof once equality has been introduced in the next chapter. Nonetheless, a direct argument should be possible. We would be grateful to any reader who can provide one.

Finally, what about the critical issues of soundness and completeness for our axiom systems? Soundness is, as usual, straightforward. One shows each axiom is valid (with respect to the appropriate semantics), and that the rules of inference preserve validity. It follows that each provable formula must also be valid. We leave this to you as an exercise.

Completeness is much more difficult and as we noted earlier, we have decided to omit completeness proofs here, and simply give a few references to the literature. The most readable reference is (Hughes and Cresswell, 1996), with (Garson, 1984) the most thorough. We note an important caveat how-ever. Hughes and Cresswell (1996) point out that their completeness proof does not work for varying domain axiom systems containing axiom B, where completeness means with respect to symmetric models. They cite a claim by Fine that a completeness proof for such a case is possible, but apparently a proof has not been published. Perhaps this is the source of our difficulties in showing that the assumption of the Barcan schema should allow the deriv-ation of the Converse Barcan formula, in the presence of axiom schema B. It is an interesting but rather technical point, which does not affect the main thrust of this book.

EXERCISES

EXERCISE 6.4.1.   Show soundness for varying domain axiomatic **K**.

CHAPTER SEVEN

EQUALITY

7.1. CLASSICAL BACKGROUND

When we say that two people own the same car, we might mean either they own the same *type* of car, e.g., a Honda, or they are joint owners of a single car. In the former case, *same* means *qualitative identity* or *equivalence*; in the latter case, it means *quantitative or numerical identity*, or, as we shall call it, *equality*. Equality is actually a special case of equivalence, as we shall see after we have introduced the notion of an *equivalence relation*.

DEFINITION 7.1.1. [Equivalence Relation] A relation $\sim$ on a set of objects S is an *equivalence relation* iff for all $x, y \in S$

1. $x \sim x$, i.e., $\sim$ is *reflexive*
2. $x \sim y$ implies $y \sim x$, i.e., $\sim$ is *symmetric*
3. $x \sim y$ and $y \sim z$ together imply $x \sim z$, i.e., $\sim$ is *transitive*

An equivalence relation $\sim$ on S determines a *partition* of the set S into disjoint subsets called *equivalence classes* (this will be proved in Section 7.6). Each element in an equivalence class bears $\sim$ to every element in that class, and to none outside—its elements are equivalent to one another. For example, if the relation is *same manufacturer* and we are talking about the set S of cars, $\sim$ partitions S into the set of Hondas, the set of Toyotas, of Fords, of Fiats, and so forth. *Equality* is also an equivalence relation: it is the *smallest* equivalence relation, so that each one of the equivalence classes is a singleton, i.e., each contains one element.

In classical first-order logic with equality, we have a two-place relation $x = y$ customarily governed by two conditions. The first,

$$(7.1) \quad x = x$$

explicitly gives us reflexivity. The second is commonly called *Leibniz's Law*:

$$(7.2) \quad (x = y) \supset (\varphi(x) \supset \varphi(y)).$$

This yields both symmetry and transitivity by judicious substitutions for the schematic letter φ. We leave this as an exercise. This assures that $=$ is an equivalence relation.

Whether equality has been captured by this axiomatization is open to question. In second-order logic, equality is definable: $x = y$ is taken to mean

$(\forall \varphi)(\varphi(x) \equiv \varphi(y))$. That is, x is equal to y iff they have *all* their properties in common. In first-order logic, however, we cannot speak of *all* properties. Formula (7.2) is a schema with φ replaceable by the available predicates of the theory. And the available predicates might not be sufficient to discern distinct objects.

Suppose, for example, that the only predicates in the theory are "x is a dog" and "x is a bird." Suppose further that we have as an axiom that no dog is a bird and conversely. This theory cannot distinguish among birds because there is no property any one bird has that any other bird lacks. (Likewise, it cannot distinguish among dogs.) Since the theory cannot distinguish among objects that belong to the same species, it is possible to consistently interpret $x = y$ as the equivalence relation x *is an element of the same species as* y. And this equivalence relation is *not* equality. The problem is the lack of expressiveness of our language—there are not enough predicates.

All of the usual principles governing equality are obtainable without necessarily interpreting $=$ as equality. In particular *The Indiscernibility of Identicals* is valid in many classical first-order models:

$$(7.3) \quad (x = y) \supset (\varphi(x) \equiv \varphi(y)).$$

It is enough to interpret $=$ as an equivalence relation that cannot distinguish between two objects, x and y, in the domain of the model that happen to share all properties *definable* in the model. Since such properties may not be *all* properties, interpreting $=$ to be something less than full equality can suffice.

There are first-order models of classical logic in which $=$ is interpreted as standard equality. This is, of course, the normal or intended interpretation; accordingly, a model in which $=$ is taken to be equality is said to be a *normal* model. But, as we have just noted, there are first-order models of classical logic in which $=$ is interpreted in a way other than standard equality. So, we cannot claim that we have captured equality completely with the two conditions (7.1) and (7.2). But we have, in fact, captured equality in the sense that a formula is true in all normal models if and only if it is true in all models allowing the kind of non-standard reading of $=$ discussed above. We give an example of how to convert a non-standard model into a normal one in Section 7.7.

We carry over to the modal setting these ideas about equality from classical logic. But the subject of equality has caused much grief for modal logic. Many of the problems that have been posed for modal logic, and which have struck at the heart of the coherence of modal logic, stem from apparent violations of the Indiscernibility of Identicals. We will discuss some of the problematic cases.

We begin with one of the most famous puzzles concerning equality and modal logic.

EXERCISE 7.1.1.   Show that = is symmetric and transitive, using (7.1) and (7.2).

7.2. FREGE'S PUZZLE

The need for an equality predicate in classical logic is clear. But further, in modal logic, we want to distinguish necessary equalities, e.g., that π is the ratio of the circumference of a circle to its diameter, from merely contingent ones, e.g., that George Washington is the first American president. Various philosophical puzzles, however, have stood in the way of this clear-sighted appraisal.

Perhaps the best known of the puzzles was introduced by Frege (1892):

Equality gives rise to challenging questions which are not altogether easy to answer. ... $a = a$ and $a = b$ are obviously statements of differing cognitive value; $a = a$ holds *a priori* and, according to Kant, is to be labeled analytic, while statements of the form $a = b$ often contain very valuable extensions of our knowledge and cannot always be established *a priori*. The discovery that the rising sun is not new every morning, but always the same, was one of the most fertile astronomical discoveries. Even to-day the identification of a small planet or a comet is not always a matter of course. Now if we were to regard equality as a relation between that which the names 'a' and 'b' designate, it would seem that $a = b$ could not differ from $a = a$ (i.e. provided $a = b$ is true). A relation would thereby be expressed of a thing to itself but to no other thing. (p. 56-7)

This, of course, is the well-known morning star/evening star puzzle. If we go far enough back in time, we reach a point when it was not known that the morning star and the evening star are the same. Still, to say that a and b are equal is to say they are the same thing, and consequently whatever is true of a is true of b, and conversely. Now, the morning star and the evening star are, in fact, equal. And certainly, the ancients knew the morning star was the morning star. Then it should follow that the ancients knew the evening star was the morning star. But they didn't.

Wittgenstein (1922) succumbed to this puzzle completely: " ... to say of *two* things that they are identical is nonsense, and to say of *one* thing that it is identical with itself is to say nothing at all." (§5.5303) He persuaded himself that there were no informative equalities, and so no need for the equality predicate:

Identity of object I express by identity of sign, and not by using a sign for identity. Difference of objects I express by difference of signs. (§5.53)

But Wittgenstein's notation is inadequate. We cannot prove the Pythagorean Theorem in his symbolism. We cannot even express it!

Frege (1879) succumbed only partially. He was persuaded by the puzzle that equality could not relate objects. But he appreciated the utility of equality, and attempted to account for the informative cases by taking equality to relate the names of the objects they designate. This solution is unsatisfactory and Frege came to abandon it. Originally he failed to distinguish clearly between what one is talking about and what one is saying about it, a distinction elaborated in his celebrated sense/reference theory announced in Frege (1892). (See (Mendelsohn, 1982) for further discussion.)

Frege (1892) diagnosed the puzzle as resting on the false assimilation of sense [Sinn] and reference [Bedeutung]. The two names "morning star" and "evening star" refer to the same object—the planet Venus—but do so in different ways, and so differ in meaning. Accordingly, the proposition that the morning star is equal to the morning star differs from the proposition that the morning star is equal to the evening star. It should come as no surprise that the ancients knew the one proposition to be true without knowing the other to be true.

But Frege's solution provides no relief for modal versions of the puzzle. In fact, it seems to exacerbate them. The simplest sentence saying one can substitute equals for equals in a modal context is

$$(x = y) \supset \Box(x = y). \tag{7.4}$$

If we interpret $\Box$ as "the ancients knew," and a and b as the morning and evening star respectively, then $(a = b) \supset \Box(a = b)$ is an instance of (7.4). Since $a = b$ is factually true, Modus Ponens brings us to the problem immediately. A number of philosophers and logicians have found this result unpalatable. As a result, they have rejected (7.4), and since (7.4) is valid in first-order modal logic—we will actually demonstrate (7.4) in Section 7.3—they have rejected quantified modal logic as well.

In the process of providing a reasonably effective solution to the puzzle about equality for *classical* logic, Frege's sense/reference solution renders equality precarious in *modal* logic, for it provides every reason to reject (7.4). Even though the objects designated by a and b are one and the same, these names need not have the same meaning. The proposition that $a = a$, which is trivially true, would then be distinct from the proposition that $a = b$, which is not. And so, the fact that $a = b$ is true would provide no reasonable assurance that $\Box(a = b)$ is true.

Frege held that when a name occurs inside a modal context, it is used to speak about its meaning, not its designation. So, to say that the ancients knew the morning star is the evening star is not to speak about the celestial body nor to ascribe a property to that celestial body. It is to speak about the meanings of the words, or, alternatively, about the proposition expressed by $a = b$. In other words, Frege's understanding of the modal operator is *de dicto*. He has no *de re* reading. But it is the *de re* reading that is crucial for first-order modal logic, and, in particular, for the validation of (7.4).[18] By effectively precluding a *de re* reading for modal statements, his approach to the problem of equality has (in yet another way) impeded the development of modal logic.

Russell (1905) provides the most satisfying solution to Frege's puzzle, although the full impact of his solution was not understood until a half-century later. Speaking a bit anachronistically, Russell held that for genuine proper names, the *Sinn* is the *Bedeutung*, but for definite descriptions and names that are disguised definite descriptions, the two are distinct. More importantly, he introduced a scope distinction for definite descriptions that, when suitably elaborated by Smullyan (1948), enabled it to formalize the *de re/de dicto* distinction. This is the path we follow in this book. On our view, the puzzle about informative identities arises because of the interplay of modal operators and singular terms. Our solution to the puzzle will await the introduction of predicate abstraction machinery, starting in Chapter 9. But there is no puzzle at this stage about (7.4) which contains only *variables*. If x and y are the very same object, and x has the property of being necessarily identical with x, then y has that very same property. The puzzle arises when we replace these variables with singular terms.

EXERCISES

EXERCISE 7.2.1.   Suppose a and b are not names or definite descriptions, but demonstratives such as "this" and "that." Can $a = a$ differ in cognative value from $a = b$? Discuss.

---

[18] With his *Sinn/Bedeutung* distinction, Frege—despite the attitude we described earlier in Section 1.2—has long been thought to provide the philosophical underpinnings for a semantics for modal logic (as was carried out, for example, by his most famous pupil Carnap (1947)). Ironically, Quine's criticisms of modal logic, as detailed in Chapter 4, appear to be heavily influenced by Frege's semantics for modal contexts.

EXERCISE 7.2.2.   "Frege's puzzle about equality is not really about equality. It is about meaning." Argue for or against this claim.

EXERCISE 7.2.3.   Wittgenstein suggested eliminating the equality sign.
How impaired would the resulting language be, if impaired at all?

7.3. THE INDISCERNIBILITY OF IDENTICALS

The Indiscernibility of Identicals, (7.3), is among the most fundamental of
logical principles. To deny it is to suppose that $x = y$ and yet that x has a
property y lacks, or conversely. Recall, x and y are not *names*, they are variables which have *objects* as values. Denying the Indiscernibility of Identicals
for objects is to suppose that the object x both has and lacks a particular
property. In other words, *The Indiscernibility of Identicals* is on a par with
the venerable *Law of Noncontradiction*, $\neg(\exists x)(\varphi(x) \wedge \neg\varphi(x))$.

Closely related to the Indiscernibility of Identity, (7.3), and frequently
conflated with it, is a *Substitution Principle*:

(7.5)   If a and b denote the same thing, then $\varphi(a)$ and $\varphi(b)$
          have the same truth value

where a and b are any two singular terms. Unlike (7.3), the *Substitution Principle* is easily shown to be false; the morning star/evening star example we
just discussed shows this. But counterexamples to the substitution principle in
no way provide counterexamples to the Indiscernibility of Identicals. On the
contrary, counterexamples like the one just described are due to the behavior
of names and descriptions. (7.3), however, holds in a first-order language (like
the one we are considering) which contains neither names nor descriptions,
but only variables.

Take φ in (7.3) to be $\square(x =)$, giving us

(7.6) $(x = y) \supset (\square(x = x) \equiv \square(x = y))$

Since $x = x$ is valid, by *Necessitation* we also have $\square(x = x)$. (7.6),
therefore, yields

(7.7) $(x = y) \supset \square(x = y)$

just as we promised in Section 7.2. (For a semantical version of this argument,
see Example 7.4.2.)

Formula (7.7) appears to fly in the face of the clear intuition that some
identities are necessary while others are only contingent. Frege's original

puzzle about the morning star and the evening star has long been taken to be just such an example of this distinction: *the morning star = the morning star* seems to be a necessary truth, but *the morning star = the evening star* is not. Once again, these complications are the result of having introduced names or definite descriptions into the language. If we look to formulas with variables only, we have the following situation. $x = x$ is logically true, and so $\Box(x = x)$ is true. $x = y$ is not logically true; but if it is true, $\Box(x = y)$ is also true. Every logical truth is a necessary truth, but there are some necessary truths that are not logical truths. But then, we have a way of making the distinction intuition wanted. While $x = y$ is, if true, necessary, just like $x = x$, it is not a logical truth. The distinction to be drawn here is not between necessary and contingent identities, but between identities that are logically— and so necessarily—true and identities that are not logically true, but are necessarily true nonetheless. Kripke (1979a) made the distinction in terms of truths that are necessary and known *a priori* and truths that are necessary and known *a posteriori*.

We have seen that if x and y are the same, that fact is necessary, i.e. $(x = y) \supset \Box(x = y)$. But what if x and y are distinct? Will their distinctness also be necessary? We take the answer to be "yes." Two objects cannot, even if circumstances change, become one. This is a different issue than whether two *names* can sometimes designate distinct objects, and sometimes the same object. Names can. Distinct objects remain distinct, no matter what.

The Indiscernibility of Identicals is closely involved with the problem of *transworld identity*, one of the sticking points in the development of modal logic. In the actual world, Julius Caesar crossed the Rubicon and marched on Rome. This is, however, a contingent property of the man. It is possible that he didn't cross the Rubicon and march on Rome, and this is to say that there is another possible world in which he didn't. But how can the Julius Caesar in this world be the very same Julius Caesar as the one in the other world, when a property the Julius Caesar in this world has, the Julius Caesar in another world lacks? This problem does not depend at all upon the role of names or descriptions, but only on the object itself and its identity conditions.

But this is not a violation of The Indiscernibility of Identicals. Julius Caesar has the property of having crossed the Rubicon in this world. But there are other worlds in which he does not have the property of having crossed the Rubicon. We do not speak simply of his having a property, but only of his having a property relative to a world. We are familiar with the problem (and the solution) in the case of time. Julius Caesar was less than 4 feet tall *when he was two years old*; Julius Caesar was more than 4 feet tall *when he was eighteen years old*. We need only change the idiom to temporal worlds. Julius Caesar was less than 4 feet tall *in the temporal world in which he was two years old*; Julius Caesar was more than 4 feet tall *in the temporal world in which he was eighteen years old*. It is the same person: he had one height at age 2 and a different one at age 18.

A somewhat different problem about transworld identity is this: How can we tell which object in another world is the same as an object in this world, especially when, in another world, the object has different local properties? The problem here is not a logical one, i.e., whether an object can exist in more than one possible world having distinct characteristics in the two worlds. The problem is one of identifying an object in another world as the same object as one that is in the actual world.

This problem has some familiar intuition in the case of temporal worlds. We see a picture of grandmother's class when she was in second grade: which one in the picture is grandmother? Of course in the picture, grandmother is not an old lady but a little girl, and she has few of the visible characteristics in that picture that she has as an old lady. We can not always answer the question. Usually we need someone who remembers what she looked like as a young girl, and we rely on her ability to pick her out again after all these years. Whether or not we are ultimately successful in identifying the correct individual as grandmother, we have some understanding of the problem and some idea of how we would go about solving it.

However no such story seems appropriate for the alethic case. We do not have pictures of objects in other possible worlds, nor do we have anything like the access to them that we have for temporal worlds. Moreover, it is unlikely that just such a problem would arise in the alethic case. It is not as if we need to find out who, in another possible world, is the Julius Caesar that is in this one. Rather, the situation would seem to be whether Julius Caesar could have a certain property, i.e., whether there is a world in which he has that property. Picking a world, and trying to determine whether he has that property in that world, seems to put the cart before the horse.

Must an object retain any of its properties in every possible world? Apparently, the object must retain the property of being identical with itself. But need there be any substantial properties the object must have in every possible world? It does not seem likely.

There are two views in the philosophical literature on the nature of objects, and of the relation between objects and their properties. On one view, the *bundle theory*, an object is considered to be its collection of properties. On the other view, the *bare particular view*, the object itself is what remains after one peels away all of its properties.[19] Despite initial appearances, we

---

[19] We cannot resist a quote here, from *Peer Gynt* by Henrik Ibsen (1867). Peer is summarizing his history, recounting the things that have made him him, while he peels an onion and compares its layers to episodes in his life.

do not favor the bare particular theory. There is no reason to suppose that there is anything left if we peel off all of its properties. In fact, it is not clear what is meant by peeling off all of its properties. For, although Bill Clinton *is* a Democrat, in another possible world he is a Republican. This property, *possibly being a Republican* is as much a property of Bill Clinton as is the property of *actually being a Democrat*. So, as we go from world to world, we do not peel off any of his properties. As a result, it is not the bare particular view that most closely characterizes our treatment of objects in modal logic but, interestingly enough, the bundle theory.

Kripke emphasizes that possible worlds are stipulated. Typically we identify a given object, say Julius Caesar, and consider what *he* would be like in another possible world. To be sure, unless we have some substantial essential qualities for the individual, just about anything we hold true of the man is false in some possible world. But our task is not to see which object in that other world is Julius Caesar and try to identify that object via the properties it has; on the contrary, we already have the object picked out as Julius Caesar, and our task is to consider what *he* is like in another possible world. One can be too gripped by the problem of identifying him. That is not the problem. We have already identified him, and we hold him constant as we consider possible alterations.

Enough informal discussion. It is time to begin the formal semantics.

EXERCISES

EXERCISE 7.3.1. "Suppose we have a statue of Achilles that is made of bronze. The statue is melted down and reshaped into a statue of Hector. Then, at one time, this bronze equaled the statue of Achilles, but at a later time this bronze is not equal to a statue of Achilles. So there are contingent identities." Discuss.

"There lies the outermost layer, all torn; that's the shipwrecked man on the jolly boat's keel. Here's the passenger layer, scanty and thin; and yet in its taste there's a tang of Peer Gynt. Next underneath it the gold-digger ego; the juice is all gone—if it ever had any. ... What an enormous number of swathings! Isn't the kernel soon coming to light? I'm blest if it is! To the innermost center it's nothing but swathings."

7.4. THE FORMAL DETAILS

From now on, one of the two-place relation symbols is designated to represent the equality relation. We use "=" for this relation symbol, and we systematically use *infix* notation, writing the familiar $x = y$ instead of $= (x, y)$. We want the equality symbol to designate the equality relation. What we will do is restrict ourselves to those models in which it does so. For what we do in this chapter, varying domain models or constant domain models serve equally well. Likewise a particular choice of $\mathbf{K}, \mathbf{T}, \mathbf{D}, \dots$ does not matter much either. Leave the choice open unless there is some reason to be specific. Let us say the default is varying domain $\mathbf{K}$, with other choices indicated as appropriate.

DEFINITION 7.4.1. [Normal Models] A model $\mathcal{M} = \langle \mathcal{G}, \mathcal{R}, \mathcal{D}, \mathcal{I} \rangle$ is called *normal* provided, for each $\Gamma \in \mathcal{G}$ we have that $\mathcal{I}(=, \Gamma)$ is the equality relation on $\mathcal{D}(\mathcal{M})$.

That is, in a normal model, at each world the equality relation symbol is interpreted to be the actual equality relation on the domain of the model. (Note that the interpretation of "=" is thus the same from world to world.) Symbolically, at a world Γ of a normal model $\mathcal{M}$,

$$\mathcal{M}, \Gamma \Vdash_v (x = y) \Longleftrightarrow v(x) = v(y).$$

EXAMPLE 7.4.2. We show the formula $[(x = y) \supset \Box(x = y)]$ is valid in all normal models.
Let $\mathcal{M} = \langle \mathcal{G}, \mathcal{R}, \mathcal{D}, \mathcal{I} \rangle$ be a model, with $\Gamma \in \mathcal{G}$. We want to show

$$\mathcal{M}, \Gamma \Vdash_v [(x = y) \supset \Box(x = y)]$$

where v is an arbitrary valuation in $\mathcal{M}$.
Well, suppose $\mathcal{M}, \Gamma \Vdash_v (x = y)$. Since $\mathcal{M}$ is a normal model, we have $v(x) = v(y)$. Now, let Δ be any member of $\mathcal{G}$ such that $\Gamma \mathcal{R} \Delta$. Since $v(x) = v(y)$, it follows that $\mathcal{M}, \Delta \Vdash_v (x = y)$. Since Δ was arbitrary, we have $\mathcal{M}, \Gamma \Vdash_v \Box(x = y)$, and this establishes the claim.

In a sense, what makes this formula valid is the fact that in our semantics, free variables range over objects, members of possible world domains, and not over names for objects. Thus if x and y are given the same object as value, that fact persists independently of a choice of possible world. In Chapter 9 we will introduce language machinery for *naming* objects, and we will see the situation becomes much more complex then.
Incidentally, it follows easily from this example that we also have the validity of $(\forall x)(\forall y)[(x = y) \supset \Box(x = y)]$. A similar remark applies to the exercises below.

EXERCISES

EXERCISE 7.4.1. Show the validity of the following in normal modal models.

1. $(x = x)$.
2. $(x = y) \supset (y = x)$.
3. $(x = y \wedge y = z) \supset x = z$.
4. $(x = y) \supset (\Phi(x) \equiv \Phi(y))$ where $\Phi(x)$ is a formula in which y does not occur, and $\Phi(y)$ is the result of substituting occurrences of y for free occurrences of x in $\Phi(x)$. (Note that since y does not occur in $\Phi(x)$, y is substitutable for x since there are no quantifiers of the form $(\forall y)$.)

EXERCISE 7.4.2. Show the validity of the following in normal modal models.

1. $\Diamond(x = y) \supset (x = y)$.
2. $\neg(x = y) \supset \Box\neg(x = y)$.

7.5. TABLEAU EQUALITY RULES

Tableau rules to deal with equality are simplicity itself. There are two fundamental features of equality. First, a thing is equal to itself, (7.1). And second, a thing is equal *only* to itself. This second principle is at the heart of "substituting equals for equals," (7.2). If $x = y$, anything we can say about x can also be said about y because, after all, it is the same thing as saying it about x. Formally what we do is add to the tableau rules of earlier chapters two additional rules, embodying these basic facts about equality. It is the choice of earlier rules that determines which modal logic we are using, and whether it is varying or constant domain. The equality rules are the same no matter what specific modal logic we work with.

We begin with a rule incorporating the idea that things equal themselves.

DEFINITION 7.5.1. [Reflexivity Rule] If p is a parameter and σ is a prefix, both of which occur on the tableau branch, then $\sigma (p = p)$ can be added to the end of the branch. Briefly,

$$\frac{}{\sigma (p = p)}$$

Next, a rule that covers the substitutivity of equals for equals.

DEFINITION 7.5.2. [Substitutivity Rule] Let $\Phi(x)$ be a formula in which x occurs free, let $\Phi(p)$ be the result of substituting occurrences of the parameter p for all free occurrences of x in $\Phi(x)$, and similarly for $\Phi(q)$. If $\sigma_1\,(p = q)$ and $\sigma_2\,\Phi(p)$ both occur on a tableau branch, $\sigma_2\,\Phi(q)$ can be added to the end. Briefly,

$$\frac{\sigma_1\,(p = q)}{\sigma_2\,\Phi(q)}\,\sigma_2\,\Phi(p)$$

As a matter of fact, the rule above can be restricted to atomic substitutions, and we still can prove completeness of the resulting system. We do not do this here, however. Before giving an example of a proof, there are a couple of tableau abbreviations that are useful.

First, we can simply close a branch that contains $\sigma\,\neg(p = p)$, because we can always add $\sigma\,(p = p)$ by the Reflexivity Rule, and then close the branch in the usual way.

Second, the Substitutivity Rule allows a *left-right* replacement (if we have $p = q$ with some prefix, we can replace p with q). But a *right-left* replacement is also possible. That is, we have a derived rule

$$\frac{\sigma_1\,(p = q)}{\sigma_2\,\Phi(p)}\,\sigma_2\,\Phi(q)$$

We can take this as abbreviating the following longer sequence of steps:

$$\begin{array}{ll} \sigma_1\,(p = q) & 1. \\ \sigma_2\,\Phi(q) & 2. \\ \sigma_1\,(p = p) & 3. \\ \sigma_1\,(q = p) & 4. \\ \sigma_2\,\Phi(p) & 5. \end{array}$$

In this, 3 is by the Reflexivity Rule. Then if we consider the formula $(x = p)$, item 3 is this with x replaced with p, and item 4 is this with x replaced with q; consequently 4 follows from 1 and 3 by the (left-right) Substitutivity Rule. And then 5 follows from 4 and 2, by Substitutivity again.

EXAMPLE 7.5.3.   Here is a (varying domain **K**) proof of $(\forall x)(\forall y)[(x = y) \supset \Box(x = y)]$.

$$1 \quad \neg(\forall x)(\forall y)\left[(x = y) \supset \Box(x = y)\right] \quad 1.$$
$$1 \quad \neg(\forall y)\left[(p_1 = y) \supset \Box(p_1 = y)\right] \quad 2.$$
$$1 \quad \neg\left[(p_1 = q_1) \supset \Box(p_1 = q_1)\right] \quad 3.$$
$$1 \quad (p_1 = q_1) \quad 4.$$
$$1 \quad \neg\Box(p_1 = q_1) \quad 5.$$
$$1.1 \neg(p_1 = q_1) \quad 6.$$
$$1.1 \neg(q_1 = q_1) \quad 7.$$

Item 2 is from 1 by an Existential Rule, as is 3 from 2; 4 and 5 are from 3 by a Conjunctive Rule; 6 is from 5 by a Possibility Rule; 7 is from 4 and 6 by the Substitutivity Rule; now the tableau is closed because of 7.

EXERCISES

EXERCISE 7.5.1.   By the *closure* of a formula is meant the result of prefixing the formula with universal quantifiers for the free variables of the formula (other than parameters). For example, the closure of $[(x = y) \supset \Box(x = y)]$ is $(\forall x)(\forall y)[(x = y) \supset \Box(x = y)]$.
   Give varying domain **K** tableau proofs of closures of the formulas in Exercises 7.4.1 and 7.4.2.

7.6. TABLEAU SOUNDNESS AND COMPLETENESS

As usual, we must establish that the new tableau rules produce a system that proves exactly the sentences valid in normal models. Soundness is easy, while completeness is by a variation of an argument originally introduced by Gödel in his proof of the completeness of an axiomatization of classical logic.

Soundness

In Section 2.5 we proved soundness for propositional tableau systems, and in Section 5.3 we extended the proof to first-order modal logics. The central fact was: if a tableau branch extension rule is applied to a satisfiable tableau, the result is another satisfiable tableau. Once this was shown, soundness followed immediately by a simple argument. If X is provable it must be valid for, if not, we could derive a contradiction as follows. Suppose X is provable but not valid. Since X is not valid, $1 \neg X$ must be satisfiable. Then the construction of a proof of X begins with a satisfiable tableau, from which only satisfiable tableaus can result. So the proof of X must itself be a satisfiable tableau. Since it is also closed, we have an impossible situation.

This argument, based on preservation of satisfiability, still applies, but we must redefine satisfiability to take equality into account. For this section we use satisfiability as given in Definition 5.3.1 with the added condition: *models must be normal*. This extra condition does not invalidate our earlier satisfiability preservation results—this is a simple matter to check. So all that is left is to show that the equality rules of this chapter also preserve satisfiability.

We gave two tableau rules for equality: the Reflexivity Rule and the Substitutivity Rule. The Reflexivity Rule says we can add $\sigma\,(p = p)$ to a branch, $\mathcal{B}$, provided both σ and p already occur on it. Now, suppose the set of sentences on branch $\mathcal{B}$ is satisfiable in the *normal* model $\mathcal{M} = \langle \mathcal{G}, \mathcal{R}, \mathcal{D}, \mathcal{I}\rangle$ with respect to the valuation v, using the prefix mapping θ. Since σ occurs on $\mathcal{B}$ and the branch is satisfiable, $\theta(\sigma)$ is defined and is some member of $\mathcal{G}$. Similarly $v(p)$ is defined and is some member of the domain of the model $\mathcal{M}$. Since the model is normal, the interpretation of the equality symbol at each world is the equality relation, so $\langle v(p), v(p)\rangle \in \mathcal{I}(=, \theta(\sigma))$ and thus $\mathcal{M}, \theta(\sigma) \Vdash_v p = p$ holds. This says the branch $\mathcal{B}$ remains satisfiable after $\sigma(p = p)$ is added. We have shown the Reflexivity Rule preserves tableau satisfiability, provided normal models are used.

Showing the Substitutivity Rule preserves satisfiability is similar, and this is sufficient to establish soundness.

Completeness

As usual, completeness is shown in the contrapositive direction. We show that if a sentence Φ is not provable using the tableau rules, including those of this Chapter, then Φ can be falsified in some normal model. To be specific, we use the varying domain **K** rules. Adapting the argument to other logics is straightforward. The argument continues from where that of Chapter 5 left off. Recall how that proof went. We gave a systematic tableau construction procedure; then we showed that if the systematic procedure failed to produce a proof, a countermodel could be extracted from the resulting, generally infinite, tableau. Our first task now is to extend the systematic procedure to include the equality rules. Then we examine the countermodel that results from a failed tableau. In general it won't be normal, but we will see that it does meet certain conditions that allow us to convert it into a normal model. The method may seem complex, but the essence of it is, in fact, quite simple. It is carried through for a particular example in Section 7.7, and you may want to look at that if you find the presentation here heavy going.

Systematic Construction Procedure Additions We gave a systematic tableau construction procedure for varying domain **K** in Section 5.3. We now extend that to take the equality rules into account. And this is quite a simple business; we just add a few extra steps to the description of stage $n + 1$. According to the description in Section 5.3, we select a prefixed formula F on a branch and do certain things to it, depending on the form of F. Now in addition to what we previously did, we also do the following immediately afterward:

1. If a new prefix σ has been introduced to a branch, add to the branch end $\sigma\ p = p$ for every parameter p occurring on the branch.
2. If a new parameter p has been introduced to a branch, add to the branch end $\sigma\ p = p$ for each prefix σ occurring on the branch.
3. For each $\sigma\ p = q$ and for each $\sigma'\ \Phi(p)$ that occur on the branch, add $\sigma'\ \Phi(q)$ to the branch end, provided this prefixed formula does not already occur on the branch.

These complete the additions to the systematic tableau construction procedure.

Suppose the sentence Φ is not provable (allowing the equality rules as above). Then a systematically constructed tableau for $1\ \neg\Phi$ will not close. Select an open branch $\mathcal{B}$ from the unclosed systematic tableau for it. As we observed in Chapter 5, this branch meets certain conditions because of the systematic construction. One example of such a condition is: If $\sigma\ \neg\neg X$ is on $\mathcal{B}$, so is $\sigma\ X$; the full list of conditions can be found in Section 5.3. Because of the items added above to the systematic procedure, we also have that the following conditions are met.

Additional Branch Conditions For an open branch $\mathcal{B}$ of a systematically constructed tableau:

1. If σ and p both occur on $\mathcal{B}$, then $\sigma\ p = p$ is on $\mathcal{B}$.
2. If $\sigma\ p = q$ and $\sigma'\ \Phi(p)$ are on $\mathcal{B}$, so is $\sigma'\ \Phi(q)$.

Next, construct a model $\mathcal{M}$ from branch $\mathcal{B}$, exactly as in Section 5.3. In this model the possible worlds are the prefixes used on $\mathcal{B}$, the domains are sets of parameters, and each prefixed formula on $\mathcal{B}$ is satisfied in $\mathcal{M}$ with respect to any valuation v_0 mapping each parameter to itself, using the assignment θ mapping each prefix to itself. Consequently the unprovable formula Φ is falsified in this model, at world 1. In addition, because of the systematic construction, this model meets the following special conditions.

Equality Conditions In the model $\mathcal{M}$, for v_0 mapping parameters to themselves, for parameters p, p_i, q, q_i:

1. For $\sigma \in \mathscr{G}$ and each $p \in \mathscr{D}(\mathscr{M})$, $\mathscr{M}, \sigma \Vdash_{v_0} p = p$.
2. Suppose R is a k-place relation symbol. If $\mathscr{M}, \sigma_1 \Vdash_{v_0} p_1 = q_1, \ldots ,$
   $\mathscr{M}, \sigma_k \Vdash_{v_0} p_k = q_k$ and $\mathscr{M}, \sigma' \Vdash_{v_0} R(p_1, \ldots , p_k)$, then $\mathscr{M}, \sigma' \Vdash_{v_0}$
   $R(q_1, \ldots , q_k)$.
3. If $\mathscr{M}, \sigma \Vdash_{v_0} p = q$ holds for some world σ, it holds for every world σ'.

Item 1 is immediate from the first of the Additional Branch Conditions. For item 2, recall that in the Model Construction of Chapter 5 atomic formulas were taken to be true only if the tableau branch explicitly said they were to be true. Thus if $\mathscr{M}, \sigma' \Vdash_{v_0} R(p_1, \ldots , p_k)$, it must be that $\sigma' R(p_1, \ldots , p_k)$ is on $\mathscr{B}$. Similarly for $\mathscr{M}, \sigma_i \Vdash_{v_0} p_i = q_i$. Then item 2 follows from Additional Branch Condition 2 concerning substitutivity. Item 3 is a consequence of items 1 and 2 by the following argument. Suppose $\mathscr{M}, \sigma \Vdash_{v_0} p = q$ and σ' is some world other than σ. By item 1, $\mathscr{M}, \sigma' \Vdash_{v_0} p = p$, and then using the substitutivity property embodied in item 2, $\mathscr{M}, \sigma' \Vdash_{v_0} p = q$.

Now we make use of these Equality Conditions to convert $\mathscr{M}$ into a *normal* model that will also falsify the unprovable sentence Φ.

First, we define a relation on the domain of the model $\mathscr{M}$ as follows. For members $p, q \in \mathscr{D}(\mathscr{M})$ (that is, for parameters), call p and q *equivalent* if $\mathscr{M}, \sigma \Vdash_{v_0} p = q$ for some (any) $\sigma \in \mathscr{G}$, where as usual, $v_0(p) = p$ for all parameters. That is, p and q are equivalent if the model $\mathscr{M}$ "thinks" they are equal. We denote this by $p \sim q$.

This notion of equivalence is an equivalence relation in the technical sense of Definition 7.1.1. Recall, this means we have the following properties: reflexivity ($p \sim p$), symmetry ($p \sim q$ implies $q \sim p$), and transitivity ($p \sim q$ and $q \sim r$ imply $p \sim r$). We check this for transitivity, the other two conditions are easier.

Suppose $p \sim q$ and $q \sim r$. Then there are $\sigma, \sigma' \in \mathscr{G}$ such that $\mathscr{M}, \sigma \Vdash_{v_0} p = q$ and $\mathscr{M}, \sigma' \Vdash_{v_0} q = r$. Now we use item 2 of the Equality Conditions. Since $\mathscr{M}, \sigma' \Vdash_{v_0} q = r$ we can replace occurrences of q with occurrences of r. Since $\mathscr{M}, \sigma \Vdash_{v_0} p = q$, we get $\mathscr{M}, \sigma \Vdash_{v_0} p = r$, and hence $p \sim r$.

A common technique in mathematics is to replace a set on which an equivalence relation is defined by the collection of *equivalence classes*. We do so here.

DEFINITION 7.6.1. [Equivalence Class] For each $p \in \mathscr{D}(\mathscr{M})$, by $\overline{p}$ we mean $\{q \in \mathscr{D}(\mathscr{M}) \mid p \sim q\}$.

Thus $\overline{p}$ is the set of all parameters that are equivalent to p. Informally we can think of this as the set of parameters the model $\mathscr{M}$ can not distinguish from p. This is called the *equivalence class of p*. Equivalence classes have certain fundamental properties.

Equivalence Class Properties For the relation $\sim$:

1. Every member of $\mathcal{D}(\mathcal{M})$ is in some equivalence class. In particular, $p \in \overline{p}$.
2. No member of $\mathcal{D}(\mathcal{M})$ is in more than one equivalence class. In particular, if $\overline{p}$ and $\overline{q}$ have a member in common, $\overline{p} = \overline{q}$.
3. $p \sim q$ if and only if $\overline{p} = \overline{q}$.

Here are the reasons for these assertions.

Item 1 is the easiest. Since equivalence is reflexive, $p \sim p$, and it follows that $p \in \overline{p}$.

For item 2, suppose $r \in \overline{p}$ and $r \in \overline{q}$. We show $\overline{p} \subseteq \overline{q}$; the converse inclusion is similar. So, suppose $x \in \overline{p}$; we must show $x \in \overline{q}$. Since $r \in \overline{p}$, $r \sim p$, and since $r \in \overline{q}$, $r \sim q$. Since $r \sim p$, by symmetry $p \sim r$. Then since $r \sim q$, by transitivity $p \sim q$. Now, $x \in \overline{p}$, so $x \sim p$. Since $p \sim q$, by transitivity again, $x \sim q$, so $x \in \overline{q}$.

Finally, for item 3. In one direction, suppose $\overline{p} = \overline{q}$. By item 1, $p \in \overline{p}$, so $p \in \overline{q}$, and hence $p \sim q$. In the other direction, suppose $p \sim q$. Then $p \in \overline{q}$ and since $p \in \overline{p}$, the equivalence classes $\overline{p}$ and $\overline{q}$ have a member in common, hence $\overline{p} = \overline{q}$ by item 2.

New Model Construction We construct a new model $\mathcal{M}' = \langle \mathcal{G}', \mathcal{R}', \mathcal{D}', \mathcal{I}' \rangle$ from the old model $\mathcal{M}$ as follows. Set $\mathcal{G}' = \mathcal{G}$ and $\mathcal{R}' = \mathcal{R}$. Take for the domain of the model the collection of equivalence classes of the domain of $\mathcal{M}$, using the equivalence relation $\sim$. Set $\overline{t} \in \mathcal{D}'(\sigma)$ provided some member of the equivalence class $\overline{t}$ is in $\mathcal{D}(\sigma)$. And finally, for a k-place relation symbol R, set $\langle \overline{t_1}, \dots, \overline{t_k} \rangle \in \mathcal{I}'(R, \sigma)$ provided $\langle t_1, \dots, t_k \rangle \in \mathcal{I}(R, \sigma)$.

The definition of $\mathcal{I}'$ needs comment, because it is not immediately clear that it actually is a definition. Say, for example, that P is a one-place relation symbol. Then according to the definition, $\langle \overline{t} \rangle \in \mathcal{I}'(P, \sigma)$ just in case $\langle t \rangle \in \mathcal{I}(P, \sigma)$. But if we could have $\overline{t_1} = \overline{t_2}$, while $\langle t_1 \rangle \in \mathcal{I}(P, \sigma)$ but $\langle t_2 \rangle \notin \mathcal{I}(P, \sigma)$, then the definition of $\mathcal{I}'$ would be ambiguous. Fortunately, this cannot happen. If $\overline{t_1} = \overline{t_2}$ then $t_1 \sim t_2$, so $\mathcal{M}, \sigma' \Vdash_{v_0} t_1 = t_2$, for some σ'. If we also have $\langle t_1 \rangle \in \mathcal{I}(P, \sigma)$ then $\mathcal{M}, \sigma \Vdash_{v_0} P(t_1)$, so $\mathcal{M}, \sigma \Vdash_{v_0} P(t_2)$ by item 2 of the Equality Conditions, and so $\langle t_2 \rangle \in \mathcal{I}(P, \sigma)$.

We thus have a well-defined model $\mathcal{M}' = \langle \mathcal{G}', \mathcal{R}', \mathcal{D}', \mathcal{I}' \rangle$. We want to show two things about it, which will be sufficient to finish the entire completeness proof.

Normality The model $\mathcal{M}'$ is normal.

By definition of $\mathcal{I}'$, $\langle \overline{p}, \overline{q} \rangle \in \mathcal{I}'(=, \sigma)$ if and only if $\langle p, q \rangle \in \mathcal{I}(=, \sigma)$, if and only if $\mathcal{M}, \sigma \Vdash_{v_0} p = q$. But by definition of $\sim$, this is equivalent to $p \sim q$, and this is equivalent to $\overline{p} = \overline{q}$ by Equivalence Class Property 3.

Key Fact For each valuation v in $\mathcal{M}$, define a corresponding valuation v' in $\mathcal{M}'$ by setting, for each variable x, $v'(x)$ is the equivalence class containing $v(x)$. That is, $v'(x) = \overline{v(x)}$. We claim: for each formula X, for each $\sigma \in \mathcal{G}$, and for each valuation v in $\mathcal{M}$:

$$\mathcal{M}, \sigma \Vdash_v X \iff \mathcal{M}', \sigma \Vdash_{v'} X$$

Before showing the Key Fact, we verify two elementary, but technical, facts about valuations in $\mathcal{M}$ and in $\mathcal{M}'$, so as not to interrupt the main argument later.

FACT ONE Suppose that in $\mathcal{M}$, w is an x-variant of v at σ. Then in $\mathcal{M}'$, w' is an x-variant of v' at σ.

PROOF OF FACT ONE Suppose w is an x-variant of v at σ. That is, v and w agree on all variables except possibly x, and further, $w(x) \in \mathcal{D}(\sigma)$. If y is a variable other than x, $v'(y) = \overline{v(y)} = \overline{w(y)} = w'(y)$, so v' and w' also agree on all variables except possibly x. Since $w'(x) = \overline{w(x)}$ then $w(x) \in w'(x)$. By definition of $\mathcal{D}'$, an equivalence class is in $\mathcal{D}'(\sigma)$ provided some member of it is in $\mathcal{D}(\sigma)$. Since $w(x) \in \mathcal{D}(\sigma)$, it follows that $w'(x) \in \mathcal{D}'(\sigma)$. Thus w' is an x-variant of v' at σ.

FACT TWO Suppose that in $\mathcal{M}'$, u is an x-variant of v' at σ. Then there is an x-variant w of v at σ in $\mathcal{M}$ such that $w' = u$.

PROOF OF FACT TWO Suppose u is an x-variant of v' at σ in $\mathcal{M}'$. We define a valuation w in $\mathcal{M}$ as follows. On variables z other than x, set $w(z) = v(z)$; we still must define $w(x)$. Since u is an x-variant of v' at σ, $u(x) \in \mathcal{D}'(\sigma)$. By definition of $\mathcal{D}'$, for this to happen some member of the equivalence class $u(x)$ must be in $\mathcal{D}(\sigma)$. Choose one such member and set $w(x)$ to be it. This completes the definition of a valuation w in $\mathcal{M}$. We next show it has the properties we want.

By definition, $w(x) \in u(x)$. Then the equivalence classes $w'(x) = \overline{w(x)}$ and $u(x)$ have the member $w(x)$ in common, so $w'(x) = u(x)$ by Equivalence Class Property 2. Also, if z is a variable other than x, $w(z) = v(z) \in v'(z) = u(z)$ (since u and v' differ only on x). Then for $z \neq x$, $w'(z) = u(z)$, since these equivalence classes have $w(z)$ in common. It follows that for the valuation w we constructed, $w' = u$. And further, by construction, w itself is an x-variant of v at σ.

Now we turn to the Key Fact itself. This fundamental connection between models $\mathcal{M}$ and $\mathcal{M}'$ is shown by an inductive argument.

Induction Hypothesis Assume the Key Fact holds for all valuations v, all worlds σ, and all formulas Y that are less complex than X.

We show that, under this hypothesis, the result also holds for X. The argument has several cases, depending on the form of X.

ATOMIC Say X is $R(z_1, \ldots, z_k)$. This case does not actually use the induction hypothesis. By definition of truth for atomic formulas, $\mathcal{M}, \sigma \Vdash_v R(z_1, \ldots, z_k)$ is equivalent to $\langle v(z_1), \ldots, v(z_k) \rangle \in \mathcal{I}(R, \sigma)$. By definition of $\mathcal{I}'$, this is equivalent to $\langle v(z_1), \ldots, v(z_k) \rangle \in \mathcal{I}'(R, \sigma)$. This is equivalent to $\langle v'(z_1), \ldots, v'(z_k) \rangle \in \mathcal{I}'(R, \sigma)$, by definition of the valuation v'. Finally this is equivalent to $\mathcal{M}', \sigma \Vdash_{v'} R(z_1, \ldots, z_k)$.

PROPOSITIONAL CASES If X is a negation, conjunction, disjunction, etc., the argument is straightforward and is left to you.

MODAL CASES Say X is $\Diamond Y$, where by the induction hypothesis, the Key Fact holds for Y. $\mathcal{M}, \sigma \Vdash_v \Diamond Y$ is equivalent to $\mathcal{M}, \tau \Vdash_v Y$ where τ is some member of $\mathcal{G}$ for which $\sigma \mathcal{R} \tau$. By the induction hypothesis, $\mathcal{M}, \tau \Vdash_v Y$ is equivalent to $\mathcal{M}', \tau \Vdash_{v'} Y$. Since $\mathcal{R}' = \mathcal{R}$, this is equivalent to $\mathcal{M}', \sigma \Vdash_v \Diamond Y$.

The case where X is $\Box Y$ is similar and is omitted.

QUANTIFIER CASES We only consider the existential case since the other is similar; say X is $(\exists x)Y$, where the Key Fact holds for Y. We give the two directions of the argument separately.

Suppose first that we have $\mathcal{M}, \sigma \Vdash_v (\exists x)Y$. Then we have $\mathcal{M}, \sigma \Vdash_w Y$ where w is some x-variant of v at σ. By the induction hypothesis this implies $\mathcal{M}', \sigma \Vdash_{w'} Y$, and by Fact One, w' is an x-variant of v' at σ. Then $\mathcal{M}', \sigma \Vdash_{v'} (\exists x)Y$.

For the other direction, suppose we have $\mathcal{M}', \sigma \Vdash_{v'} (\exists x)Y$. Then there is a valuation u, in $\mathcal{M}'$, that is an x-variant of v' at σ, such that $\mathcal{M}', \sigma \Vdash_u Y$. By Fact Two, there is an x-variant w of v in $\mathcal{M}$ such that $w' = u$. Since $w' = u$, $\mathcal{M}', \sigma \Vdash_{w'} Y$. By the induction hypothesis, $\mathcal{M}, \sigma \Vdash_w Y$. And since w is an x-variant of v at σ, $\mathcal{M}, \sigma \Vdash_v (\exists x)Y$.

This completes the inductive argument, and establishes the Key Fact.

THEOREM 7.6.2. [Tableau Completeness With Equality] If the sentence Φ is valid in all normal varying domain first-order **K** models, Φ has a tableau proof using the varying domain **K** rules and the equality rules.

Proof Suppose Φ has no tableau proof. Apply the systematic tableau construction procedure, including the additions to incorporate the equality rules. It cannot produce a closed tableau, so an open branch $\mathcal{B}$ is generated. Using the method described in Chapter 5, produce a model $\mathcal{M}$ with prefixes as possible worlds, in which all the formulas on the branch are satisfied. In particular, since $1 \neg \Phi$ begins the branch $\mathcal{B}$, $\mathcal{M}, 1 \nVdash_{v_0} \Phi$ (where v_0 maps parameters to themselves).

Now, use the New Model Construction given above to produce a normal model $\mathcal{M}'$. It follows from the Key Fact that $\mathcal{M}', 1 \nVdash_{v'_0} \Phi$, and so Φ can be falsified in a normal model. ∎

We have dealt with only a single case: varying domain **K**. But the other cases are just small and tedious variations on this one. We stop here.

7.7. AN EXAMPLE

The completeness proof for tableaus with equality is full of details, and it is easy to lose the big picture. In this section we present the general ideas through an example. Specifically, we attempt to give a varying domain **K** proof of

$$[(\forall x)\Box(\exists y)(x = y) \wedge (\exists x)P(x)] \supset [\Diamond(\exists x)P(x) \supset (\exists x)\Diamond P(x)]$$

and from our failure, we construct a normal model falsifying it.

The failed tableau is given in Figure 6. We have not constructed it fully using our systematic procedure since the result is infinite, but we still have done enough to allow the extraction of a counter-model from it.

If we use the methods of Section 5.3, we can construct a falsifying model, call it $\mathcal{M}$, with two worlds, 1 and 1.1, given schematically in Figure 7.

As usual, we just do at each world what the tableau branch "tells" us to do. For example, since item 16 of the tableau is $1\, P(q_{1.1})$, at world 1 we interpret P to hold of $q_{1.1}$. We leave it to you to verify that the open branch of the tableau in Figure 6 is satisfied in the model $\mathcal{M}$ of Figure 7. Note that the model is *not* normal, because at world 1.1 we have $p_1 = q_{1.1}$ (because of item 15 of the tableau) though p_1 and $q_{1.1}$ are different objects.

The domain of model $\mathcal{M}$ is $\{p_1, p_{1.1}, q_{1.1}\}$. As specified in the previous section, we define a relation $\sim$ on this set by taking $x \sim y$ to hold if some world of the model above "thinks" $x = y$. So we have $p_1 \sim p_1$, $p_{1.1} \sim p_{1.1}$, $q_{1.1} \sim q_{1.1}$, $p_1 \sim q_{1.1}$, and $q_{1.1} \sim p_1$. Then we form equivalence classes, where $\overline{x} = \{y \mid y \sim x\}$. This gives us just two distinct classes.

$$\overline{p_1} = \overline{q_{1.1}} = \{p_1, q_{1.1}\}$$
$$\overline{p_{1.1}} = \{p_{1.1}\}$$

$$1 \quad \neg\{[(\forall x)\Box(\exists y)(x = y) \land (\exists x)P(x)] \supset$$
$$[\Diamond(\exists x)P(x) \supset (\exists x)\Diamond P(x)]\} \quad 1.$$
$$1 \quad (\forall x)\Box(\exists y)(x = y) \land (\exists x)P(x) \quad 2.$$
$$1 \quad \neg[\Diamond(\exists x)P(x) \supset (\exists x)\Diamond P(x)] \quad 3.$$
$$1 \quad (\forall x)\Box(\exists y)(x = y) \quad 4.$$
$$1 \quad (\exists x)P(x) \quad 5.$$
$$1 \quad \Diamond(\exists x)P(x) \quad 6.$$
$$1 \quad \neg(\exists x)\Diamond P(x) \quad 7.$$
$$1 \quad P(p_1) \quad 8.$$
$$1 \quad \Box(\exists y)(p_1 = y) \quad 9.$$
$$1 \quad \neg\Diamond P(p_1) \quad 10.$$
$$1.1 \quad (\exists x)P(x) \quad 11.$$
$$1.1 \quad (\exists y)(p_1 = y) \quad 12.$$
$$1.1 \neg P(p_1) \quad 13.$$
$$1.1 \quad P(p_{1.1}) \quad 14.$$
$$1.1 \quad p_1 = q_{1.1} \quad 15.$$
$$1 \quad P(q_{1.1}) \quad 16.$$
$$1.1 \neg P(q_{1.1}) \quad 17.$$
$$1 \quad p_1 = p_1 \quad 18.$$
$$1 \quad p_{1.1} = p_{1.1} \quad 19.$$
$$1 \quad q_{1.1} = q_{1.1} \quad 20.$$
$$1.1 \quad p_1 = p_1 \quad 21.$$
$$1.1 \quad p_{1.1} = p_{1.1} \quad 22.$$
$$1.1 \quad q_{1.1} = q_{1.1} \quad 23.$$
$$1.1 \quad q_{1.1} = p_1 \quad 24.$$

Items 2 and 3 are from 1 by a Conjunctive Rule, as are 4 and 5 from 2, and 6 and 7 from 3; 8 is from 5 by an Existential Rule; 9 is from 4 by a Universal Rule, as is 10 from 7; 11 is from 6 by a Possibility Rule; 12 is from 9 by a Necessity Rule, as is 13 from 10; 14 is from 11 by an Existential Rule, as is 15 from 12; 16 is from 8 and 15 by the Substitutivity Rule, as is 17 from 13 and 15; 18, 19, 20, 21, 22, and 23 are by the Reflexivity Rule; and 24 is from 15 and 21 by the Substitutivity Rule.

Figure 6.   A Failed Tableau Using Equality

Now we define a second model, $\mathcal{M}'$, based on the first one. It is given schematically in Figure 8. The worlds and the accessibility relation are the same as in $\mathcal{M}$. The domain of the model $\mathcal{M}'$ is the set of equivalence classes, $\{\overline{p_1}, \overline{p_{1.1}}\}$. We say an equivalence class is in the domain of a world in $\mathcal{M}'$ if some member of it is in the domain of that world in the model $\mathcal{M}$. Then $\overline{p_1}$

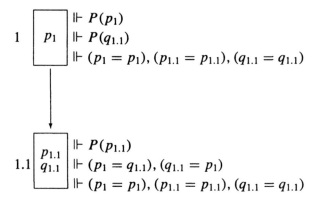

$$\Vdash P(p_1)$$
$$\Vdash P(q_{1.1})$$
$$\Vdash (p_1 = p_1), (p_{1.1} = p_{1.1}), (q_{1.1} = q_{1.1})$$

$$\Vdash P(p_{1.1})$$
$$\Vdash (p_1 = q_{1.1}), (q_{1.1} = p_1)$$
$$\Vdash (p_1 = p_1), (p_{1.1} = p_{1.1}), (q_{1.1} = q_{1.1})$$

Figure 7. A Non-Normal Counter Model $\mathcal{M}$

is in the domain of world 1 in $\mathcal{M}'$ since p_1 is a member of $\overline{p_1}$, and is in the domain of 1 in $\mathcal{M}$. Likewise $\overline{p_1}$ is in the domain of world 1.1 in $\mathcal{M}'$ since $q_{1.1}$ is a member, and $q_{1.1}$ is in the domain of world 1.1 in the model $\mathcal{M}$. Similarly $\overline{p_{1.1}}$ turns out to be in the domain of world 1.1 in $\mathcal{M}'$, but not in the domain of world 1.

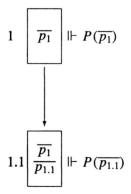

Figure 8. A Normal Counter Model $\mathcal{M}'$

We interpret P in the new model $\mathcal{M}'$ by saying it holds of an equivalence class, at a world, if it holds of its members, at that world, in the model $\mathcal{M}$. Thus, for instance, P holds of $\overline{p_1}$ at world 1 in $\mathcal{M}'$ since P holds of both p_1 and $q_{1.1}$ at world 1 in the model $\mathcal{M}$.

The interpretation of the relation symbol $=$ is equality, so we do not indicate it in the diagram of Figure 8.

In the model $\mathcal{M}'$, $\overline{p_1}$ is common to both worlds. This is essential for the formula $(\forall x)\square(\exists y)(x = y)$ to be true at world 1. But we leave it to you to verify that this new, normal, model $\mathcal{M}'$ does indeed falsify the formula we failed to prove.

EXERCISES

EXERCISE 7.7.1.   Give a varying domain **K** proof, using the equality rules, of the following sentence.

$$(\forall x)\square(\exists y)(x = y) \supset [(\exists x)\Diamond P(x) \supset \Diamond(\exists x)P(x)]$$

EXERCISE 7.7.2.   Construct a normal, varying domain, **K** model in which the following sentence is falsified.

$$(\forall x)\square(\exists y)(x = y) \supset [(\forall x)\square P(x) \supset \square(\forall x)P(x)]$$

Do so by attempting a tableau proof as we did above, and extracting a counter-model from it.

CHAPTER EIGHT

EXISTENCE AND ACTUALIST QUANTIFICATION

We saw, in Chapter 4, that two basic kinds of quantification were natural in first-order modal logics: possibilist and actualist. Possibilist quantifiers range over what *might* exist. This corresponds semantically to constant domain models where the common domain is, intuitively, the set of things that could exist. We also saw that in the possibilist approach we could introduce an existence primitive (Section 4.8) and relativize quantifiers to it, permitting the possibilist approach to paraphrase the actualist version.

Now it is time to consider things from the other side. In this chapter we take the actualist quantifier as basic. Our models are varying domain, and we think of the domain of each world as what actually exists at that world. At each world quantifiers range over the domain of that world only. We will see that this allows us to *define* an existence formula, rather than taking one as primitive. Further, we will be able to formulate single-formula versions of the Barcan and the Converse Barcan schemes. And this, in turn, will allow us to use varying domain machinery to reason in constant domain settings. In other words, the actualist quantifier allows us to talk in a possibilist mode. Incidentally, we will see that monotonic and anti-monotonic semantics can be captured by this approach as well.

The upshot of this and our earlier work in Chapter 4 is that a choice between the possibilist and the actualist quantifier cannot be decided on formal grounds. With good will on both sides, a possibilist philosopher and an actualist philosopher can talk meaningfully to each other—it probably happens all the time. It will take a bit more work to get each to speak meaningfully to a classical philosopher. The hurdle in this case is that the modal logicians believe it is meaningful to speak of things that do not exist, whereas classically, this has been thought to be incoherent. After presenting some of the formal details, we will rehearse the issues and arguments surrounding this difficult issue.

8.1. TO BE

The basic semantics of this chapter is varying domain—frames have a domain function that assigns to each possible world a domain of quantification for that world, and different worlds are allowed to have different domains. (We also assume all models are normal—the equality symbol is interpreted by the

equality relation on the domain of the frame. We will not say this each time.)
The value a valuation assigns to a free variable, x, in a varying domain model
must be something in the domain of the *model*. That is, it must be something
in the domain of *some* world, but at any given world that value may or may
not exist. If the value assigned to x exists at world Γ, that is, if it is in the
domain of Γ, quantifiers at Γ have that value in their range, and if the value
assigned to x does not exist at Γ, that value is out of quantifier range. This
means we can make use of the well-known formula $(\exists y)(y = x)$ to express
the existence of the value of x at a world. The following Proposition makes
this idea precise; the proof is simple, and is left to you.

PROPOSITION 8.1.1. Let $\mathcal{M} = \langle \mathcal{G}, \mathcal{R}, \mathcal{D}, \mathcal{I} \rangle$ be a normal, varying domain
model, and let v be a valuation. Then $\mathcal{M}, \Gamma \Vdash_v (\exists y)(y = x)$ if and only if
$v(x) \in \mathcal{D}(\Gamma)$, that is, if and only if $v(x)$ exists at world Γ.

This Proposition says we can use $(\exists y)(y = x)$ to express the existence of
x. The formula plays a fundamental role, so we introduce special notation for
it.

DEFINITION 8.1.2. [E(x)] For a variable x, $\mathsf{E}(x)$ abbreviates $(\exists y)(y = x)$,
where y is a variable distinct from x.

$\mathsf{E}(x)$ is true at a world of a model with respect to a valuation if and only if
the value assigned to x exists at that world. Now that we can explicitly assert
existence, certain peculiarities of varying domain semantics can be dealt with
naturally and easily. For instance, the classically valid formula $(\forall x)\Phi(x) \supset$
$\Phi(z)$ is not valid in varying domain modal semantics. At a world Γ of a
varying domain model, the value assigned to the free variable z in $\Phi(z)$ might
not exist, and so not be in the range of the quantifier $(\forall x)$. Then, even though
everything that exists at Γ makes $\Phi(x)$ true, the value assigned to z need not.
But if we knew the value assigned to z did, in fact, exist at Γ, things would
be different. The formula $[(\forall x)\Phi(x) \wedge \mathsf{E}(z)] \supset \Phi(z)$ *is* valid. We leave it to
you to check this. Thus we retain universal instantiation for those variables
that actually designate existent objects. This may be unfamiliar because in
classical logic variables are *assumed* to designate existent objects, so $\mathsf{E}(z)$ is
always satisfied and consequently is omitted.

There is a family of logics called *free logics*, whose characteristic feature is that variables need not designate (Bencivenga, 1986). The formal
machinery for varying domain modal logic, in fact, gives us a free logic.
(There are more things in heaven and earth, Horatio, than are dreamt of by
your quantifiers.)

EXERCISE 8.1.1.   Prove Proposition 8.1.1.

8.2. TABLEAU PROOFS

Varying domain quantifier rules were given in Chapter 5, and rules for equality were given in Chapter 7. Since quantification and equality is all that is involved in $E(z)$, we need no new tableau machinery to deal with formulas that contain it.

EXAMPLE 8.2.1.   Here is a varying domain proof of $(\forall z)E(z)$. Since quantifiers intuitively range over what exists, this states the obvious fact that everything that exists, exists. As it happens, modal issues are not involved in this example.

$$1\ \neg(\forall z)E(z)\quad 1.$$
$$1\ \neg(\forall z)(\exists y)(y = z)\quad 2.$$
$$1\ \neg(\exists y)(x = p_1)\quad 3.$$
$$1\ \neg(p_1 = p_1)\quad 4.$$

In this, 2 is 1 unabbreviated; 3 is from 2 by an Existential rule; 4 is from 3 by a Universal rule. Now closure is by a (derived) equality rule.

There are a couple of derived tableau rules that often simplify the use of $E(x)$ in tableau construction. The varying domain universal quantification rule only permits the use of parameters having the same subscript as the prefix—that is, from $\sigma\ (\forall x)\Phi(x)$ we can get $\sigma\ \Phi(p_\sigma)$, but not $\sigma\ \Phi(p_\tau)$ where $\tau \neq \sigma$. However, if we "know" p_τ exists at σ, we should be able to conclude $\sigma\ \Phi(p_\tau)$ after all. The following makes this precise.

DEFINITION 8.2.2.   [Parameter Existence Rule] A varying domain tableau branch containing $\sigma\ E(p_\tau)$ and $\sigma\ (\forall x)\Phi(x)$ can be extended with $\sigma\ \Phi(p_\tau)$. (There is a similar rule for branches containing $\sigma\ \neg(\exists x)\Phi(x)$.)

Here is the justification for this derived rule. Suppose, on a tableau branch, we have $\sigma\ E(p_\tau)$ and also $\sigma\ (\forall x)\Phi(x)$. Then $\sigma\ \Phi(p_\tau)$ can be added, by the following sequence of legitimate moves.

$$\sigma\ E(p_\tau)\quad 1.$$
$$\sigma\ (\forall x)\Phi(x)\quad 2.$$
$$\sigma\ (\exists y)(y = p_\tau)\quad 3.$$
$$\sigma\ q_\sigma = p_\tau\quad 4.$$
$$\sigma\ \Phi(q_\sigma)\quad 5.$$
$$\sigma\ \Phi(p_\tau)\quad 6.$$

Item 3 is item 1 unabbreviated; 4 is from 3 by an existential rule where q_σ is a new parameter; 5 is from 2 by a universal rule; 6 is from 4 and 5 by Substitutivity.

Here is another useful derived tableau rule. We leave its proof as an exercise. The intuition is that, while free variables might have nonxistent objects as values, the rules for parameters in varying domain tableau proofs ensure they always designate existents.

DEFINITION 8.2.3. [Parameter NonExistence Rule] A branch of a varying domain tableau containing $\sigma\ \neg E(p_\sigma)$ closes.

EXERCISES

EXERCISE 8.2.1. Justify the Parameter NonExistence derived rule.

EXERCISE 8.2.2. Show that the formula $[(\forall x)\Phi(x) \wedge E(z)] \supset \Phi(z)$ is valid in all varying domain models.

EXERCISE 8.2.3. Give a varying domain **K** proof of the following sentence.

$$[(\forall x)\Diamond E(x) \wedge (\exists x)\Box P(x)] \supset \Diamond(\exists x)P(x)$$

EXERCISE 8.2.4. Attempt to give a varying domain **K** proof of the following sentence.

$$(\exists x)\Box P(x) \supset \Diamond(\exists x)P(x)$$

From the failed attempt, construct a counter-model.

8.3. THE PARADOX OF NONBEING

We have been speaking about things that do not exist. Eventually we will show that the existence of unicorns, for example, can be formally denied in our models using definite descriptions and an existence predicate. But over the centuries philosophers have found speaking of nonexistence to be very problematic. In this and the next few sections we discuss the issues that are involved.

Puzzles and conundrums abound if one tries either to assert or to deny that something exists. The oldest among them finds its inspiration in some fragments of Parmenides, who cautioned against saying "what is not":

Come now and I will tell thee—listen and lay my word to heart—the only ways of inquiry that are to be thought of: one, that [That which is] is, and it is impossible for it not to be, is the Way of Persuasion, for Persuasion attends on Truth.

Another, that It is not, and must needs not be—this, I tell thee is a path that is utterly undiscernible; for thou couldst not know that which is not—for that is impossible—nor utter it.

For it is the same thing that can be thought and that can be.

What can be spoken of and thought must be; for it is possible for it to be, but it is not possible for "nothing" to be. These things I bid thee ponder; for this is the first way of inquiry from which I hold thee back. (Cornford, 1957, fragments 2,3,6 ll.1-3)

The ancient Greeks used "is" to express predication, to express identity, to express existence, and even to express truth; "is not," accordingly, could mean that a thing lacked a (or any) property, that it was not identical (with something), that it failed to exist, or that it was false. The ambiguity of his injunction rendered Parmenides's remarks exceedingly rich and evocative, stunning and provocative: a goldmine for philosophical excavation. The great craftsman Plato could not resist applying his logical spade to this archaeological dig, painstakingly separating out layers of different meanings in the sediment that the pressures of thought and speech had forced together. His results he recorded as a dialogue between Socrates and the mysterious Eleatic Stranger in the *Sophist*.

In that dialogue, Plato distinguished the "is" of predication from the "is" of identity, and even, some commentators would add, the "is" of existence. He determined that Parmenides was wrong if taken to urge that one can neither assert nor believe what is not, in the sense that one can neither assert nor believe false propositions. But he found an impassable kernel in Parmenides's injunction: speaking and thinking about nonexistent objects. Quine characterized the predicament in which he landed *Plato's Beard*: "historically," Quine says, "it has proved tough, frequently dulling the edge of Occam's Razor."(Quine, 1948, p. 2) Here are some highlights from Plato's discussion.

The Eleatic Stranger says "that the term 'what is not' must not be applied to anything that exists; ... and since it cannot be applied to what exists, neither can it properly be applied to 'something.' " From which he concludes that "to speak of what is not 'something' is to speak of no thing at all." He echoes Parmenides:

Must we not even refuse to allow that in such a case a person is *saying* something, though he may be speaking of nothing? Must we not assert that he is not even saying anything when he sets about uttering the sounds 'a thing that is not'?

Matters get worse. "[T]he nonexistent reduces even one who is refuting its claims to such straits that, as soon as he sets about doing so, he is forced to contradict himself," continues the Stranger. For when he spoke of the nonexistent as "*being* a thing not to be uttered or spoken of or expressed," he applied the term 'being' to it; moreover, "in speaking of it as 'a thing not to be expressed or spoken of or uttered,' [he] was ... using language as if referring to a single thing." And so he ends plaintively:

In that case there is nothing to be said for me. I shall be found to have had the worst of it, now and all along, in my criticism of the nonexistent.

Parmenides, then, appears vindicated: we are reduced to babbling or incoherence or self-contradiction if we try to speak about the utterly nonreal.

Thus we come to a genuine philosophical paradox, *The Paradox of Non-Being*: to say of a thing that is not that it is not is to say that it is something, viz., a thing that is not, and so it is. Here is a modern formulation due to Cartwright (1960):

To deny the existence of something—of unicorns, for example—we must indicate what it is the existence of which is being denied; and this requires that unicorns be referred to or mentioned; the negative existential must be about them. But things which do not exist cannot be referred to or mentioned; no statement can be about them. So, given that we have denied their existence, unicorns must after all exist. The apparently true negative existential is thus either false or not really a statement at all; and, since the argument applies as well in any other case, we seem forced to conclude that there are no true negative existentials. (p. 21)

The conclusion of the argument is unacceptable. But the argument is valid. From the two premises,

(A) If an individual (call him "John") denies the existence of something, then John refers to what he says does not exist;

(B) Things which do not exist cannot be referred to or mentioned; no statement can be about them.

the following conclusion is drawn:

(C) If John denies the existence of something, then what John says does not exist does exist.

This argument is logically straightforward. It requires little more than transitivity of implication. We need only to rephrase premise (B) to read

(D) If John refers to or mentions something, Ks, then Ks exist.

Then, as an instance of (D), taking "Ks" to be "what John says does not exist," we have

(E) If John refers to or mentions what he says does not exist, then what John says does not exist does exist.

And, from (A) and (E), by transitivity, we reach the desired conclusion (C).

The argument can be generalized. By parallel reasoning, we can show that one cannot think or believe anything about something that does not exist. And this is an equally unpalatable conclusion. Responses to the paradox have sorted themselves into two groups. Berlin (9 50) called them *Inflationists* and *Deflationists*. Inflationists accept premise (A) and deny premise (B); Deflationists accept premise (B) and deny premise (A). (Our semantics adopts the Inflationist position.)

8.4. DEFLATIONISTS

The Deflationists are the dominant group in twentieth century Anglo-American philosophy. They accept premise (B) of Section 8.3 but reject premise (A): existence is a necessary condition for reference, but no such semantic relation comes into play when denials of existence are made.

The main impetus for the Deflationist view stems from Russell (1905). We shall discuss Russell's analysis of names and definite descriptions in Chapter 12. But the flavor of his solution to the Paradox of NonBeing can be given here.

Before Russell, one would have thought that a sentence like

(8.1)   Winged horses exist

would, like

(8.2)   Winged horses fly

be treated as a subject/predicate sentence: (8.1) would be analyzed in such a way that the term "Winged horses" refers to some things and the sentence says about these things that they have the property of existence. By the same token,

(8.3)   Winged horses do not exist

would be treated as the negation of a subject/predicate sentence: (8.3) would be analyzed in such a way that the term "Winged horses" refers to some things and the sentence says about them that they lack the property of existence. So (8.1) would be expressed in a first-order language as

(8.4) $(\exists x)(x$ is a winged horse $\wedge\ x$ exists$)$

and (8.3) would be expressed as

(8.5) $(\exists x)(x$ is a winged horse $\wedge\ \neg x$ exists$)$

But, on Russell's view, this is wrong. Neither (8.1) nor (8.3) involves any reference to things *per se*. There are no winged horses, and so there are no such objects to refer to. Instead, (8.1) says there are things of a certain sort and (8.3) denies that there are things of a certain sort. So, the proper first-order interpretations of (8.1) and (8.3) are, respectively,

(8.6) $(\exists x)(x$ is a winged horse$)$.

and

(8.7) $\neg(\exists x)(x$ is a winged horse$)$.

What about a sentence like

(8.8)   Pegasus exists

which contains a singular term that *prima facie* refers to an object? Russell's view is that the name in (8.8) is not really serving as a genuine referential term and so the sentence should not be assigned the logical form of an atomic sentence, $F(a)$; rather we are to replace the name by a description or, as Quine (1948) later suggested, simply create one,

(8.9) $(\exists x)(x$ pegasizes$)$

where *pegasizes* or *is-identical-with-Pegasus* is an artificial predicate, true only of Pegasus.

The Deflationist *reductio* entitles us to conclude that either it cannot be the case that "Winged horses" is the subject of (8.1) or it cannot be the case that "exists" is the predicate. The Russellian Deflationist, however, appears to infer something stronger, namely, that "Winged horses" is not the subject and "exists" is not the predicate. This is unnecessarily strong. Let us look at the four possible analyses of (8.1)

1. "Winged horses" is subject; "exists" is predicate.
2. "Winged horses" is subject; "exists" is not predicate.
3. "Winged horses" is not subject; "exists" is predicate.
4. "Winged horses" is not subject; "exists" is not predicate.

The *reductio* only eliminates 1. Item 2 has no plausibility because one cannot have a subject without a predicate, so it too can be eliminated. This leaves 3 and 4. "Winged horses" cannot be the subject, whether it is true or false that "exists" is the predicate. So, there is little reason to be found in The Paradox of NonBeing for rejecting "exists" as a predicate.

Unfortunately, discussions of the Paradox of NonBeing have become intertwined with the notorious Ontological Argument for the existence of God. There is a widely entrenched belief that, to avoid the conclusion of the Ontological Argument, one must deny that *existence* is a property of objects. The Paradox of NonBeing has been viewed as confirming the logical correctness of this view and taking it one step further. Indeed, it has become standard to argue that "Winged horses" cannot be the subject of (8.1) *because* "exists" is not the predicate. We quote from the very influential explanation and defense of Russell's theory from Ryle (1932):

Since Kant, we have, most of us, paid lip service to the doctrine that 'existence is not a quality' and so we have rejected the pseudo-implication of the ontological argument: 'God is perfect, being perfect entails being existent, God exists.' For if existence is not a quality, it is not the sort of thing that can be entailed by a quality.

But until fairly recently it was not noticed that if in 'God exists' 'exists' is not a predicate (save in grammar), then in the same statement 'God' cannot be (save in grammar) the subject of predication. The realization of this came from examining negative existential propositions like 'Satan does not exist' or 'Unicorns are nonexistent'. If there is no Satan, then the statement 'Satan does not exist' cannot be about Satan in the way in which 'I am sleepy' is about me. Despite appearances the word 'Satan' cannot be signifying a subject of attributes. (p. 42)

But Ryle is wrong. Kant (1781) denied that "exists" was a predicate that enlarged the concept; he never denied that it was a predicate.

The idea of translating (8.1) into first-order logic as (8.6) appears to have been proposed first in the modern literature by Frege (1884): in his vocabulary, (8.1) is a statement about a concept, for it says that something falls under the concept *winged horse*.[20] Frege's analysis has been echoed by many others who followed him, and it has been supposed that the logical work of the grammatical predicate "exists" is exhausted by the existential quantifier,

---

[20] Frege never provided an analysis for *singular* existentials like (8.8). Carnap (1947) introduced the notion of an *individual concept* to this end; Quine (1948), working within the alternate tradition of handling singular terms devised by Russell (1905), invented the artificially constructed predicate used in (8.9).

∃. There are two ways of understanding the view just expressed. First, "exists" is not a predicate; an existential assertion is to be viewed as an existentially quantified sentence following the translation scheme just indicated (i.e., replace (8.1) by (8.6), replace (8.8) by (8.9)). Second, whether or not "exists" is a predicate, the translation indicated is appropriate and exhaustive of the content of an existential claim.

But neither of these views is correct. In the first place, there is no reason why we cannot include a predicate "x exists" in a classical first-order language if we hold that it is a predicate true of everything. Denials of existence, of course, turn out to be false. But in a very important sense, they should: classically everything exists, and so it is false that there are things that don't exist. Where the situation gets tricky, as in our use of names and descriptions in later chapters, we need a more complicated logical story about the role of these expressions. Russell's, among others, will do. But, nothing in this more complicated story requires that we abandon the idea that "exists" is a predicate.

So, there is little reason on the Deflationist view to deny that "exists" is a predicate. Is this predicate, however, redundant in classical first-order logic? More precisely, is the work done by "exists" exhausted by the existential quantifier? The answer is "No." For there is no way of saying "Something exists" or "Everything exists" unless there is a predicate—either primitive or defined—available to do the work of "exists": it will make no sense simply to use the quantifiers.

Our discussion of the Deflationist response to the Paradox of NonBeing has focused on the idea that in denying that something exists, we need refer to it. But we have still to look at the other assumption in the Deflationist position, viz., that we can only refer to things that exist.

8.5. PARMENIDES' PRINCIPLE

Premise (**B**), from Section 8.3, carries the burden in the argument. We call it *Parmenides' Principle*, and we repeat it for convenience.

PARMENIDES' PRINCIPLE Things which do not exist cannot be referred to or mentioned; no statement can be about them.

Parmenides' Principles is widely espoused. Plato, as we just saw, accepted it. So too did Frege (1884) and Russell (1905). More recently, Strawson (1950) says that one cannot refer to fictional or imaginary creatures. Searle (1968) takes "Whatever is referred to must exist" as a basic axiom of the theory of reference. And Quine (1948) holds that a singular term denotes if, and only if, the term can be replaced by a bound variable.

Still, Parmenides' Principle is not obviously true. In fact, on our ordinary understanding of the key terms "exist" and "refer to," it cannot plausibly be defended. For we do assert, for example, that Santa Claus lives at the North Pole, that Pegasus was captured by Bellerophon, that phlogiston does not exist. In each case, we talk about, respectively, Santa Claus, Pegasus, Bellerophon, and phlogiston—entities which no one today believes to exist, yet communication and intelligibility are not impaired. We can perfectly well identify what we are talking about. Surely, nobody confuses Santa Claus with either the archangel Gabriel or the novelist Marcel Proust, yet we all recognize that Santa Claus, also known as Saint Nick, is none other than Kris Kringle.

Ordinary language considerations aside, Parmenides' Principle is objectionable at a more profound level. Wittgenstein (1922), in an oft-quoted remark, warned of attempting to circumscribe the conceivable:

... in order to be able to set a limit to thought, we should have to find both sides of the limit thinkable (i.e., we should have to be able to think what cannot be thought). (p. 3)

Parmenides' Principle, which appears to speak about what cannot be spoken about, commits just such a blunder. This is precisely the problem the Eleatic Stranger puts his finger on in the passage from the *Sophist* quoted in Section 8.3.

Let's go through the problem slowly. In the original form as Cartwright phrased it, the condition on reference was "Things which do not exist cannot be referred to or mentioned; no statement can be about them." Stated this way, (B) is self-defeating: the very making of the claim falsifies what is being claimed. Consider the second half of (B), which says that no statement can be about things which do not exist. If no statement can be about things which do not exist, then, in particular, (B) cannot be about things which do not exist. But certainly, premise (B) tells us what no statement can be about: things which do not exist. The same self-defeating quality also infects the first part of (B). (B) says that the expression "things which do not exist" cannot refer to things which do not exist. Yet (B) does succeed, via this expression, in indicating what it is that cannot be referred to. And so, as we argued before, premise (B) refers to things which do not exist.

There does, however, seem to be a simple way out of this bind. Admit that (B) is ill-constructed and give it up. Instead of specifying what cannot be referred to, rephrase the condition so as to specify what can be referred to. This can be done easily. Recall the conditional form of premise (B) which we used for testing the validity of the paradox:

(D)   If John refers to or mentions something, Ks, then Ks exist.

(D) is not, in any straightforward sense, about things which do not exist, and there does not seem to be any reference to things which do not exist; So, (D) is not self-defeating like (B). And, most important, (D) says just what was intended in (B), namely, that existence is a necessary condition for referring. The dilemma we posed in the previous paragraph has, therefore, been sidestepped: we do not have to specify what cannot be referred to, and so we do not have to refer to things which do not exist.

   This solution, however, is illusory. Consider the following instantiation of (D):

(F)   If John refers to or mentions apples, then apples exist.

Among the statements governed by the condition laid down in (F) is (F) itself. For, if (F) is true, then since in the very claiming of (F), apples are referred to, apples exist. In other words, (F) is self-referential (a property it shares with (B)). Self-reference is not of itself pernicious. Although (F) has no untoward consequences, another instantiation of (D), namely,

(G)   If John refers to or mentions things which do not exist, then things which do not exist do exist

has the contradictory consequence:

Things which do not exist do exist,

since (G) itself makes a reference to things which do not exist, (D), then, fares even worse than (B), for whereas (B) was merely self-defeating, (D) implies a patent contradiction.

   In denying that certain things can be referred to, we explicitly refer to them, so that the very claiming of premise (B) is its own falsification. The Deflationist solution to the Paradox, which essentially involves accepting (B), cannot even get off the ground.[21]

   The more promising approach, then, would be to admit that one can speak about things that do not exist. This Inflationist solution, which has been the object of derision for so many years, therefore merits another look.

---

[21] A standard strategy here is to ascend to the "formal mode"—to speak about words rather than what the words are about—but the efficacy of this maneuver is suspect. If the formal mode version lacks the self-defeating character of the material mode version, it is open to the criticism that it is not equivalent, as it is supposed to be, to its material mode counterpart.

8.6. INFLATIONISTS

The author of *The Principles of Mathematics*, Bertrand Russell, famously urged: "[W]hat does not exist must be something; or it would be meaningless to deny its existence; and hence we need the concept of being as that which belongs to the nonexistent." Russell explains:

Being is that which belongs to every conceivable term, to every possible object of thought—in short, to everything that can possibly occur in a proposition, true or false, and to all such propositions themselves. Being belongs to whatever can be counted. If A be any term that can be counted as one, it is plain that A is something, and therefore that A is. 'A is not' must always be either false or meaningless. For if A were nothing, it could not be said not to be; 'A is not' implies that there is a term whose being is denied, and hence that A is. Thus unless 'A is not' be an empty sound, it must be false—whatever it may be, it certainly is. Numbers, the Homeric gods, chimeras, and four-dimensional spaces all have being, for if these were not entities of a kind, we could make no propositions about them. Thus being is a general attribute of everything, and to mention anything is to show that it is. (Russell, 1938, p. 449)

Russell allows us to deny the existence of a thing; it is *being* we cannot deny. *Being* is the more general attribute that belongs to everything; *existence* holds of only some beings.

Russell, therefore accepts premise (A) of the Paradox (Section 8.3): he agrees that in order to deny the existence of an object, we must refer to it. But he rejects premise (B), supplanting it instead with

(H)   Things which do not have being cannot be referred to or mentioned; no statement can be about them.

Existence is not a necessary condition for referring, but *Being* (something) is. There is an analogue of the original paradox for the Inflationist, the upshot of which is that there can be no true, meaningful denials of *Being*.

Russell's inflationism appears crass and dogmatic by comparison with the more subtle views of Alexius Meinong, the Austrian philosopher from whom he took his cue. To be sure, Meinong (1889) expressed his contempt for this prejudice for the actual:

[T]he totality of what exists, including what has existed and will exist, is infinitely small in comparison with the totality of the objects of knowledge. This fact easily goes unnoticed, probably because the lively interest in reality which is part of our nature tends to favor that exaggeration which finds the non-real a mere nothing ... or, more precisely, which finds the non-real to be something for which science has no application or at least no application of any worth. (p. 82)

And his own position is specifically provocative:

Those who like paradoxical modes of expression could very well say: 'There are objects of which it is true to say that there are no such objects'. (p. 83)

But his suggestion is no mere terminological trick to inflate the universe. Meinong spoke of objects as beyond *Being* and *NotBeing*, a point embodied in his Principle of the Independence of *Being-so* [Sosein] from *Being* [Sein]:

[T]he *Sosein* of an Object is not affected by its *Nichtsein*. The fact is sufficiently important to be explicitly formulated as the principle of the independence of *Sosein* from *Sein*. The area of applicability of this principle is best illustrated by consideration of the following circumstance: the principle applies, not only to Objects which do not exist in fact, but also to Objects which could not exist because they are impossible. Not only is the much heralded gold mountain made of gold, but the round square is as surely round as it is square. (p. 82)

His goal, then, is not the overpopulated unruly slum of a universe Quine (1948) ridiculed. Meinong's point is that an object can have properties even if it does not exist—indeed, even if it could not exist. Whether or not an object has such-and-such properties is *independent* of whether it exists. And this is just the semantic principle we have chosen in dealing with varying domain semantics: $F(x)$ is assigned a truth value at a world in which the value of x does not exist. We do appear to be able to identify and re-identify things that do not exist—Pegasus, Sherlock Holmes, Santa Claus. These do not appear to be utterly unreal. It is not as if one is faced with emptiness, a void: there is something on which to focus our attention and about which we can converse intelligently.

Meinong was a student of Brentano. He was influenced by Brentano's view that all mental attitudes are directed to objects—in some cases, to objects that do not exist. Meinong's task was to characterize the "intentional objects" Brentano had posited as that to which the mind is directed. An individual who is thinking of Pegasus is thinking of something. To be sure, Pegasus does not exist; but the thought is directed toward him, and so he must be something. I can entertain the thought (as all, by and large, admit) that the great racehorse Man-O-War had wings: here, my thought is directed at a nonexistent state of affairs, namely, Man-O-War's having wings. Why can I not similarly entertain the thought that Pegasus ran in the Kentucky Derby? Here my thought is also directed at a nonexistent state of affairs, namely, Pegasus's having run in the Kentucky Derby. Meinong's proposal is that referring is intentional. Unlike the verb "hit," where if it is true that John hit Jim, Jim must exist, John can refer to Jim even though Jim does not exist. "Refer," then, is like "worship," "assume," "postulate": the object of the verb need not exist.

Russell appears to be invoking the following principle.

Russell's Schema: If '*A*' is a meaningful singular term, *A* must be.

Essence does not entail *Existence*, as one tradition would have it, but it does entail *Being* on this view. But Meinong invokes something a bit stronger, an *Unrestricted Comprehension Schema for Singular Terms*, which we might put as follows.

Meinong's Schema: If 'the *A*' is a meaningful singular term, the *A* must be *A*.

Routley (1980) calls this "the characterization postulate." So stated, the comprehension schema is highly dubious.

Meinong's position has come down to us by and large as a prime example of philosophical foolishness, a target of derision ever since Russell regained his "robust sense of reality" and turned on it. It has found very few adherents. (Although Parsons (1985) and Routley (1980) have attempted to resuscitate the view.) What we have tried to do in these few paragraphs is make the motivation clearer. The problem Meinong addressed is real, and requires real solutions. It cannot be swept under the carpet. But the particular solution he offered has insuperable difficulties, which Russell himself pointed out.

Russell (1905) claimed that it violated the Law of Noncontradiction. According to Meinong, any significant denial of existence requires that that which is denied existence *be* something. Consider the claim that there is no least prime number in the open interval (19, 23). This assertion is meaningful, even true; so there must be a least prime number in the interval (19, 23), else it would make no sense to deny its existence. But which number could this be? The only integers in the interval are 20, 21 and 22, and none of these is prime. We appear to be committed to saying that there is a prime number in the interval (19, 23) . . . and also that there is no prime number in the interval (19, 23). To be sure, Meinong is aware of this result: he himself notes that the round square is as much round as it is square, which is to say that it is as much round as it is not round. How are we to make sense of this? Impossible objects are objects that cannot possibly be—and therefore cannot possibly be so or not be so.

Meinong, when faced with this difficulty, remained steadfast in his position and believed he could defend it. Russell, who pointed out the difficulty to him, was convinced this philosophical battle was over. He went over the hill and joined the Deflationists. But this, as we have seen, is fraught with its own difficulties.

We adopt a position less radical than either of these two men, while retaining the flavor of the Meinongian approach. We allow only possible objects into our formal model. So we reject both Russell's and Meinong's schemas. We distinguish terms that designate nonexistent objects, like "the golden mountain," from terms that fail to designate at all, like "the round square." But the discussion of this distinction will have to await the introduction of names and descriptions into our logical vocabulary, and the machinery of predicate abstraction, beginning in the next chapter.

EXERCISES

EXERCISE 8.6.1.   "There is nothing inflationary about inflationism, since inflationism postulates no more existents than does deflationism." Discuss.

8.7. UNACTUALIZED POSSIBLES

Our discussion, so far, has shown that the question whether "exists" is a predicate is actually several questions that are often not clearly distinguished:

1. Is "exists" a *predicate* in either the logical or grammatical sense?
2. Is "exists" a *universal* predicate, i.e., is it true of everything?
3. Is "exists" a *definable* predicate, i.e., is it expressible in terms already available in the theory?

We summarize the answers given by the actualist and possibilist philosophers in Figure 9. Whatever the differences between the two positions, each is committed to the idea of things that don't exist, but could, and we want to discuss this in the present section.

Question	Actualist	Possibilist
1: Predicate	Yes	Yes
2: Universal	Yes	No
3: Definable	Yes	No

Figure 9.   Actualist and Possibilist Existence

Quine (1948) presents the following famous criticism of the fictitious philosopher Wyman (a stand-in for Meinong) who believes in *unactualized possibles*:

Wyman's slum of possibles is a breeding ground for disorderly elements. Take, for instance, the possible fat man in that doorway; and, again, the possible bald man in that doorway. Are they the same possible man, or two possible men? How do we decide? How many possible men are there in that doorway? Are there more possible thin ones than fat ones? How many of them are alike? Or would their being alike make them one? Are no *two* possible things alike? Is this the same as saying that it is impossible for two things to be alike? Or, finally, is the concept of identity simply inapplicable to unactualized possibles? But what sense can be found in talking of entities which cannot meaningfully be said to be identical with themselves and distinct from one another? These elements are well-nigh incorrigible. By a Fregean therapy of individual concepts, some effort might be made at rehabilitation; but I feel we'd do better simply to clear Wyman's slum and be done with it. (p. 4)

We do not wish to defend the Meinongian view that the very meaningfulness of a singular term requires that there *be* an object the singular term stands for. We do not wish to endorse this reason for introducing unactualized possibles. But, we do use the notion of unactualized possibles in our construction of varying domain semantics, and we need to say something to defend the coherence of the idea.

The proper response to Quine's story is that the doorway is empty. There is nothing in the doorway. There is no fat man in the doorway. There is no thin man in the doorway. There is no possible fat man in the doorway. There is no possible thin man in the doorway. There is just nothing in the doorway.

It makes no sense to place unactualized possibles inside the actual world. That is the point of saying that they are unactualized.

Wyman thinks that there is no *actual* man in the doorway but that there is a *possible* one in the doorway. What does this mean? It could mean two distinct things. On the one hand, it could mean there is a man who is actual and who is not in the doorway, but who, in another possible world, is in the doorway. Julius Caesar is not in the doorway; but in another possible world, he is. On the other hand, it could mean that there is a possible man, e.g., Sherlock Holmes—an element in the domain of another possible world but not the actual one—who, in that world, is in the doorway. In neither case, however, do we have *in this world* anyone in the doorway.

Suppose we consider "the future man in the doorway." (Of course, it is always open to someone to say that this is an improper description insofar as it most likely fails to pick out a unique individual. Ignore this for the following discussion.) We might mean the actual man who in the future is in the doorway; we might mean the man who is not actual now but will be in the future who is in the doorway. In each case, there is no man in the doorway but there will be; in the first case, it is a man who exists now, but in the second case, it is a man who will exist in the future.

The temporal analogy gives a better handle on the question "How many?" In the first case, that depends upon which men will in the future appear in the doorway. It is doubtful whether we can make that determination now, but in the future, long after everyone who is now alive is dead, such a determination can be made. We go back to the year 1998 and see which men alive in 1998 subsequently appeared in the doorway. The question is understandable and can in principle be answered. How about with the second case? Here we have no clear answer. The problem is that we never reach a point at which we look back and count. There is always another moment in time. It could be, the correct answer is that there is an unlimited number of such men.

An unactualized possible does not exist in this world. But this does not mean that objects cannot exist in more than one possible world. We can speak of what *this* individual is like in another world. Similarly, we can speak of what individual in another world is in *this* doorway. It is not as if we have a shadowy individual in *this* doorway (Quine's "possible fat man in the doorway"). Once again, in the actual world, there is nothing in the doorway. In another world, there is.

EXERCISES

EXERCISE 8.7.1.   Construct a varying domain **K** model with distinct objects a and b so that, at a particular world Γ, a and b satisfy exactly the same formulas, provided those formulas do not contain $=$, and such that at Γ, $E(x)$ is true if the value of x is a, and false if the value of x is b.

EXERCISE 8.7.2.   The solution to Quine's puzzle works best for varying domain semantics. Is there a comparable story to be told for constant domain semantics?

8.8. BARCAN AND CONVERSE BARCAN, AGAIN

We have concluded the discussion of philosophical issues that began with Section 8.3. Now we turn to more formal matters, concerning the behavior of models and tableaus.

Monotonicity (objects that exist continue to do so in alternate worlds) and anti-monotonicity (no additional objects exist in alternate worlds) are important existence assumptions. These were investigated in Section 4.9. The two together give us locally constant domain models (Definition 4.9.9), and these have the same set of validities as constant domain models (Proposition 4.9.10). That is, we can think of varying domain semantics, restricted to monotonic, anti-monotonic models, as essentially constant domain semantics.

Monotonicity and anti-monotonicity are semantic conditions, but we saw in Section 4.9 that they correspond to something syntactic as well. Monotonicity of a frame is equivalent to the validity of all instances of the Converse

Barcan formula, while anti-monotonicity is equivalent to the validity of all instances of the Barcan formula. But these syntactic versions have their drawbacks: there are infinitely many instances of both the Barcan and the Converse Barcan schemas. Now, however, we have a richer language than we had in Chapter 4—equality has been added—and it turns out that these infinitely many instances can be replaced by single formulas. Although these formulas may not be sentences—they may involve free variables—they correspond to intuition quite closely.

We begin with the Converse Barcan formula, for which there are actually two equivalents, one of which does not involve a free variable. Semantically, the Converse Barcan formula corresponds to monotonicity. Monotonicity says that whatever exists continues to do so in all alternative worlds. At each world the actualist quantifier ranges over what exists at that world. So using our existence formula (which is defined in terms of equality) we can express monotonicity very simply with the single sentence $(\forall x)\Box E(x)$.

There is an alternative way of capturing monotonicity syntactically. Since it involves free variables, it is somewhat less handy than the sentence we just gave, but it provides some additional insights. The idea is that instead of expressing existence at the present world using the universal quantifier, as we did above, we make use of the defined existence formula. Here is our second way of capturing monotonicity: $E(x) \supset \Box E(x)$.

The following theorem says we have succeeded in what we set out to do: express monotonicity syntactically.

THEOREM 8.8.1. Let $\mathcal{F} = \langle \mathcal{G}, \mathcal{R}, \mathcal{D} \rangle$ be a varying domain frame. The following are equivalent:

1. $\mathcal{F}$ is monotonic.
2. The Converse Barcan formula is valid in every model based on $\mathcal{F}$.
3. $E(x) \supset \Box E(x)$ is valid in every normal model based on $\mathcal{F}$.
4. $(\forall x)\Box E(x)$ is valid in every normal model based on $\mathcal{F}$.

Proof The equivalence of items 1 and 2 was shown earlier in Chapter 4, as Proposition 4.9.6. The equivalence of item 1 and item 3 is rather straightforward, and we leave it as an exercise. That item 2 implies item 4 is easy: one instance of the Converse Barcan formula is $\Box(\forall x)E(x) \supset (\forall x)\Box E(x)$, but we also have the validity of $\Box(\forall x)E(x)$ (in Example 8.2.1 we gave a tableau proof of $(\forall z)E(z)$, and the validity of its necessitation follows). Showing that item 4 implies item 2 is most easily done using tableaus, and we leave it as an exercise as well. ∎

There is an important observation to be made at this point. Given a frame, the interpretation of the equality symbol in all normal models based on this frame will be the same since the equality symbol is always interpreted by the equality relation on the frame. It follows that a formula involving only the equality relation is valid in *some* model based on a frame if and only if it is valid in *every* model based on that frame. This is the case for $E(x) \supset \Box E(x)$ and $(\forall x)\Box E(x)$, and is a most important feature.

Next we turn to the Barcan formula. This time we have only a single equivalent. The Barcan formula corresponds to anti-monotonicity: anything that exists at a world alternative to this one also exists here. The open formula $\Diamond E(x) \supset E(x)$ expresses this very nicely. And the following says formally that it does so. Part of it was proved in Chapter 4 as Proposition 4.9.8. The rest is left as an exercise.

THEOREM 8.8.2. Let $\mathcal{F} = \langle \mathcal{G}, \mathcal{R}, \mathcal{D} \rangle$ be a varying domain frame. The following are equivalent:

1. $\mathcal{F}$ is anti-monotonic.
2. The Barcan formula is valid in every model based on $\mathcal{F}$.
3. $\Diamond E(x) \supset E(x)$ is valid in every normal model based on $\mathcal{F}$.

EXERCISES

EXERCISE 8.8.1.   Show the equivalence of items 1 and 3 in Theorem 8.8.1.

EXERCISE 8.8.2.   Give a varying domain tableau proof of the following.

$$(\forall x)\Box E(x) \supset [\Box(\forall x)\Phi(x) \supset (\forall x)\Box\Phi(x)]$$

EXERCISE 8.8.3.   Show the equivalence of items 1 and 3 in Theorem 8.8.2.

8.9. USING VALIDITIES IN TABLEAUS

Almost all tableau proofs we have given have been direct—they did not make use of assumptions. In Section 2.4 we did discuss how to use local and global assumptions in propositional tableau proofs, but for the most part this has played little role so far. Now such ideas become particularly useful because, as we have seen in the previous section, there are formulas that capture monotonicity and anti-monotonicity, and these are important semantic notions. We only need *global* assumptions in tableaus now, but we will need to make use of open formulas, which hitherto have played no role in tableaus. For convenience, we start from the beginning here—review of Section 2.4 would be nice, but is not necessary.

DEFINITION 8.9.1. [Using a Closed Global Assumption] To use a *closed* formula Φ as a global assumption in a tableau proof the rule is: $\sigma\ \Phi$ can be added to any branch, for any prefix σ that occurs on the branch.

This means we can have a version of tableaus incorporating *monotonicity* by using the varying domain rules and taking $(\forall x)\Box E(x)$ as a global assumption.

EXAMPLE 8.9.2. Here is a tableau proof, using the varying domain **K** rules, of $\Box(\exists x)\Diamond A(x) \supset \Box\Diamond(\exists x)A(x)$, *taking* $(\forall x)\Box E(x)$ *as an assumption.* Thus in effect it is a monotonic proof of the formula.

$$
\begin{array}{lll}
1 & \neg[\Box(\exists x)\Diamond A(x) \supset \Box\Diamond(\exists x)A(x)] & 1. \\
1 & \Box(\exists x)\Diamond A(x) & 2. \\
1 & \neg\Box\Diamond(\exists x)A(x) & 3. \\
1.1 & \neg\Diamond(\exists x)A(x) & 4. \\
1.1 & (\exists x)\Diamond A(x) & 5. \\
1.1 & \Diamond A(p_{1.1}) & 6. \\
1.1.1 & A(p_{1.1}) & 7. \\
1.1.1 & \neg(\exists x)A(x) & 8. \\
1.1 & (\forall x)\Box E(x) & 9. \\
1.1 & \Box E(p_{1.1}) & 10. \\
1.1.1 & E(p_{1.1}) & 11. \\
1.1.1 & \neg A(p_{1.1}) & 12.
\end{array}
$$

Items 2 and 3 are from 1 by a Conjunctive Rule; 4 is from 3 by a Possibility Rule; 5 is from 2 by a Necessity Rule; 6 is from 5 by an Existential Rule; 7 is from 6 by a Possibility Rule; 8 is from 4 by a Necessity Rule; 9 is our assumption; 10 is from 9 by a Universal Rule; 11 is from 10 by a Necessity Rule; and 12 is from 8 and 11 by the derived Parameter Existence Rule.

Using an *open* formula as a global assumption is almost as easy. Remember, a free variable can represent anything in the domain of the model.

DEFINITION 8.9.3. [Using an Open Global Assumption] To use an open formula $\Phi(x)$ as a global assumption in a tableau proof the rule is: $\sigma\ \Phi(p_\tau)$ can be added to any branch, for any prefix σ and any parameter p_τ that occur on the branch.

Now we can get the effect of anti-monotonicity by using the varying domain rules and taking $\Diamond E(x) \supset E(x)$ as an assumption.

EXAMPLE 8.9.4. Here is a tableau proof, using the varying domain **K** rules, of an instance of the Barcan formula, $(\forall x)\Box A(x) \supset \Box(\forall x)A(x)$, taking $\Diamond E(x) \supset E(x)$ as an assumption. In effect, it is a tableau verification of part of Theorem 8.8.2.

$$
\begin{array}{ll}
1 & \neg[(\forall x)\Box A(x) \supset \Box(\forall x)A(x)] \quad 1. \\
1 & (\forall x)\Box A(x) \quad 2. \\
1 & \neg\Box(\forall x)A(x) \quad 3. \\
1.1 & \neg(\forall x)A(x) \quad 4. \\
1.1 & \neg A(p_{1.1}) \quad 5. \\
1 & \Diamond E(p_{1.1}) \supset E(p_{1.1}) \quad 6.
\end{array}
$$

Items 2 and 3 are from 1 by a Conjunctive Rule; 4 is from 3 by a Possibility Rule; 5 is from 4 by an Existential Rule; 6 is our global assumption;

At this point the tableau branches, using item 6. The left branch contains $1 \neg\Diamond E(p_{1.1})$, from which we get $1.1 \neg E(p_{1.1})$, and this branch closes immediately using the Parameter NonExistence derived rule. The right branch continues as follows.

$$
\begin{array}{ll}
1 & E(p_{1.1}) \quad 7. \\
1 & \Box A(p_{1.1}) \quad 8. \\
1.1 & A(p_{1.1}) \quad 9.
\end{array}
$$

Item 8 follows from items 2 and 7 using the Parameter Existence derived rule; 9 follows from 8 by a Necessity Rule.

Finally, by using both $(\forall x)\Box E(x)$ and $\Diamond E(x) \supset E(x)$ as global assumptions, varying domain tableau machinery allows us to construct what are, in effect, constant domain proofs. We do not recommend this—the constant domain tableau rules we gave earlier are simpler to use. The point is simply that varying domain rules can be made to do constant domain work, just as constant domain rules can simulate varying domain arguments. On purely formal grounds, neither version has primacy.

EXERCISES

EXERCISE 8.9.1. Give a varying domain **K** proof of

$$\Box(\forall x)(\exists y)\Diamond R(x, y) \supset (\forall x)\Box\Diamond(\exists y)R(x, y)$$

using $(\forall x)\Box E(x)$ as a global assumption.

8.10. ON SYMMETRY

This section consists of a small remark, but it is of some technical interest. If the accessibility relation of a frame is symmetric, monotonicity and anti-monotonicity are equivalent—either implies the other. To say monotonicity and anti-monotonicity are equivalent is to say each instance of the Barcan formula is derivable from some instances of the Converse Barcan formula, and each instance of the Converse Barcan formula is derivable from some instances of the Barcan formula.

If we use the Barcan/Converse Barcan equivalents involving $E(x)$, the equivalence is straightforward to verify. Recall that to say a frame has a symmetric accessibility relation is equivalent to saying that all instances of the schema $X \supset \Box \Diamond X$ are valid in it, or equivalently, all instances of $\Diamond \Box X \supset X$ are valid. Now we have the following informal argument (formalizable axiomaticallly, however).

> 1. $\Diamond E(x) \supset E(x)$ anti-monotonicity
> 2. $\Box \Diamond E(x) \supset \Box E(x)$ necessitation
> 3. $E(x) \supset \Box E(x)$ using $X \supset \Box \Diamond X$

Thus assuming a syntactic equivalent to anti-monotonicity, plus a syntactic equivalent to symmetry, we can derive a syntactic equivalent to monotonicity. The other direction is equally easy.

Using the Barcan and Converse Barcan formulas directly is more work. Each instance of the Barcan formula is implied by a corresponding instance of the Converse Barcan formula, assuming symmetry. We give this as a tableau exercise below, Exercise 8.10.1. In the other direction, each instance of the Converse Barcan formula is implied by an instance of the Barcan formula as a global assumption, assuming symmetry. This is Exercise 8.10.2 below. We leave it as an open problem, how this can be established if we do not make use of equality.

EXERCISES

EXERCISE 8.10.1. Let B be the formula $(\forall x)\Box A(x) \supset \Box(\forall x)A(x)$ (an arbitrary instance of the Barcan formula). Let CB be the formula

$$\Box(\forall y)[(\forall x)\Box A(x) \supset \Box A(y)] \supset (\forall y)\Box[(\forall x)\Box A(x) \supset \Box A(y)]$$

(an instance of the Converse Barcan formula). Give a varying domain **B** proof of $CB \supset B$.

EXERCISE 8.10.2.   Give a varying domain **K** proof of the Converse Barcan formula

$$(\exists x)\lozenge P(x) \supset \lozenge(\exists x)P(x)$$

using the (open) Barcan formula

$$\lozenge(\exists x)(x = y) \supset (\exists x)\lozenge(x = y)$$

as a global assumption.

CHAPTER NINE

TERMS AND PREDICATE ABSTRACTION

Most presentations of first-order classical logic allow constant and function symbols to appear in formulas. This is the appropriate point for us to introduce them into our treatment of first-order modal logic. But doing so brings some unexpected complications with it. On the other hand, the rewards are great. The material that follows is, perhaps, the central part of this book.

9.1. WHY CONSTANTS SHOULD NOT BE CONSTANT

Classically, constant symbols can represent mathematical objects, like the numbers 3 or π, but they can also represent things like Ghengis Khan or the concept of goodness—in fact, any object that might reside in the domain of a classical model. Similarly, function symbols represent mathematical functions like *sin* or *cos*, and also functions like *the-mother-of*. In a classical model whose domain is a set of people (including each person's mother), a particular function symbol can represent the function that assigns, to each person in the domain, that person's mother. See almost any treatment of classical logic for the formal details—(Fitting, 1996a), for instance.

Suppose we pick a constant symbol, c, with the informal intention that it mean *the tallest person in the world*. Of course c will designate different people at different times. Classically we regard the expression as incomplete, and we eliminate any ambiguity by completing it to something like *the tallest person in the world on January 1, 1900*. Let us use the constant symbol c' for the completed meaning. The designation of the constant c' is independent of the here and now. Let's see how this works in a modal setting, reading $\Box$ temporally, to mean *at all future times*, and thinking of possible worlds as instants of time. The tallest person at one instant of time might not be the same person as the tallest person at another instant of time, and so the incomplete c might very well designate different people at different instants of time. But the completed constant c' designates the same person at all times—whoever was tallest on January 1, 1900. Constant symbols that designate the same object in all possible worlds are called *rigid designators*. In our example, c' is a rigid designator, but c is non-rigid. (In computer science, non-rigid designators are sometimes called *flexible*.)

Classically, such ambiguity is frowned upon and ambiguities are eliminated from logic. In modal talk, however, the ambiguity of nonrigid constants is actually quite desirable. For instance, consider the sentence,

> Someday, somebody will be taller than the tallest person
> in the world.

A natural reading of this sentence is

> The person that c designates at some future time is taller
> than the person that c designates at the present time.

And this is to read the constant expression in the original sentence as non-rigid, *the tallest person in the world* **now**, so that it can designate different objects at different times.

Here is another example where non-rigidity is desirable. Suppose we read the modal operator $\Box$ epistemically as "I know that." We want to create a natural model for the situation expressed by

> I know the world population is at least 5 billion, but I do
> not know if it is smaller than 6 billion.

For the epistemic model, we take the set of possible worlds, $\mathcal{G}$, to consist of all states of affairs that are compatible with my present knowledge. Let the constant symbol c designate the size of the world population. In the present world, c designates whatever number corresponds to the actual world population. In every possible world in $\mathcal{G}$, c designates a number that is at least 5 billion, but in some possible worlds the number that c designates is also greater than 6 billion. Then (assuming arithmetic operations have their usual interpretation) the sentence

$$\Box(c > 5 \cdot 10^9) \wedge \neg\Box(c < 6 \cdot 10^9)$$

should be true at the member of $\mathcal{G}$ corresponding to the actual state of affairs.

But the ambiguity that is introduced is a bit more complicated than just allowing constant symbols to designate different things at different worlds. We can see the complications in the following formal example.

EXAMPLE 9.1.1. Let $\mathcal{M}$ be the constant domain two-world model given schematically as follows.

$$\Gamma \quad \boxed{\alpha, \beta} \qquad \mathcal{I}(P, \Gamma) = \{\alpha\}$$

$$\downarrow$$

$$\Delta \quad \boxed{\alpha, \beta} \qquad \mathcal{I}(P, \Delta) = \{\alpha\}$$

We have two worlds, Γ and Δ, both with domain $\{\alpha, \beta\}$, with Δ accessible from Γ. The relation symbol P is interpreted to be true of α (only) at both worlds. And let us interpret the constant symbol c non-rigidly, taking it to designate α at Γ and β at Δ.

Now, let v be an arbitrary valuation. What status should we ascribe to the following?

$$\mathcal{M}, \Gamma \Vdash_v \Diamond P(c) \tag{9.1}$$

There are *two* reasonable scenarios.

First we could say that, since we are asking about world Γ, and c designates α there, then what we could mean by (9.1) is $\mathcal{M}, \Gamma \Vdash_{v'} \Diamond P(x)$, where x is a variable and v' is like v except that $v'(x) = \alpha$. We do this since, at Γ, c designates α. Then since Δ is the only world accessible from Γ, this is equivalent to $\mathcal{M}, \Delta \Vdash_{v'} P(x)$, and this is *true*, since $v'(x) = \alpha \in \mathcal{I}(P, \Delta)$.

Second we could say that, since $\Diamond P(c)$ is a formula whose principal operation symbol is $\Diamond$, by (9.1) we could mean $\mathcal{M}, \Delta \Vdash_v P(c)$. But this in turn should be equivalent to $\mathcal{M}, \Delta \Vdash_{v'} P(x)$, where v' is like v except that $v'(x) = \beta$. We do this now since, at Δ, c designates β. But this is *false*, since $v'(x) = \beta \notin \mathcal{I}(P, \Delta)$.

If a constant c designates non-rigidly, then the truth value of a sentence containing it, $\Phi(c)$, will vary depending upon how we select the object designated. "The tallest building in the world is in New York City" is not true now, but it has been true in the past, though not always. This is a complication we would expect from non-rigid designation. But what the example above shows is that non-rigid designation by the constant c interacts badly with the modal operator $\Diamond$ and we can get different truth values depending on whether we take designation of the constant symbol as primary, or whether we take the modal operation of moving to an alternate world as primary. And both situations come up everyday.

Consider, for instance,

(9.2)   The morning star is the evening star, and always will be.

Here we mean that the *objects*, morning star and evening star, are identical and, consequently, will remain so at all future times. By contrast, in

(9.3)   The ancients did not know that the morning star was the evening star,

we mean that, while the morning star and the evening star are, in fact, identical, there are alternate situations compatible with the knowledge of the ancients in which the objects designated by "morning star" and by "evening star" are different. In (9.2), we treat designation as primary, after which we consider future states of affairs. In (9.3), we consider alternate states of affairs before we consider what is designated.

Examples of non-rigid constant symbols, like "the tallest person in the world," are *definite descriptions*. We will not treat definite descriptions formally until Chapter 12. For the time being, terms built up from constant and function symbols will serve as a stand-in for definite descriptions. Our formal treatment in this chapter will handle constant and function symbols, so that we wind up with a full modal analog of first-order classical logic. The formal problems that arise for non-rigid constant and function symbols will also be problems for definite descriptions, and if we present an adequate treatment of these problems now, it will make things clearer in Chapter 12. Our informal examples of readings for non-rigid constant and function symbols should be taken as anticipations of the formal treatment of definite descriptions that is to come.

Much of what follows is indifferent to whether we use a possibilist or an actualist reading of quantifiers, that is, whether we use constant or varying domain models. Systematically we will use varying domain semantics, since this includes constant domain semantics as a special case (by taking the domain function to be constant). When only one version is appropriate, we will say so.

Also, to anticipate another important issue in connection with definite descriptions, we will be distinguishing terms that designate nonexistent objects from terms that do not designate. Assuming a temporal reading, the expression "the victorious General of Waterloo" designates a nonexistent, Wellington. But the expression "the present King of France" at the present time designates no one at all. The treatment of non-designation is a separate issue from that of non-rigidity. It is more perspicuous if we deal with one problem at a time. In this chapter we show how non-rigidity can be treated formally, assuming terms always designate. In Chapter 11 we deal with non-designation.

The task of this chapter is set by the blunt lesson of Example 9.1.1. *Formal syntax drawn from that of classical logic cannot distinguish modal meanings we can readily distinguish intuitively.*

9.2. SCOPE

A modal sentence $\Box F(c)$ containing the constant c has a syntactic ambiguity that can engender a semantic ambiguity. On the one hand, we can regard the constant as primary: c designates an object in the actual world and that object is said, in every possible world, to be F. On the other hand, we can regard the modal operator as primary: in every possible world, the object designated by c in that world is said to be F. There are, in other words, two operations involved in reading the sentence. One of these is the world-shift operation, represented by the necessity operator. This operation is familiar from earlier chapters. But, designation (of an object by a constant symbol) is also an operation. These two operations do not commute, and so there is ambiguity in reading the sentence. It is, in fact, a *scope* ambiguity.

When we add a list of numbers, e.g., $3 + 4 + 5$, the order in which we proceed does not matter. We get the same result whether we first compute $3 + 4$ and then add that result to 5, or, alternatively, whether we first compute $4 + 5$ and add that result to 3. Either way, we end up with 12.

$$\overbrace{\underbrace{7 + 5}_{3 + 4 + 5}}^{12} \qquad \overbrace{3 + \underbrace{9}_{4 + 5}}^{12}$$

On the other hand, when we mix operations, as in $3 \times 4 + 5$, ordering does matter. If we compute 3×4 first and add the result to 5, we get 17. On the other hand, if we compute $4 + 5$ first and multiply the result by 3, we get 27.

$$\overbrace{\underbrace{12 + 5}_{3 \times 4 + 5}}^{17} \qquad \overbrace{3 \times \underbrace{9}_{4 + 5}}^{27}$$

The notation $3 \times 4 + 5$ is therefore ambiguous: without a convention indicating the order of operations, it can denote 17 or it can denote 27. The usual way of clarifying the notation is by using parentheses: $(3 \times 4) + 5 = 17$ and $3 \times (4 + 5) = 27$.

We are familiar with this use of parentheses in propositional logic to clarify the scope of operations in a formula. For example, we distinguish $(P \lor Q) \land R$ from $P \lor (Q \land R)$. Standard notation is designed to be free of ambiguity so that a complex formula will uniquely decompose into its atoms, guaranteeing that an assignment of truth values to the atoms yields a unique truth value to the complex. But, we have reached a point where standard notation breaks down—non-rigid constant symbols make it possible to write ambiguous formal sentences.

Aristotelian logicians identify a scope ambiguity engendered by negation in a sentence like

(9.4)   Wellington is *not* happy.

On the one hand, it could be the sentence that is negated, i.e.,

(9.5)   It is *not* the case that Wellington is happy,

or it could just be the predicate that is negated, i.e.,

(9.6)   Wellington is *not*-happy.

For the Aristotelian, (9.6) implies (9.5), but not conversely. In modern first-order logic, this distinction is not marked in the symbolism. Taking w to abbreviate "Wellington" and Hx to be the predicate "x is happy", (9.5) could be symbolized as $\neg(Hw)$ and (9.6) could be symbolized as $(\neg H)w$. Think of the first as denying a certain positive property to Wellington, and the second as ascribing a negative property to him. But in standard formal logic, both versions are represented by $\neg Hw$, and the Aristotelian distinction is lost. This is reasonable since in modern non-modal symbolic logic the distinction plays no role. Modally, however, the distinction is significant, and we will shortly introduce a better notation to represent it.

Distinctions like the Aristotelian one above were found to be significant when definite descriptions that lacked designations were considered, even without modal operators present. This led to the introduction of a formal scoping mechanism, introduced by Whitehead and Russell (1925) in connection with their celebrated Theory of Descriptions.[22] On their view,

The present King of France is *not* happy,

is ambiguous. Let us abbreviate "the present King of France" as f. Then, as above, we might represent the two readings by:

(9.7) $\neg(Hf)$

and

(9.8) $(\neg H)f$

(This, of course, is not Russell's notation.) In (9.7), we are denying that the present King of France has a certain property, namely happiness, whereas in (9.8), we are saying that the present King of France has a certain property, namely non-happiness. The logical distinction between the two shows itself when we consider a situation, like now, in which there is no King of France. If we assume a nonexistent being has no properties, the first should be true since it denies the King has a property; and the second should be false, since it says he has a property.

For Russell, scope was a logical factor when the description failed to designate an existent, but even if it did designate an existent, scope was still

---

[22] Russell (1905) originally introduced a distinction between *primary* and *secondary* occurrences of a description to handle such modal contexts. The full *scope* treatment was not presented in any detail until Whitehead and Russell (1925).

a factor in non-truth-functional contexts. This is precisely the situation that affects us, for modal contexts are non-truth-functional.

In Russell's terminology, in (9.7) the description f has *small* or *narrow* scope, and in (9.8), it has *large* or *wide* or *broad* scope. This is a distinction somewhat analogous to the *de re/de dicto* distinction. We will see that such distinctions are not really general enough to account for all the cases of interest, and a more general scoping mechanism is needed.

Earlier, we identified the medieval distinction between *de dicto* and *de re* necessity for formulas involving modal operators and quantifiers. We now find that the very same distinction needs to be made also for formulas involving modal operators and constants. The sentence

The number of planets is necessarily greater than 7

can be read *de dicto* as

(9.9)   It is necessarily true that the number of planets is greater than 7.

and *de re* as

(9.10)   The number of planets is such that it is necessarily greater than 7

In (9.9) it is the proposition [*dictum*] that is said to be necessarily true. In (9.10), on the other hand, it is the thing [*res*], i.e., the number of the planets, that is said to have a certain property necessarily, the property of *being greater than 7*.

The *de re/de dicto* distinction covers the modals and propositional attitude constructions generally. For example, upon meeting someone of whom one has only seen photographs, one might say

I thought you were taller than you are.

This can be taken in either of the following ways:

(9.11)   I thought your height was greater than your height.

(9.12)   Your height, n, is such that I thought you were taller than n.

In (9.11), the *de dicto* reading, I attribute to myself a thought that is patently false; in (9.12), the *de re* reading, I do no such thing. Under most circumstances, it is (9.12) that is intended, not (9.11).

Temporal language also invites a *de re/de dicto* distinction. Consider

The prime minister will be a Laborite.

This could be taken to mean either of the following:

(9.13)   It will be the case that the prime minister is a Laborite,

or

(9.14)   The prime minister is such that he will be a Laborite.

In (9.14) we are speaking of the current prime minister; in (9.13) we are speaking about whoever will be the prime minister in the future, not necessarily the same person as the one who occupies the position now. On the *de dicto* reading, (9.13), one moves to a future world to determine the reference of the singular term and says of that individual that he is a Laborite. On the *de re* reading, (9.14), one picks out the individual who is prime minister in this world and says of him that there is a future world in which he is a Laborite.

EXERCISES

EXERCISE 9.2.1.   Give an example illustrating the *de re/de dicto* distinction for the epistemic case. Similarly for the epistemic case.

9.3. PREDICATE ABSTRACTION

Informally, classical first-order formulas represent *predicates*. A formula such as $\Phi(x)$, with one free variable, defines in each model a certain class of objects—loosely the members of the domain of that model for which $\Phi(x)$ is true. There is a harmless ambiguity in the notation that generally goes unnoticed. Suppose $\Phi(x)$ is the formula $P(x) \wedge Q(x)$. We take this formula to express the conjunction of two simpler predicates, P and Q; but, alternatively, we could have taken it to express the application of a compound predicate $[P \wedge Q]$. We will not try to make the distinction any clearer since, under any plausible interpretation, the results will be the same.

In a modal setting, however, things are quite different. We saw in Example 9.1.1 that $\Diamond P(c)$ gives different results if we think of it as the "possible-P" predicate applied to c, or if we think of it as asserting that the P predicate applied to c is possible. Formulas can no longer be thought of as representing predicates, pure and simple. Rather, a representation of a predicate can be *abstracted* from a formula. This is the purpose of the device of predicate abstraction. We make a distinction between a formula,

$\Phi(x)$,

and the predicate abstracted from it,

$$\langle \lambda x.\Phi(x) \rangle.$$

In part, this is by analogy with the *lambda-calculus*, in which a distinction is made between an *expression* like $x + 3$, and the *function abstracted from it*, $\langle \lambda x.x + 3 \rangle$. Once the formal machinery is in place we will see, as promised, that

$$\langle \lambda x.(P(x) \wedge Q(x)) \rangle(c)$$

and

$$\langle \lambda x.P(x) \rangle(c) \wedge \langle \lambda x.Q(x) \rangle(c)$$

do indeed behave alike. On the other hand,

$$\langle \lambda x.\Diamond P(x) \rangle(c)$$

and

$$\Diamond \langle \lambda x.P(x) \rangle(c)$$

are quite different, with behaviors corresponding to the two readings that arose in Example 9.1.1.

We have been abbreviating formulas like $(P(x) \wedge Q(x))$ by $P(x) \wedge Q(x)$, omitting outer parentheses. We will often apply a similar device within predicate abstracts, abbreviating $\langle \lambda x.(P(x) \wedge Q(x)) \rangle(c)$ by $\langle \lambda x.P(x) \wedge Q(x) \rangle(c)$. No ambiguity arises, and since parentheses multiply when predicate abstraction is involved, clarity is often enhanced by parenthesis omission.

All the basic ideas of predicate abstraction were introduced into modal logic by Stalnaker and Thomason (1968), and continued in (Thomason and Stalnaker, 1968). Bressan (1972) gave an extensive development, involving a treatment of higher-order modal logic, but the work is complex, and it has not had the influence it deserves. Further modal applications of predicate abstraction appear in (Fitting, 1972a; Fitting, 1973; Fitting, 1975). Then there was a long gap, until recent work in (Fitting, 1991; Fitting, 1993; Fitting, 1996b). This is probably the first book to make use of predicate abstraction in modal logic since (Bressan, 1972).

9.4. ABSTRACTION IN THE CONCRETE

Most of the machinery of modal models remains exactly as it was in Chapter 4. There are specific modifications and additions to allow predicate abstraction. We say what these are, beginning with syntax issues.

Syntax

We start with the notion of a term—a standard item in classical first-order logic.

Just as with relation symbols, we assume we have available an infinite list of *one place function symbols*, f_1^1, f_2^1, f_3^1, ... , an infinite list of *two place function symbols*, f_1^2, f_2^2, f_3^2, ... , an infinite list of *three place function symbols*, f_1^3, f_2^3, f_3^3, ... , and so on. We will generally be informal in our notation, and use f, g, or something similar for a function symbol, with its arity determined from context. We also assume we have an infinite list of *constant symbols*, c_1, c_2, c_3, ... (Constant symbols are sometimes taken to be function symbols of arity 0.) We will use c, d, and the like, informally as constant symbols.

DEFINITION 9.4.1. [term] The collection of *terms* is specifiec by the following rules.

1. Every variable (including parameters) is a term.
2. Every constant symbol is a term.
3. If f is an n-place function symbol, and t_1, ... , t_n are terms, then $f(t_1, \ldots , t_n)$ is a term.

For example, if f is a 3-place function symbol, g is a 2-place function symbol, x and y are variables, and c and d are constant symbols, then

$$f(x, g(c, x), g(y, g(c, d)))$$

is a term. You might try showing this, using the definition.

We keep the definition of atomic formula exactly as it was in Chapter 4. An atomic formula is an expression of the form $R(x_1, \ldots , x_n)$, where R is an n-place relation symbol and $x_1, \ldots , x_n$ are variables. (Classically, arbitrary terms are allowed to appear in atomic formulas where we have only allowed variables. As we saw above, this leads to ambiguous readings, and we avoid it modally.)

Finally, our earlier definition of *formula* continues to be used, but with the addition of the following.

DEFINITION 9.4.2. [First-Order Modal Formulas, continued] The following clauses are added to Definition 4.1.2:

6A. If Φ is a formula and v is a variable then $\langle \lambda v.\Phi \rangle$ is a *predicate abstract*; the free variable occurrences of $\langle \lambda v.\Phi \rangle$ are those of Φ except for occurrences of v.

6B. If $\langle \lambda v.\Phi \rangle$ is a predicate abstract and t is a term, $\langle \lambda v.\Phi \rangle(t)$ is a formula; the free variable occurrences of $\langle \lambda v.\Phi \rangle(t)$ are those of $\langle \lambda v.\Phi \rangle$ together with all variable occurrences in t.

For example, the following is a formula, and exactly one variable occurrence in it is free: $(\forall y)\langle \lambda x.P(x) \rangle(f(x, y))$. Try showing this.

It is intended that $\langle \lambda v.\Phi \rangle$ be thought of as the predicate "abstracted" from the formula Φ. Then $\langle \lambda v.\Phi \rangle(t)$ should be read, *the object designated by t has the property* $\langle \lambda v.\Phi \rangle$.

Frequently iterated predicate abstraction comes up. It makes formula reading simpler to introduce a modified notation for this. From now on, we often abbreviate a formula like

$$\langle \lambda x_1.\langle \lambda x_2.\langle \lambda x_3.\Phi \rangle(t_3) \rangle(t_2) \rangle(t_1)$$

with the simpler expression

$$\langle \lambda x_1, x_2, x_3.\Phi \rangle(t_1, t_2, t_3)$$

and similarly for other formulas involving iterated predicate abstraction. Of course, when behavior in models, or in proofs, is an issue, it is the unabbreviated version we must work with.

Semantics

We have added constant and function symbols to the language, so we must give meaning to them in models. Not surprisingly, constant symbols should designate objects, and function symbols should designate functions. Since we allow both to be non-rigid, reference is allowed to change from world to world.

DEFINITION 9.4.3. [Non-Rigid Interpretation] $\mathcal{I}$ is a *non-rigid interpretation* in an augmented frame $\mathcal{F} = \langle \mathcal{G}, \mathcal{R}, \mathcal{D} \rangle$ if $\mathcal{I}$ meets the conditions of Definition 4.6.2 (assigning relations to relation symbols) and also meets the following conditions.

1. To each constant symbol c, and to each $\Gamma \in \mathcal{G}$, $\mathcal{I}$ assigns some member of the domain of the frame. That is, $\mathcal{I}(c, \Gamma) \in \mathcal{D}(\mathcal{F})$.
2. To each n-place function symbol f, and to each $\Gamma \in \mathcal{G}$, $\mathcal{I}$ assigns some n-ary function from the domain of the frame to itself. That is, $\mathcal{I}(f, \Gamma) : \mathcal{D}(\mathcal{F})^n \to \mathcal{D}(\mathcal{F})$.

It can, of course, happen that the reference of a term at a world of a model might not be something in the domain of that world. We are explicitly allowing terms to designate nonexistent objects, though the objects designated must exist at *some* world. (Notice that this becomes moot if we are dealing with a constant domain model, for then existence anywhere is existence everywhere.)

As an example, consider the definite description "the first President of the United States." (As we noted earlier, we will treat definite descriptions formally in Chapter 12. For now, intuition will suffice.) Using the natural temporal reading, the definite description designates George Washington: it designates him now, in this world, even though he does not now exist.

DEFINITION 9.4.4. [Non-Rigid Model] A *non-rigid model* is a structure $\mathcal{M} = \langle \mathcal{G}, \mathcal{R}, \mathcal{D}, \mathit{l} \rangle$ where $\langle \mathcal{G}, \mathcal{R}, \mathcal{D} \rangle$ is an augmented frame and l is a non-rigid interpretation in it.

We used constant and function symbols to build up terms, so next we must assign designations to them in models. Terms generally contain not only function and constant symbols, but also free variables. Meanings are assigned to function and constant symbols by the interpretation of the model, while values are assigned to variables by valuations. Consequently, to specify a designation for a term, both must be taken into account.

DEFINITION 9.4.5. [Term Evaluation] Let $\mathcal{M} = \langle \mathcal{G}, \mathcal{R}, \mathcal{D}, \mathit{l} \rangle$ be a non-rigid model, let $\Gamma \in \mathcal{G}$, and let v be a valuation in $\mathcal{M}$. We associate a value at Γ with each term t, denoted $(v \star \mathit{l})(t, \Gamma)$, as follows.

1. If x is a free variable (possibly a parameter), $(v \star \mathit{l})(x, \Gamma) = v(x)$.
2. If c is a constant symbol, $(v \star \mathit{l})(c, \Gamma) = \mathit{l}(c, \Gamma)$.
3. If f is an n-place function symbol,

$$(v \star \mathit{l})(f(t_1, \ldots, t_n), \Gamma) = \mathit{l}(f, \Gamma)((v \star \mathit{l})(t_1, \Gamma), \ldots, (v \star \mathit{l})(t_n, \Gamma))$$

Part 3 of this Definition is a little complicated, but the underlying ideas are straightforward. Suppose each of the terms $t_1, \ldots, t_n$ has been assigned a value at Γ, say $\alpha_1, \ldots, \alpha_n$ respectively. The interpretation l associates some n-ary function with the function symbol f at Γ, say $\mathit{l}(f, \Gamma) = \theta$. Then the value assigned to the term $f(t_1, \ldots, t_n)$ at Γ is $\theta(\alpha_1, \ldots, \alpha_n)$, the result of applying the designation of f at Γ to the designations of $t_n, \ldots, t_n$, at Γ.

Finally, the main event.

DEFINITION 9.4.6. [Truth in a Non-Rigid Model] Let $\mathcal{M} = \langle \mathcal{G}, \mathcal{R}, \mathcal{D}, \mathit{l} \rangle$ be a non-rigid model. The definition of $\mathcal{M}, \Gamma \Vdash_v X$ is exactly as in Definition 4.6.7, with the addition of one more clause:

9. $\mathcal{M}, \Gamma \Vdash_v \langle \lambda x.\Phi \rangle(t) \iff \mathcal{M}, \Gamma \Vdash_w \Phi$, where w is the x-variant of v such that $w(x) = (v \star \mathit{l})(t, \Gamma)$.

That is, $\langle \lambda x.\Phi \rangle(t)$ is true at Γ if Φ turns out to be true at Γ when we assign to x whatever it is that t designates at the world Γ.

EXAMPLE 9.4.7.   We return to Example 9.1.1, applying the machinery we have now developed. Earlier we specified an interpretation of the constant symbol c informally. Now we make it formal by adding the following conditions on the interpretation $\mathcal{I}$.

$$\mathcal{I}(c, \Gamma) = \alpha$$
$$\mathcal{I}(c, \Delta) = \beta$$

We repeat our earlier picture of the situation.

$$
\begin{array}{lll}
\Gamma & \boxed{\alpha, \beta} & \mathcal{I}(P, \Gamma) = \{\alpha\} \\
& \downarrow & \\
\Delta & \boxed{\alpha, \beta} & \mathcal{I}(P, \Delta) = \{\alpha\}
\end{array}
$$

Based on this, here are a few simple calculations.

$$
\begin{aligned}
(v \star \mathcal{I})(c, \Gamma) &= \mathcal{I}(c, \Gamma) \\
&= \alpha
\end{aligned}
$$

and

$$
\begin{aligned}
(v \star \mathcal{I})(c, \Delta) &= \mathcal{I}(c, \Delta) \\
&= \beta
\end{aligned}
$$

According to Definition 9.4.6, $\mathcal{M}, \Gamma \Vdash_v \langle \lambda x.\Diamond P(x) \rangle(c)$ is equivalent to $\mathcal{M}, \Gamma \Vdash_w \Diamond P(x)$, where $w(x) = (v \star \mathcal{I})(c, \Gamma)$, which is α by the calculation above. Since Δ is the only world in the model accessible from Γ, this in turn is equivalent to $\mathcal{M}, \Delta \Vdash_w P(x)$, which is equivalent to $w(x) \in \mathcal{I}(P, \Delta)$. Since indeed, $\alpha \in \mathcal{I}(P, \Delta)$, $\langle \lambda x.\Diamond P(x) \rangle(c)$ is *true* at Γ.

Next, $\mathcal{M}, \Gamma \Vdash_v \Diamond \langle \lambda x.P(x) \rangle(c)$ is equivalent to $\mathcal{M}, \Delta \Vdash_v \langle \lambda x.P(x) \rangle(c)$. By the added condition of Definition 9.4.6, this is equivalent to $\mathcal{M}, \Delta \Vdash_w P(x)$ where $w(x) = (v \star \mathcal{I})(c, \Delta) = \beta$. This is further equivalent to $w(x) \in \mathcal{I}(P, \Delta)$, which is not the case since $\beta \notin \mathcal{I}(P, \Delta)$. Thus $\Diamond \langle \lambda x.P(x) \rangle(c)$ is *false* at Γ.

Predicate abstraction allows us to separate the ambiguous formula $\Diamond P(c)$ into two distinct formulas, $\langle \lambda x.\Diamond P(x) \rangle(c)$ and $\Diamond \langle \lambda x.P(x) \rangle(c)$, and these are, indeed, not equivalent. The example above shows that $\langle \lambda x.\Diamond P(x)\rangle(c) \supset \Diamond\langle \lambda x.P(x)\rangle(c)$ is not valid. Exercise 9.4.2 asks you to show the converse is not valid either.

EXERCISES

EXERCISE 9.4.1.   Suppose the variable y is substitutable for x in $\Phi(x)$. Show the validity of the formula $\langle \lambda x.\Phi(x)\rangle(y) \equiv \Phi(y)$. Assume the semantics is first-order **K**, with varying domains.

EXERCISE 9.4.2.   For this question the intended semantics is that of first-order **K**, with constant domains, allowing predicate abstraction.

1. Show there is a model in which $\Diamond\langle \lambda x.P(x)\rangle(c) \supset \langle \lambda x.\Diamond P(x)\rangle(c)$ is not valid.
2. Show that $(\forall y)\Diamond\langle \lambda x.P(x)\rangle(y) \supset \langle \lambda x.\Diamond P(x)\rangle(c)$ is valid in all such models.

EXERCISE 9.4.3.   Show the validity, in all varying domain **K** models allowing predicate abstraction, of the following.

1. $\langle \lambda x.(\Phi \supset \Psi)\rangle(t) \equiv (\langle \lambda x.\Phi\rangle(t) \supset \langle \lambda x.\Psi\rangle(t))$.
2. $\langle \lambda x.\neg\Phi\rangle(t) \equiv \neg\langle \lambda x.\Phi\rangle(t)$.

9.5. READING PREDICATE ABSTRACTS

Take the following example. Suppose the date is January 1, 1902. The current President of the United States, Theodore Roosevelt, is 43. Now, the sentence

(9.15) George knows the President is at least 35.

has two interpretations. First, George might know that Roosevelt is at least 35. On this reading, (9.15) implicitly binds the term "President" to the person Theodore Roosevelt so that it is taken to be an assertion specifically about *him*. (George might have known Theodore Roosevelt from childhood, and know he is at least 35 without even knowing he is President.) Second, George might have in mind the requirement in the U.S. constitution that a President be at least 35. In this case George is not specifically referring to Theodore Roosevelt—indeed, he might not even know who the current President is. Formalizing these two versions involves different bindings.

Take □ to be "George knows"; take $A(x)$ to be "x is at least 35"; and take p to be a non-rigid constant symbol "President of the United States." For the formula

$$\langle \lambda x. \Box A(x) \rangle (p)$$

to be true at this particular time, the person currently designated by p must satisfy the condition $\Box A(x)$. Rather awkwardly then, we could read the formal sentence as, "It is true of the person who is now the President of the United States, that George knows the person's age is at least 35." The more colloquial reading is, "George knows, of the current President, that he is at least 35." Placing □ inside a predicate abstract generally can be read using an "of" phrasing.

On the other hand, the formula

$$\Box \langle \lambda x. A(x) \rangle (p)$$

can be read, mechanically, as "George knows that whoever is the President of the United States, the person is at least 35." The more colloquial reading is "George knows that the President is at least 35." This time we used a "that" reading.

Here is another example. I once had a dog named Benedict who, late in life, had a leg amputated. Suppose you, a stranger to the dog, saw him. You directly see a dog with three legs, and so you know that the dog you are looking at has three legs. That dog is Benedict, and so you know, of my dog Benedict, that he has three legs. But suppose I never told you the name of the dog. Then while you know, *of* Benedict, that he has three legs, you don't know the truth of the assertion *that* Benedict has three legs.

This is a distinction we make every day, but generally it is implicit in our conversation and phrasing. It is part of the unspoken background. When we must put it into words, the of/that phrasing used above is probably the best we can do. The reason it seems a little awkward is simply that so often we never do put it explicitly into words—we rely on context to clarify. Nonetheless, when formalizing, such distinctions must appear.

In this text we adopt a notational device, *predicate abstraction*, which will enable us to make the relevant distinctions we have discussed, including the *de re*/*de dicto* distinction, readily apparent as scope distinctions.

In discussing Russell's scope distinction, we had identified an ambiguity in the English sentence

The King of France is not happy

that depended upon whether we regarded the "not" as operating on the whole sentence or just on the predicate. Using predicate abstract notation, we can symbolize the two readings as follows:

(9.16) $\neg\langle\lambda x.H(x)\rangle(f)$

and

(9.17) $\langle\lambda x.\neg H(x)\rangle(f)$

(9.16) denies *that* the claim Hf is true; (9.17) says *of* f, that *he is not H*. (Compare the versions above, using predicate abstraction, with the unofficial versions (9.7) and (9.8).)

Predicate abstraction notation allows us to make a similar distinction for the temporal examples (9.13) and (9.14) from Section 9.2. We distinguish

(9.18) $\mathbf{F}\langle\lambda x.L(x)\rangle(m)$

from

(9.19) $\langle\lambda x.\mathbf{F}L(x)\rangle(m)$

In the first, we attribute *being a Laborite* to a future Prime Minister; in the second, we apply the property of *being a future Laborite* to the present Prime Minister.

Predicate abstraction also disambiguates another problem which might arise when we have a relational expression which can be analyzed in different ways. Take the example

(9.20) Caesar killed Caesar

(9.20) can be analyzed in two ways:

— The two-place relation "x killed y" is applied the ordered pair $<$Caesar, Caesar$>$
    $\langle\lambda x.\langle\lambda y.x$ killed $y\rangle(\text{Caesar})\rangle(\text{Caesar})$
— The one-place predicate "x killed x" is applied to Caesar
    $\langle\lambda x.x$ killed $x\rangle(\text{Caesar})$

These represent the two ways in which a predicate can be abstracted from (9.20). In the latter case we have what is known as a reflexive predicate. In the former case, we do not. The distinction is lost when we see the final sentence, "Caesar killed Caesar."

EXERCISES

EXERCISE 9.5.1.   Predicate abstraction is tricky when we have more than one variable involved. Consider the sentence, "Necessarily Tolstoy authored *War and Peace*." Using predicate abstraction and the necessity operator, $\Box$, formalize this sentence in as many ways as you can. Then explain whether the different formalizations express different readings or not.

ABSTRACTION CONTINUED

In this chapter we combine the machinery of predicate abstraction with other fundamental machinery, such as equality and an existence predicate. And we develop tableau systems that take predicate abstracts into account.

10.1. EQUALITY

In a *normal* model the relation symbol "=" is interpreted to be the equality relation on the domain of the model. The relation is thus the same from world to world. We investigated this notion in Chapter 7, before we included constant and function symbols in the language. Now it is time to see how equality, predicate abstracts, and non-rigid terms interact.

The first, fairly obvious, result says that at world Γ the predicate abstract $\langle \lambda x, y.(x = y) \rangle$ holds for terms t_1 and t_2 just in case both t_1 and t_2 evaluate to the same object at Γ.

PROPOSITION 10.1.1. In a normal model $\mathcal{M} = \langle \mathcal{G}, \mathcal{R}, \mathcal{D}, \mathit{I} \rangle$, for two terms t_1 and t_2, where t_2 does not contain x,

$$\mathcal{M}, \Gamma \Vdash_v \langle \lambda x, y.(x = y) \rangle (t_1, t_2) \Leftrightarrow (v \star \mathit{I})(t_1, \Gamma) = (v \star \mathit{I})(t_2, \Gamma)$$

Proof In the following, v' is the x-variant of v such that $v'(x) = (v \star \mathit{I})(t_1, \Gamma)$, and v'' is the y-variant of v' such that $v''(y) = (v' \star \mathit{I})(t_2, \Gamma)$.

$$
\begin{aligned}
&\mathcal{M}, \Gamma \Vdash_v \langle \lambda x, y.(x = y) \rangle (t_1, t_2) && \Longleftrightarrow \\
&\mathcal{M}, \Gamma \Vdash_v \langle \lambda x.(\lambda y.(x = y))(t_2) \rangle (t_1) && \Longleftrightarrow \\
&\mathcal{M}, \Gamma \Vdash_{v'} \langle \lambda y.(x = y) \rangle (t_2) && \Longleftrightarrow \\
&\mathcal{M}, \Gamma \Vdash_{v''} (x = y) && \Longleftrightarrow \\
&v''(x) = v''(y) && \Longleftrightarrow \\
&v'(x) = (v' \star \mathit{I})(t_2, \Gamma) && \Longleftrightarrow \\
&(v \star \mathit{I})(t_1, \Gamma) = (v' \star \mathit{I})(t_2, \Gamma) && \Longleftrightarrow \\
&(v \star \mathit{I})(t_1, \Gamma) = (v \star \mathit{I})(t_2, \Gamma) &&
\end{aligned}
$$

The last step is justified because t_2 does not contain x, and so v and v' behave alike as far as t_2 is concerned. ∎

We introduce special notation for the equality abstract holding between terms. Sometimes it simplifies the appearance of formulas, though we will be sparing of its use.

DEFINITION 10.1.2. [Term Equality] We use $t \approx u$ as an abbreviation for $\langle \lambda w, z.z = w \rangle(u, t)$, where w and z are new variables.

Recall from Chapter 7 that

$$(x = y) \supset \Box(x = y) \tag{10.1}$$

is valid, where x and y are free variables. This most decidedly contrasts with

$$(t \approx u) \supset \Box(t \approx u) \tag{10.2}$$

where t and u are terms involving function and constant symbols. This is *not* generally valid. It is easy to create counterexamples, taking t and u to be constant symbols. We leave this to you.

The difference between (10.1) and (10.2) is a simple one. In (10.1), $x = y$ asserts that the *objects* that are the values of x and y are the same. But in (10.2), $t \approx u$ asserts that the terms t and u *designate* the same object, which is quite a different thing. If x and y are the same object, that object is, and always will be, one, and hence $\Box(x = y)$ follows. But if t and u happen to designate the same object, they might fail to do so under other circumstances, so $\Box(t \approx u)$ does not follow. In fact, $\Box(t \approx u)$ expresses a notion considerably stronger than that of simple equality—it has the characteristics of *synonymy*.

With term equality notation available, the Proposition above can be restated in a somewhat clearer form.

COROLLARY 10.1.3.   In a normal model $\mathcal{M} = \langle \mathcal{G}, \mathcal{R}, \mathcal{D}, \mathcal{I} \rangle$, for two terms t_1 and t_2,

$$\mathcal{M}, \Gamma \Vdash_v t_1 \approx t_2 \Leftrightarrow (v \star \mathcal{I})(t_1, \Gamma) = (v \star \mathcal{I})(t_2, \Gamma)$$

Now we discuss several representative sentences involving equality and predicate abstraction, some of which are valid and some of which are not. The actual choice of modal logic does not matter.

We saw in Chapter 7 that

$$(x = y) \supset \Box(x = y)$$

is a valid open formula. Here we want to consider how the formula behaves when, instead of free variables, we have terms and predicate abstracts involved. The complication is that there are a number of ways of abstracting on

$$\Box(x = y)$$

Each of the following can be obtained.

1. $\langle \lambda x, y.\Box(x = y)\rangle$
2. $\Box \langle \lambda x, y.x = y\rangle$
3. $\langle \lambda x.\Box \langle \lambda y.x = y\rangle\rangle$
4. $\langle \lambda y.\Box \langle \lambda x.x = y\rangle\rangle$

These do not all behave alike.

EXAMPLE 10.1.4. For any terms t and u, where u does not contain x, the sentence

$$\langle \lambda x, y.(x = y)\rangle(t, u) \supset \langle \lambda x, y.\Box(x = y)\rangle(t, u)$$

is valid.

Suppose $\mathcal{M}, \Gamma \Vdash_v \langle \lambda x, y.(x = y)\rangle(t, u)$, where $\mathcal{M} = \langle \mathcal{G}, \mathcal{R}, \mathcal{D}, \mathcal{I}\rangle$ is a normal model. Then $(v \star \mathcal{I})(t, \Gamma) = (v \star \mathcal{I})(u, \Gamma)$ by Proposition 10.1.1. We must show that we also have $\mathcal{M}, \Gamma \Vdash_v \langle \lambda x, y.\Box(x = y)\rangle(t, u)$. Following through the definitions, this means we must show we have $\mathcal{M}, \Gamma \Vdash_{v'} \Box(x = y)$, where v' agrees with v on all variables except x and y, and $v'(x) = (v \star \mathcal{I})(t, \Gamma)$, and $v'(y) = (v \star \mathcal{I})(u, \Gamma)$.

To show this, let Δ be an arbitrary world accessible from Γ. We must show we have $\mathcal{M}, \Delta \Vdash_{v'} (x = y)$, which in turn says we must have $(v' \star \mathcal{I})(x, \Delta) = (v' \star \mathcal{I})(y, \Delta)$. Since x and y are variables, this reduces to showing that $v'(x) = v'(y)$, but in fact, by the preceding paragraph, this equality does hold.

The example above can be read as: consider the objects that t and u denote; if those objects are equal, they are necessarily equal. This is not really about non-rigid denotation—it is about the things denoted. Equal objects are equal, no matter what. More colloquially, it can be read: if it is true *of* t and u that they are equal, then it is also true *of* them that they are necessarily equal.

Example 10.1.4 concerned abstracts of type 1 in the list above. The next example shows that if abstracts of type 2 are involved, the result is *not* always valid. This is important, since the two versions are easily confused.

EXAMPLE 10.1.5. Where a and b are constant symbols, the sentence

$$\langle \lambda x, y.(x = y)\rangle(a, b) \supset \Box \langle \lambda x, y.(x = y)\rangle(a, b)$$

is *not* valid.

Consider the following schematically presented model $\mathcal{M}$ (which, incidentally, is constant domain).

$$\Gamma \quad \boxed{\alpha, \beta} \qquad \begin{array}{l} \mathit{l}(a, \Gamma) = \alpha \\ \mathit{l}(b, \Gamma) = \alpha \end{array}$$

$$\Delta \quad \boxed{\alpha, \beta} \qquad \begin{array}{l} \mathit{l}(a, \Delta) = \alpha \\ \mathit{l}(b, \Delta) = \beta \end{array}$$

We have $\mathcal{M}, \Gamma \Vdash_v \langle \lambda x, y.(x = y)\rangle(a, b)$ (for any v) because this reduces to $(v \star \mathit{l})(a, \Gamma) = (v \star \mathit{l})(b, \Gamma)$, and this is correct since $\mathit{l}(a, \Gamma) = \mathit{l}(b, \Gamma) = \alpha$.

Next, if we had $\mathcal{M}, \Gamma \Vdash_v \Box\langle \lambda x, y.(x = y)\rangle(a, b)$, we would also have $\mathcal{M}, \Delta \Vdash_v \langle \lambda x, y.(x = y)\rangle(a, b)$, which reduces to $(v \star \mathit{l})(a, \Delta) = (v \star \mathit{l})(b, \Delta)$. But this is not the case since $\mathit{l}(a, \Delta) = \alpha$, $\mathit{l}(b, \Delta) = \beta$, and $\alpha \neq \beta$.

This example can be read: if a and b designate the same object, then necessarily they designate the same object. Read this way, we should not be surprised at invalidity. Just because a and b designate the same thing today doesn't mean they must do so tomorrow. Again, read colloquially, we have the invalidity of: if it is true *that* a and b are equal, then it is necessarily true *that* a and b are equal.

The two examples above have considerable significance for the morning star/evening star puzzle. Let a and b be the non-rigid designators "the morning star," and "the evening star," respectively, and read $\Box$ as "the ancients knew." There is no doubt that in the actual world, the morning star is the evening star, $a \approx b$. Unabbreviated,

(10.3) $\quad \langle \lambda x, y.(x = y)\rangle(a, b)$.

Now by Example 10.1.4, the sentence

(10.4) $\quad \langle \lambda x, y.(x = y)\rangle(a, b) \supset \langle \lambda x, y.\Box(x = y)\rangle(a, b)$

is valid. It follows from (10.3) and (10.4) that

(10.5) $\quad \langle \lambda x, y.\Box(x = y)\rangle(a, b)$

is true in the actual world. This is, in fact, correct. The phrases "morning star" and "evening star" designate the same object, and the ancients certainly knew *of* any object that it is identical with itself.

If the sentence

(10.6) $\langle \lambda x, y.(x = y) \rangle (a, b) \supset \Box \langle \lambda x, y.(x = y) \rangle (a, b)$

were also valid, then from the true (10.3) we could conclude

(10.7) $\Box \langle \lambda x, y.(x = y) \rangle (a, b)$.

But Example 10.1.5 shows that the sentence (10.6) is simply not a valid one. So the problematic conclusion (10.7), which can be read "The ancients knew *that* the morning star and the evening star are equal," cannot be drawn.

EXAMPLE 10.1.6. This example is more complex. The sentence

$$\left[\langle \lambda x. \Diamond \langle \lambda y.x = y \rangle (p) \rangle (a) \wedge \langle \lambda x. \Box S(x) \rangle (a) \right] \supset$$
$$\Diamond \langle \lambda x.S(x) \rangle (p)$$

is valid. A concrete instance of this sentence will sound familiar to you. Note our careful usage of "of" and "that" in what follows. We give the modal operators a temporal reading: $\Box X$ is read, "X is and will always be the case," and $\Diamond X$ is read, "X is or will be the case some time in the future." Suppose the possible world in which we evaluate the sentence is the world of 1850. Finally, suppose that a is "Abraham Lincoln," p is "the President of the United States," and $S(x)$ is "x sees the start of the American Civil War." Under this reading,

$\langle \lambda x. \Diamond \langle \lambda y.x = y \rangle (p) \rangle (a)$

says, "It is true, of Abraham Lincoln, that he is or will be President of the United States," more succinctly, "Abraham Lincoln will someday be President of the United States." Likewise,

$\langle \lambda x. \Box S(x) \rangle (a)$

reads, "It is true now, and will always be true, of Abraham Lincoln, that he sees the start of the Civil War." More colloquially, "Abraham Lincoln sees the start of the Civil War." The conclusion we can draw,

$\Diamond \langle \lambda x.S(x) \rangle (p)$

reads "It is the case or will happen that the President of the United States sees the start of the Civil War."

We leave it to you as an exercise to verify the validity of this sentence. In Section 10.6 we provide a tableau proof of it.

EXAMPLE 10.1.7. Suppose you tell me, "Today is Bastille day," and I respond, "I didn't know that. I know that Bastille day is always July 14, but I didn't know today was July 14." Let us see how we formalize this small dialogue.

We give $\Box X$ the epistemic reading "I know that X." We let b be "Bastille day," t be "today," and j be "July 14." These three constant symbols are non-rigid, for there is no assurance that any will designate the same day every time, and consequently, their designation can vary with different states of my knowledge.

Now, you tell me that today is Bastille day,

$$\langle \lambda x, y.(x = y) \rangle (t, b).$$

From this it follows, by Example 10.1.4, that

$$\langle \lambda x, y.\Box(x = y) \rangle (t, b),$$

i.e., I know, *of* today and of Bastille day, they are the same. And this is trivial. They are in fact the same, and I know of each object that it is self-identical.

I claimed to know *that* Bastille day is always July 14, that is, I asserted

$$\Box \langle \lambda x, y.(x = y) \rangle (b, j).$$

But I also said I did not know, of today, that it was July 14, that is,

$$\langle \lambda x.\neg\Box\langle \lambda y.(x = y) \rangle (j) \rangle (t).$$

This purported to explain my not knowing, of today, that it was Bastille day, that is,

$$\langle \lambda x.\neg\Box\langle \lambda y.(x = y) \rangle (b) \rangle (t).$$

Now, in fact, the sentence

$$\left[\Box \langle \lambda x, y.(x = y) \rangle (b, j) \wedge \langle \lambda x.\neg\Box\langle \lambda y.(x = y) \rangle (j) \rangle (t) \right] \supset$$
$$\langle \lambda x.\neg\Box\langle \lambda y.(x = y) \rangle (b) \rangle (t)$$

is valid. We leave it to you to verify this. Alternatively, it too can be proved using tableau rules.

EXAMPLE 10.1.8. In 1556 Charles V, Holy Roman Emperor, resigned his offices and spent the remainder of his life near the monastery of San Yuste in Spain repairing clocks. One can imagine Charles, around 1550, planning these events and musing "Someday the Emperor won't be the Emperor."

We give $\Box$ the same temporal reading as in Example 10.1.6. Also, let c be "Charles V," and e be "Holy Roman Emperor."

Charles is certainly not asserting $\langle\lambda x, y.\Diamond\neg(x = y)\rangle(c, e)$, which says *of* he, Charles, and he, the Emperor, that they will at some time be different individuals. For in that case he would be uttering a contradiction: it's negation, $\neg\langle\lambda x, y.\Diamond\neg(x = y)\rangle(c, e)$, is a valid sentence. Rather, Charles is asserting $\langle\lambda x.\Diamond\langle\lambda y.\neg(x = y)\rangle(e)\rangle(c)$, i.e., it is true, of Charles, that he has the property of someday being different than the Emperor. In this case we do not have a validity to deal with—there is nothing logically certain about Charles' assertion. But the sentence is *satisfiable*—there is a way of making it true in a model. It could happen.

EXERCISES

EXERCISE 10.1.1.   Determine the status of the following sentence:

$$\langle\lambda x, y.(x = y)\rangle(a, b) \supset \langle\lambda x.\Box\langle\lambda y.(x = y)\rangle(a)\rangle(b).$$

That is, either show it is valid, or produce a counter-model.

EXERCISE 10.1.2.   Show the validity of $(x = y) \equiv (x \approx y)$ where x and y are variables.

EXERCISE 10.1.3.   Show the validity of the sentence of Example 10.1.6.

EXERCISE 10.1.4.   Show the validity of the sentence of Example 10.1.7.

EXERCISE 10.1.5.   Finish Example 10.1.8 and produce a model in which the sentence

$$\neg\langle\lambda x, y.\Diamond\neg(x = y)\rangle(c, e) \wedge \langle\lambda x.\Diamond\langle\lambda y.\neg(x = y)\rangle(e)\rangle(c)$$

is true at some possible world.

10.2. RIGIDITY

Kripke (1980) introduced the terminology of rigid designation into the modern literature on modal logic. (The notion was anticipated by Marcus (1992).) In his terminology, a singular term is a *rigid designator* if it refers to the same object in every possible world in which that object exists. A singular term is *strongly rigid* if it is rigid and the referent exists (i.e., the term designates something) in every possible world. Issues of existence complicate matters.

To keep terminology simple we depart somewhat from Kripke, and use the term *rigid* for a term that always designates the same object, whether or not the object exists. Even so, there are two closely related versions that we must distinguish.

DEFINITION 10.2.1.   Let $\mathcal{M} = \langle \mathcal{G}, \mathcal{R}, \mathcal{D}, \mathcal{I} \rangle$ be a model, and let t be a closed term. Also let v be any valuation (the actual choice won't matter, since t has no free variables).

1. We say t is *rigid* in $\mathcal{M}$ if t designates the same thing in each world of the model. More precisely, t is rigid if $(v * \mathcal{I})(t, \Gamma)$ is the same for all $\Gamma \in \mathcal{G}$.
2. We say t is *locally rigid* provided, for any $\Gamma, \Delta \in \mathcal{G}$, if $\Gamma \mathcal{R} \Delta$ then $(v * \mathcal{I})(t, \Gamma) = (v * \mathcal{I})(t, \Delta)$.

Thus rigidity refers to *all* worlds, while local rigidity only refers to *related* worlds. (The relationship between constant domain models and locally constant domain models in Chapter 4, was a similar issue, and not coincidentally.)

Rigidity and local rigidity are different things; it is easy to produce examples that show this. Nonetheless, they characterize the same logic, in the following sense.

PROPOSITION 10.2.2.   Assume Φ is a formula in which the closed term t occurs. The following are equivalent:

1. Φ is valid in all models in which t is rigid.
2. Φ is valid in all models in which t is locally rigid.

Proof In one direction the argument is simple. Suppose Φ is valid in all models in which t is locally rigid. Obviously every model in which t is rigid is also a model in which t is locally rigid. Consequently Φ is valid in all models in which t is rigid.

For the other direction, suppose Φ is not valid in some model $\mathcal{M} = \langle \mathcal{G}, \mathcal{R}, \mathcal{D}, \mathcal{I} \rangle$ in which t is locally rigid. We produce a model in which Φ is not valid, and in which t is rigid. (The construction is the same as that of Proposition 4.9.10, but we repeat it here for convenience.)

Since Φ is not valid in $\mathcal{M}$, there is some world, say $\Gamma_0 \in \mathcal{G}$, such that $\mathcal{M}, \Gamma_0 \not\Vdash_{v_0} \Phi$, for some valuation v_0.

Let us say a world $\Gamma \in \mathcal{G}$ is *relevant* to Γ_0 if there is a "path" from Γ_0 to Γ, in the following sense: there is a sequence of worlds, $\Delta_1, \Delta_2, \ldots, \Delta_n$, starting with Γ_0 (that is, $\Delta_1 = \Gamma_0$), finishing with Γ (that is, $\Delta_n = \Gamma$), and with each world related to the next (that is, $\Delta_i \mathcal{R} \Delta_{i+1}$). We allow the sequence to be of length 1 as well, so that Γ_0 is relevant to Γ_0. We use this notion to define a new model, as follows.

Let $\mathcal{G}'$ consist of all members of $\mathcal{G}$ that are relevant to Γ_0. Let $\mathcal{R}'$ be $\mathcal{R}$ restricted to the members of $\mathcal{G}'$. Likewise let $\mathcal{D}'$ be $\mathcal{D}$ restricted to members of $\mathcal{G}$ and let $\mathcal{I}'$ be $\mathcal{I}$ restricted to members of $\mathcal{G}$. This gives us a new model $\mathcal{M}' = \langle \mathcal{G}', \mathcal{R}', \mathcal{D}', \mathcal{I}' \rangle$.

Now, the following finishes the proof.

1. Each term 'behaves the same' in both models, at worlds common to both. That is, for a term u, for $\Gamma \in \mathcal{G}'$, and for a valuation v, $(v * \mathcal{I})(u, \Gamma) = (v * \mathcal{I}')(u, \Gamma)$. (We leave this as an exercise).
2. It follows that since t was locally rigid in $\mathcal{M}$, then t is rigid in $\mathcal{M}'$. (Why?)
3. Each formula also 'behaves the same' in the two models as well. That is, for a formula X, for $\Gamma \in \mathcal{G}'$, and for a valuation v, $\mathcal{M}, \Gamma \Vdash_v X$ if and only if $\mathcal{M}', \Gamma \Vdash_v X$. (We also leave this as an exercise.)
4. It follows that since Φ was not true at Γ_0 (using valuation v_0) in $\mathcal{M}$, and since Γ_0 is also a world of $\mathcal{M}'$, then Φ is not true at Γ_0 in $\mathcal{M}'$ (again using valuation v_0).

We thus have produced a model in which Φ is not valid, and in which t is rigid. This completes the proof. ∎

Now that we know rigidity and local rigidity are equivalent as far as formula validity goes, we can work with whichever is convenient. For the rest of this section we work exclusively with *local* rigidity.

Next we show that rigidity, which is a semantic concept, corresponds to a syntactic notion, and in fact this syntactic notion can be given in several versions whose equivalence is not at all obvious. We present the results for a constant symbol c for simplicity, though more complex terms behave similarly.

We have seen many examples where syntactic properties corresponded to semantic properties, but semantic properties of *frames*, not of particular models. This is another such example, with some minor complications.

DEFINITION 10.2.3. [$\mathcal{I}_0$-compatible] Let $\mathcal{F} = \langle \mathcal{G}, \mathcal{R}, \mathcal{D} \rangle$ be an augmented frame, and let $\mathcal{I}_0$ be a *partial* interpretation in this frame, defined only for the constant symbol c. That is, for each $\Gamma \in \mathcal{G}$, $\mathcal{I}_0(c, \Gamma)$ is defined, but $\mathcal{I}_0$ is undefined on other constant, function, and relation symbols. We call $\mathcal{I}_0$ a *c-interpretation*.

We say $\mathcal{I}_0$ interprets c *locally rigidly* provided, for any $\Gamma, \Delta \in \mathcal{G}$, if $\Gamma \mathcal{R} \Delta$, then $\mathcal{I}_0(c, \Gamma) = \mathcal{I}_0(c, \Delta)$.

We say a model $\mathcal{M} = \langle \mathcal{G}, \mathcal{R}, \mathcal{D}, \mathcal{I} \rangle$ that is based on the frame $\mathcal{F}$ is $\mathcal{I}_0$-compatible provided $\mathcal{I}$ and $\mathcal{I}_0$ agree on c at each world of $\mathcal{G}$.

If $\mathcal{L}_0$ interprets c locally rigidly, when we talk of $\mathcal{L}_0$-compatible models based on $\mathcal{F}$ we are really talking about a family of models in which the interpretation of c is fixed (though interpretations of other terms are not), and that interpretation is locally rigid.

PROPOSITION 10.2.4. Let $\mathcal{F} = \langle \mathcal{G}, \mathcal{R}, \mathcal{D} \rangle$ be an augmented frame, and $\mathcal{L}_0$ be a c-interpretation in this frame. The following are equivalent:

1. $\mathcal{L}_0$ interprets c locally rigidly.
2. All instances of the schema $\Box \langle \lambda x.\Phi \rangle(c) \equiv \langle \lambda x.\Box\Phi \rangle(c)$ are valid in every $\mathcal{L}_0$-compatible model based on $\mathcal{F}$.
3. All instances of the schema $\Box \langle \lambda x.\Phi \rangle(c) \supset \langle \lambda x.\Box\Phi \rangle(c)$ are valid in every $\mathcal{L}_0$-compatible model based on $\mathcal{F}$.
4. All instances of the schema $\langle \lambda x.\Box\Phi \rangle(c) \supset \Box \langle \lambda x.\Phi \rangle(c)$ are valid in every $\mathcal{L}_0$-compatible model based on $\mathcal{F}$.

Before proving this, we note its significance. Item 2 says that, for c, *de re* and *de dicto* distinctions cannot be seen. The equivalence between 1 and 2 essentially says that the lack of *de re* and *de dicto* distinctions is characteristic of rigidity. What is somewhat unexpected is the further equivalence between this and items 3 and 4. These latter say that either half of the equivalence in item 2 suffices. The deeper significance of this technical result is not understood (at least not yet, at least not by us). Exercise 10.2.4 gives further equivalent *de re/de dicto* schemas. Now we turn to the proof.

Proof Showing that item 1 implies the other three is rather easy, and we leave it as an exercise.

Now we show that if item 1 is false, so is item 3, and hence also item 2. So, suppose $\mathcal{L}_0$ does not interpret c locally rigidly. Say there are worlds, $\Gamma, \Delta \in \mathcal{G}$, with $\Gamma \mathcal{R} \Delta$, but $\mathcal{L}_0(c, \Gamma) \neq \mathcal{L}_0(c, \Delta)$. We use this information to falsify item 3.

We define an $\mathcal{L}_0$-compatible model by defining a suitable interpretation $\mathcal{L}$ as follows. On c, $\mathcal{L}$ and $\mathcal{L}_0$ agree, and on other function and constant symbols, make some arbitrary choice (it won't matter). Thus $\mathcal{L}$ and $\mathcal{L}_0$ agree on c, so the model we are constructing will be $\mathcal{L}_0$-compatible. Let P be a one-place relation symbol. For each world $\Omega \in \mathcal{G}$, set $\mathcal{L}(P, \Omega) = \{\mathcal{L}_0(c, \Omega)\}$. That is, at each world we take P to be true of exactly what c designates at that world— informally, we can read P as "is c." On other relation symbols, again make some arbitrary choice (it, too, won't matter). We thus have our $\mathcal{L}_0$-compatible model $\mathcal{M} = \langle \mathcal{G}, \mathcal{R}, \mathcal{D}, \mathcal{L} \rangle$.

We first check that we have $\mathcal{M}, \Gamma \Vdash_v \Box \langle \lambda x.P(x) \rangle(c)$ (for any v). This will be the case provided that for any $\Omega \in \mathcal{G}$ such that $\Gamma \mathcal{R} \Omega$ we have $\mathcal{M}, \Omega \Vdash_v \langle \lambda x.P(x) \rangle(c)$, and this is so because, by definition, $\mathcal{L}(c, \Omega) \in \mathcal{L}(P, \Omega)$.

We next check that we do *not* have $\mathcal{M}, \Gamma \Vdash_v \langle \lambda x.\Box P(x)\rangle(c)$. Well if we did, we would have $\mathcal{M}, \Gamma \Vdash_{v'} \Box P(x)$, where v' is the x-variant of v such that $v'(x) = \mathcal{l}(c, \Gamma)$. Further, if we had this, since $\Gamma \mathcal{R} \Delta$, we would have $\mathcal{M}, \Delta \Vdash_{v'} P(x)$, and thus $v'(x) \in \mathcal{l}(P, \Delta)$. However by definition, the only member of $\mathcal{l}(P, \Delta)$ is $\mathcal{l}_0(c, \Delta)$, but $v'(x) = \mathcal{l}_0(c, \Gamma)$, and $\mathcal{l}_0(c, \Gamma) \neq \mathcal{l}_0(c, \Delta)$.

It follows that we have

$$\mathcal{M}, \Gamma \nVdash_v \Box\langle \lambda x.P(x)\rangle(c) \supset \langle \lambda x.\Box P(x)\rangle(c)$$

And thus the failure of item 1 implies the failure of item 3. We leave item 4 to you. ∎

It is a consequence of this Proposition that items 3 and 4 are equivalent. Formally, this means each instance of schema 3 is implied by one or more instances of schema 4, and each instance of schema 4 is implied by one or more instances of schema 3. In fact, we can identify the instances needed.

Let $\Phi(x)$ be a formula in which the variable y does not appear, and let $\Phi(y)$ be the result of substituting y for free occurrences of x in $\Phi(x)$. We now define four formulas, involving the constant symbol c.

$$
\begin{aligned}
A_3 &= \Box\langle \lambda y.\langle \lambda x.\Phi(y) \supset \Phi(c)\rangle(c)\rangle(c) \supset \\
&\qquad \langle \lambda y.\Box\langle \lambda x.\Phi(y) \supset \Phi(x)\rangle(c)\rangle(c) \\
A_4 &= \langle \lambda x.\Box\Phi(x)\rangle(c) \supset \Box\langle \lambda x.\Phi(x)\rangle(c) \\
B_4 &= \langle \lambda y.\langle \lambda x.\Box(\Phi(x) \supset \Phi(y))\rangle(c)\rangle(c) \supset \\
&\qquad \langle \lambda y.\Box\langle \lambda x.\Phi(x) \supset \Phi(y)\rangle(c)\rangle(c) \\
B_3 &= \Box\langle \lambda x.\Phi(x)\rangle(c) \supset \langle \lambda x.\Box\Phi(x)\rangle(c)
\end{aligned}
$$

Notice that A_4 and B_3 are arbitrary cases of items 4 and 3 of Proposition 10.2.4, while A_3 and B_4 are related instances of items 3 and 4 respectively. Now it can be shown that $A_3 \supset A_4$ and $B_4 \supset B_3$ are both valid (in varying domain **K**, say). Showing this is a good work-out for tableaus, and we give it as Exercise 10.6.6, after tableau rules have been introduced.

Recall that both the Barcan and the Converse Barcan schemas could be replaced with single formulas, using equality. A similar thing happens here: rigidity can be captured by a single formula.

PROPOSITION 10.2.5. The constant symbol c is locally rigid in a model if and only if the following is valid in that model: $\langle \lambda y.\Box\langle \lambda x.x = y\rangle(c)\rangle(c)$.

Proof We have already seen that local rigidity is equivalent to the validity of either of the following formula schemas:

1. $\Box\langle\lambda x.\Phi\rangle(c) \supset \langle\lambda x.\Box\Phi\rangle(c)$
2. $\langle\lambda x.\Box\Phi\rangle(c) \supset \Box\langle\lambda x.\Phi\rangle(c)$

Now, $\Box\langle\lambda y.\langle\lambda x.x = y\rangle(c)\rangle(c)$ is valid, and $\Box\langle\lambda y.\langle\lambda x.x = y\rangle(c)\rangle(c) \supset \langle\lambda y.\Box\langle\lambda x.x = y\rangle(c)\rangle(c)$ is an instance of the first schema above, from which $\langle\lambda y.\Box\langle\lambda x.x = y\rangle(c)\rangle(c)$ follows. Incidentally $\langle\lambda y.\langle\lambda x.\Box(x = y)\rangle(c)\rangle(c)$ is also valid, and $\langle\lambda y.\langle\lambda x.\Box(x = y)\rangle(c)\rangle(c) \supset \langle\lambda y.\Box\langle\lambda x.x = y\rangle(c)\rangle(c)$ is an instance of the second schema above, from which $\langle\lambda y.\Box\langle\lambda x.x = y\rangle(c)\rangle(c)$ again follows. Thus either of the schemas implies the equality version.

For the other direction the desired result follows from the validity of either of the following two formulas.

$$\langle\lambda y.\Box\langle\lambda x.x = y\rangle(c)\rangle(c) \supset [\Box\langle\lambda x.\Phi\rangle(c) \supset \langle\lambda x.\Box\Phi\rangle(c)]$$
$$\langle\lambda y.\Box\langle\lambda x.x = y\rangle(c)\rangle(c) \supset [\langle\lambda x.\Box\Phi\rangle(c) \supset \Box\langle\lambda x.\Phi\rangle(c)]$$

You can check the validity of these directly, but tableau proofs are easier. We give these as Exercise 10.6.7, after tableau rules have been presented. ∎

EXERCISES

EXERCISE 10.2.1.   Finish the proof of Proposition 10.2.2. Show item 1 by induction on the complexity of term u. Also show item 3 using induction on the complexity of formula X.

EXERCISE 10.2.2.   Show that item 1 of Proposition 10.2.4 implies item 2, and hence trivially items 3 and 4.

EXERCISE 10.2.3.   Show that if item 1 of Proposition 10.2.4 is false, so is item 4.

EXERCISE 10.2.4.   Extend Proposition 10.2.4 by showing items 1 – 4 are further equivalent to

5. All instances of the schema $\Diamond\langle\lambda x.\Phi\rangle(c) \equiv \langle\lambda x.\Diamond\Phi\rangle(c)$ are valid in every $\mathcal{L}_0$-compatible model based on $\mathcal{F}$.
6. All instances of the schema $\Diamond\langle\lambda x.\Phi\rangle(c) \supset \langle\lambda x.\Diamond\Phi\rangle(c)$ are valid in every $\mathcal{L}_0$-compatible model based on $\mathcal{F}$.
7. All instances of the schema $\langle\lambda x.\Diamond\Phi\rangle(c) \supset \Diamond\langle\lambda x.\Phi\rangle(c)$ are valid in every $\mathcal{L}_0$-compatible model based on $\mathcal{F}$.

10.3. A Dynamic Logic Example

In Section 1.5 we briefly mentioned *dynamic logic*. Recall, this is a version of modal logic in which there are many modal operators, each corresponding to an action—typically, the action of executing a computer program. Since the variables used in most computer languages change their values from time to time, dynamic logic can provide us with an example of non-rigid constant symbols that is natural, but of a different kind than we have been considering.

Suppose c is a variable in the sense of computer programming, and which we will think of as a non-rigid constant symbol. Consider the assignment statement $c := c + 1$, which increments the value of c. Corresponding to this is the dynamic logic modal operator $\boxed{c := c+1}$, where the intention is that $\boxed{c := c+1}\,\Phi$ should mean: after the assignment statement $c := c + 1$ is executed, Φ will be true. For notational convenience, in what follows we will abbreviate $\boxed{c := c+1}$ as $\Box$.

To express the intended behavior of $\Box$ one might be tempted to write something like the following, where we assume arithmetic notation is interpreted in the usual way.

$$(\forall x)[(c = x) \supset \Box(c = x + 1)] \tag{10.8}$$

Informally, if the current value of c is x, after the assignment statement $c := c + 1$ has been executed, the value of c will be $x + 1$. As written, the formulation makes no use of predicate abstraction, and this omission immediately leads us to a serious problem. In *classical* first-order logic the following is valid:

$$(\forall x)[(c = x) \supset P(x)] \equiv P(c). \tag{10.9}$$

Using this, (10.8) immediately yields:

$$\Box(c = c + 1) \tag{10.10}$$

which is silly. Clearly predicate abstraction is essential.

Using predicate abstraction, (10.8) can be restated correctly as follows:

$$(\forall x)[(c \approx x) \supset \Box(c \approx x + 1)]. \tag{10.11}$$

(Recall our abbreviation $t \approx u$, which stands for $\langle \lambda w, x.z = w \rangle(u, t)$, where w and z are new variables.) But in modal logic using predicate abstraction, we have the validity of the following, which is similar to (10.9).

$$(\forall x)[(c \approx x) \supset P(x)] \equiv \langle \lambda x.P(x) \rangle(c). \tag{10.12}$$

Using (10.12), the correct (10.11) can be replaced by

$$\langle \lambda x.\Box(c \approx x + 1)\rangle(c) \tag{10.13}$$

which uses no quantification at all. This formula should be compared with the incorrect (10.10).

This example can be explored a little further yet. The three sentences

1. $\langle \lambda x.E(x)\rangle(c)$
2. $(\forall x)\Box\neg(x \approx x + 1)$
3. $\langle \lambda x.\Box(c \approx x + 1)\rangle(c)$

together imply

4. $\neg\langle \lambda x.\Box\langle \lambda y.x = y\rangle(c)\rangle(c)$

in the logic **D**. (We give this as an exercise in Section 10.6, once appropriate tableau rules have been introduced.) Sentence 1 corresponds, in computer programming terminology, to c being initialized. Sentence 2 asserts that x and $x + 1$ are never the same, which is all we need from arithmetic now. Sentence 3 captures the behavior of the assignment statement $c := c + 1$ (recall, we are thinking of $\Box$ as being the action of executing this program step). Inference is in the logic **D**, not **K**. This amounts to assuming that $c :=$ $c + 1$ is executable—there must always be accessible states that $c := c + 1$ takes us to. Finally, the consequence, 4, expresses the (obvious) fact that c is not rigid, by Proposition 10.2.5. More generally, arithmetic variables in programming languages are never rigid. It is how they differ from constants of programming languages.

10.4. RIGID DESIGNATORS

Rigidity can be characterized in a straightforward way, as we saw above. In Proposition 10.2.4 we showed that a singular term c is rigid if all instances of the following are valid.

$$\langle \lambda x.\Box\Phi(x)\rangle(c) \equiv \Box\langle \lambda x.\Phi(x)\rangle(c)$$

That is, a rigid designator is a designator for which scope does not matter. It should be clear that although we can define rigidity in terms of scope, the notion of scope is much broader: as Kripke (1979b) has argued, none of the family of distinctions—rigid/nonrigid, *de re/de dicto*, referential/attributive—can replace the scope distinction, any more than the primary/secondary distinction of Russell (1905) can. After all, we can have nesting of modal operators, say, or multi-variable predicates, and the richness of the interpretations cannot be captured by a simple on/off feature analysis.

But the notion so defined speaks about terms in a *formal language*. The question remains whether any *natural language* expressions should be represented by rigid designators when formalized. It should be noted, incidentally, it is possible that a term might be rigid under one interpretation of □ and ◊ but not under another. So, when we speak about the rigidity of natural language expressions, we have to specify the interpretation of the modal operators.

Let us, for now, suppose that we are working with the usual alethic modalities. Then it turns out that some definite descriptions are rigid, but not all. "The number of U.S. Supreme Court Justices" is non-rigid. Although there are, as a matter of fact, 9 Supreme Court Justices, there is no reason why, in other possible worlds, there couldn't be fewer justices, or more. And the nonrigidity of the expression shows itself in the fact that the two following sentences differ in truth value:

(10.14) The number of Supreme Court Justices is necessarily odd

and

(10.15) Necessarily, the number of Supreme Court Justices is odd.

The *de re* reading (10.14) is true because there are 9 Supreme Court Justices, and this number is, in every possible world, odd. The *de dicto* reading (10.15) on the other hand, is false, because the number of Supreme Court Justices can vary from world to world, and in some, that number could be even.

On the other hand, "the ratio of the circumference to the diameter of a circle" is rigid, for it designates the same number, "π," in every possible world. In this case, the two sentences

> The ratio of the circumference of a circle to its diameter is necessarily irrational

and

> Necessarily, the ratio of the circumference of a circle to its diameter is irrational.

have the same truth value.

In the case of proper names, the situation is a bit more complicated. Kripke (1980) has argued that ordinary proper names are rigid when the modalities are understood alethically. The reference of a proper name is fixed by some arbitrary procedure in which names are assigned to individuals—a baptismal ceremony, perhaps. But once the reference has been fixed, that name refers to the individual it has been assigned to, in every possible world. When we consider whether it is possible (say) for Cicero to be a modal logician, we consider whether there is a possible world in which *that person* who had been baptized "Cicero" is a modal logician.

Kripke's *causal/historical* account of proper names is to be contrasted with the *description theory* associated with Russell (1905) and Frege (1892). Russell and Frege regarded a proper name as short for some definite description. "Homer," for example, means "the author of the *Illiad*." And if these two mean the same thing, then in every possible world, they will refer to the same object. But this, Kripke has pointed out, does not seem right, because it is only a contingent property of Homer that he wrote the *Illiad*. On the description theory, however, Homer will be whoever writes the *Illiad* in that world, and since this can vary, the reference of "Homer" will vary from world to world. By contrast, Kripke's view is that the reference of a name is fixed, not by an associated description, but originally, by a "baptism," and subsequently, by the intention of the speakers of the language to maintain the reference of the original baptism.

On Kripke's view, then, where $\Box$ and $\Diamond$ are interpreted alethically, if a and b are proper names, they are rigid. From this the equivalence of $\langle \lambda x, y.\Box(x = y)\rangle(a, b)$ and $\Box\langle \lambda x, y.x = y\rangle(a, b)$ follows, and so we might write $\Box(a = b)$ as shorthand for both. So in particular, since

(10.16) Hesperus = Phosphorus

then

(10.17) $\Box$(Hesperus = Phosphorus).

This, however, reawakens the old puzzle about the morning star and the evening star and what the ancients really knew (and when). If we take a to be *Hesperus* and b to be *Phosphorus*, and $\Box$ to be *The ancients knew that*, we have the unfortunate result that, if Hesperus is Phosphorus, then the ancients knew that this was so. And, it appears, they did not. The problem, quite clearly, lies with the understanding that these names are rigid designators. We see that, although they are rigid within the context of an *alethic* reading of $\Box$, they cannot be rigid under an *epistemic* reading of $\Box$. In fact, it is hard to see whether there are any rigid designators under the epistemic reading. Kripke distinguished very clearly between what he called *metaphysical necessity* and *epistemic necessity*. If (10.16) is true, he claimed, it is metaphysically necessary; but that is not to say it is epistemically necessary. Kripke followed this line further. For him, a statement that was epistemically necessary was *a priori*, and so there are statements that are necessarily true but not *a priori* true; and by the same token, there are statements that are only contingently true, but are known *a priori*.

EXERCISES

EXERCISE 10.4.1.   Are the following two claims equivalent?

1. A rigid designator is one for which scope does not matter.
2. A rigid designator is one which is always given wide scope.

10.5. EXISTENCE

For this section actualist quantification is required—varying domain semantics is the relevant one.

As we have set things up, the value of a term at a particular world of a varying domain model may or may not exist at that world, that is, it may or may not be a member of the domain of that world. This is a situation we are all familiar with, though perhaps not in such a formal setting. Consider, for instance, a temporal model, where possible worlds are time instants. Take the domain at each instant to be the set of people existing at that particular time. Consider the function that maps each person to his or her mother. For anyone who is alive at the present, but whose mother has died, the mother-of function maps that person, who is in the domain of the present world, to someone who is not in the domain of the present world, though she is in the domain of some previous ones.

Since terms can, at particular worlds, designate objects that do not exist at those worlds, it would be useful to have syntactic machinery for specifying that a term does, in fact, designate an existent object. Recall, from Chapter 8, the formula $E(x)$, which abbreviates $(\exists y)(y = x)$. There is a free variable here, x, and so we can form a predicate abstract based on it. We introduce special notation for it.

DEFINITION 10.5.1.   [$E(t)$] By $\mathbf{E}$ we mean the predicate abstract $\langle \lambda x.E(x) \rangle$. Consequently for a term t, $\mathbf{E}(t)$ is the formula $\langle \lambda x.E(x) \rangle (t)$ or, unabbreviated, $\langle \lambda x.(\exists y)(y = x) \rangle (t)$.

This provides exactly the syntactic machinery wanted. Proposition 8.1.1, concerning E has a straightforward extension to $\mathbf{E}$.

PROPOSITION 10.5.2.   Let $\mathcal{M} = \langle \mathcal{G}, \mathcal{R}, \mathcal{D}, \mathcal{I} \rangle$ be a (normal, varying domain, non-rigid) model, let v be a valuation, and t be a term. Then $\mathcal{M}, \Gamma \Vdash_v \mathbf{E}(t)$ if and only if $(v \star \mathcal{I})(t, \Gamma)$ exists at Γ, that is, if and only if $(v \star \mathcal{I})(t, \Gamma) \in \mathcal{D}(\Gamma)$.

EXERCISE 10.5.1.   Let c be a constant symbol. Give a varying domain model showing $(\forall x)P(x) \supset \langle \lambda x.P(x)\rangle(c)$ is not valid. Also show $[(\forall x)P(x) \wedge E(c)] \supset \langle \lambda x.P(x)\rangle(c)$ is valid in all varying domain models.

EXERCISE 10.5.2.   Prove Proposition 10.5.2.

10.6. TABLEAU RULES, VARYING DOMAIN

Not surprisingly, tableau rules for predicate abstracts are different depending on whether we use the possibilist or the actualist quantifier. And the difference comes down to one simple fact. With either kind of quantifier, terms can have different values at different worlds. But, if we use the actualist quantifier—varying domain models—the value of a term at a world might not be something that exists at that world; while with the possibilist quantifier—constant domain models—the value of a term at a world must exist there, and everywhere. Differences in tableau rules all follow from these points. In this section we give the rules corresponding to varying domain semantics, and in the next section, those for the constant domain version.

You've seen the general ideas behind soundness and completeness proofs for tableau systems, and they carry over to the present setting quite well. But the proofs are somewhat involved, though they contain no fundamentally new ideas. Consequently we have decided to omit them for both the varying and constant domain versions.

When first-order tableau rules were introduced in Chapter 5 the language was extended with *parameters*. Now it must be extended once more, to deal with constant and function symbols during proofs.

DEFINITION 10.6.1.   [Extended Term] An *extended term* is like a term except that some constant and function symbols may have prefixes, which we write as subscripts. A *grounded term* is an extended term in which every constant and function symbol has a subscript.

We allow extended terms to appear in tableau proofs, in positions where terms are ordinarily allowed. Extended terms must not appear in the sentences we are attempting to prove. In this respect they have exactly the same role as parameters.

Recall that we intuitively think of prefixes as denoting possible worlds, or more loosely, as being possible worlds. Now if c is a constant symbol, think of the extended term c_σ intuitively as the object that c denotes at world σ. Likewise for a function symbol f, think of f_σ as the function that f denotes at σ. This should be enough to supply motivation for the tableau rules we give below.

EXAMPLE 10.6.2. $f_{1.1}(x, g(c_1, d))$ is an extended term. It is not grounded since the function symbol g and the constant symbol d have no subscript. The following is an extended term that is grounded: $f_{1.1}(x, g_{1.1.1}(c_1, d_{1.1.1}))$.

At one extreme, an extended term might have no subscripted constant or function symbols, in which case it is simply a term. At the other extreme everything is subscripted, and we have a grounded term. Free variables are allowed to appear in extended terms. Since parameters are a particular kind of free variable, these too can occur.

DEFINITION 10.6.3.   Suppose σ is a prefix and t is an extended term. By $t@\sigma$ we mean the grounded term that results when every unsubscripted constant and function symbol in t is given σ as its subscript.

EXAMPLE 10.6.4.   For the extended term $f_{1.1}(x, g(c_1, d))$ we have

$$f_{1.1}(x, g(c_1, d))@1.1.1 = f_{1.1}(x, g_{1.1.1}(c_1, d_{1.1.1})).$$

Now, here are the varying domain tableau rules. First, the quantifier rules are *exactly* as they were in Chapter 5. We repeat them for convenience.

DEFINITION 10.6.5.   [Universal Rules—Varying Domain] In the following, p_σ is any parameter that is associated with the prefix σ.

$$\frac{\sigma \ (\forall x)\Phi(x)}{\sigma \ \Phi(p_\sigma)} \qquad \frac{\sigma \ \neg(\exists x)\Phi(x)}{\sigma \ \neg\Phi(p_\sigma)}$$

DEFINITION 10.6.6.   [Existential Rules—Varying Domain] In the following, p_σ is a parameter associated with the prefix σ, subject to the condition that p_σ is new to the tableau branch.

$$\frac{\sigma \ (\exists x)\Phi(x)}{\sigma \ \Phi(p_\sigma)} \qquad \frac{\sigma \ \neg(\forall x)\Phi(x)}{\sigma \ \neg\Phi(p_\sigma)}$$

A key fact to note here is that in the universal rules only *parameters* are involved. Extended terms play no role. Intuitively, this is because we don't generally know if an extended term designates an existent object, and so we don't generally know if it is in the range of a quantifier. This is the specific point of difference between these rules and the constant domain ones given in the next section.

DEFINITION 10.6.7. [Abstraction Rules] For any extended term t:

$$\frac{\sigma \; \langle \lambda x.\Phi(x)\rangle(t)}{\sigma \; \Phi(t@\sigma)} \qquad \frac{\sigma \; \neg\langle \lambda x.\Phi(x)\rangle(t)}{\sigma \; \neg\Phi(t@\sigma)}$$

And finally, the equality rules are essentially as they were in Chapter 7, except that grounded terms can be used in places where, earlier, we were restricted to parameters. We state the more general versions in full.

DEFINITION 10.6.8. [Reflexivity Rule] If t is a grounded term whose subscripts all occur on the tableau branch, and σ is a prefix which occurs on the tableau branch, then $\sigma \; (t = t)$ can be added to the end of the branch. Briefly,

$$\overline{\sigma \; (t = t)}$$

DEFINITION 10.6.9. [Substitutivity Rule] Let $\Phi(x)$ be a formula (possibly containing extended terms) in which x occurs free, let t and u be grounded terms, and let $\Phi(t)$ be the result of substituting occurrences of the grounded term t for all free occurrences of x in $\Phi(x)$, and similarly for $\Phi(u)$. If $\sigma_1(t = u)$ and $\sigma_2\Phi(t)$ both occur on a tableau branch, $\sigma_2\Phi(u)$ can be added to the end. Briefly,

$$\frac{\sigma_1 \; (t = u)}{\sigma_2 \; \Phi(t)}$$
$$\overline{\sigma_2 \; \Phi(u)}$$

EXAMPLE 10.6.10. In Example 10.1.6 we promised a tableau proof of

$$\left[\langle \lambda x.\Diamond\langle \lambda y.x = y\rangle(p)\rangle(a) \wedge \langle \lambda x.\Box S(x)\rangle(a) \right] \supset \Diamond\langle \lambda x.S(x)\rangle(p).$$

Well, here it is.

$$
\begin{array}{ll}
1 & \neg\big\{\big[\langle \lambda x.\Diamond\langle \lambda y.x = y\rangle(p)\rangle(a) \wedge \langle \lambda x.\Box S(x)\rangle(a)\big] \\
& \quad \supset \Diamond\langle \lambda x.S(x)\rangle(p)\} \quad 1. \\
1 & \langle \lambda x.\Diamond\langle \lambda y.x = y\rangle(p)\rangle(a) \wedge \langle \lambda x.\Box S(x)\rangle(a) \quad 2. \\
1 & \neg\Diamond\langle \lambda x.S(x)\rangle(p) \quad 3. \\
1 & \langle \lambda x.\Diamond\langle \lambda y.x = y\rangle(p)\rangle(a) \quad 4. \\
1 & \langle \lambda x.\Box S(x)\rangle(a) \quad 5. \\
1 & \Diamond\langle \lambda y.a_1 = y\rangle(p) \quad 6. \\
1.1 & \langle \lambda y.a_1 = y\rangle(p) \quad 7. \\
1.1 & a_1 = p_{1.1} \quad 8. \\
1.1 & \neg\langle \lambda x.S(x)\rangle(p) \quad 9. \\
1.1 & \neg S(p_{1.1}) \quad 10. \\
1 & \Box S(a_1) \quad 11. \\
1.1 & S(a_1) \quad 12. \\
1.1 & S(p_{1.1}) \quad 13. \\
\end{array}
$$

In this, 2 and 3 are from 1 by a Conjunctive Rule, as are 4 and 5 from 2; 6 is from 4 by an Abstraction Rule; 7 is from 6 by a Possibility Rule; 8 is from 7 by Abstraction; 9 is from 3 by a Necessity Rule; 10 is from 9 by Abstraction; 11 is from 5 by Abstraction; 12 is from 11 by a Necessity Rule; and 13 is from 8 and 12 by Substitutivity.

Since terms can designate nonexistent objects at worlds of varying domain models, explicit existence assumptions can play an important role.

EXAMPLE 10.6.11.   We observed earlier that $(\forall x)P(x) \supset \langle \lambda x.P(x)\rangle(c)$ is not valid, where c is a constant symbol. But here is a tableau proof of $[(\forall x)P(x) \wedge E(c)] \supset \langle \lambda x.P(x)\rangle(c)$. As it happens, modal issues are not involved in this example.

$$
\begin{array}{ll}
1 \; \neg\{[(\forall x)P(x) \wedge E(c)] \supset \langle \lambda x.P(x)\rangle(c)\} & 1. \\
1 \; (\forall x)P(x) \wedge E(c) & 2. \\
1 \; \neg\langle \lambda x.P(x)\rangle(c) & 3. \\
1 \; (\forall x)P(x) & 4. \\
1 \; \langle \lambda x.(\exists y)(y = x)\rangle(c) & 5. \\
1 \; (\exists y)(y = c_1) & 6. \\
1 \; p_1 = c_1 & 7. \\
1 \; P(p_1) & 8. \\
1 \; P(c_1) & 9. \\
1 \; \neg P(c_1) & 10.
\end{array}
$$

In this, 2 and 3 are from 1, and 4 and 5 are from 2 by a Conjunction Rule (note that 5 is $E(c)$ unabbreviated); 6 is from 5 by an Abstraction Rule; 7 is from 6 by an Existential Rule (so p_1 is a parameter); 8 is from 4 by a Universal Rule; 9 is from 7 and 8 by Substitutivity; 10 is from 3 by an Abstraction Rule.

Before giving another example, we present a derived rule that simplifies tableau calculation. The universal quantification rule only allows the use of parameters, but it can be broadened. Suppose on a tableau branch we have $\sigma \, E(t)$, where t is some extended term, not necessarily a parameter. It seems reasonable that, since we "know" t designates something that exists at σ, we can assume that whatever t designates, it is in the range of a universal quantifier at σ. Then if we also have $\sigma \, (\forall x)\varphi(x)$ on the branch, we should be able to add $\sigma \, \varphi(t@\sigma)$. In fact we can, by the following sequence of legitimate moves.

$$\sigma \, E(t) \quad 1.$$
$$\sigma \, (\forall x)\varphi(x) \quad 2.$$
$$\sigma \, \langle \lambda x.(\exists y)(y = x)\rangle(t) \quad 3.$$
$$\sigma \, (\exists y)(y = t@\sigma) \quad 4.$$
$$\sigma \, p_\sigma = t@\sigma \quad 5.$$
$$\sigma \, \varphi(p_\sigma) \quad 6.$$
$$\sigma \, \varphi(t@\sigma) \quad 7.$$

Item 3 is item 1 unabbreviated; 4 is from 3 by a predicate abstraction rule; 5 is from 4 by an existential rule, so p_σ is a parameter; 6 is from 2 by a universal rule; 7 is from 5 and 6 by substitutivity.

We summarize this as the following derived rule.

DEFINITION 10.6.12. [Object Existence Derived Rule] A tableau branch containing $\sigma \, E(t)$, where t is an extended term, and containing $\sigma \, (\forall x)\varphi(x)$, can be extended with $\sigma \, \varphi(t@\sigma)$. (Similarly for $\sigma \, \neg(\exists x)\varphi(x)$.)

Now, here is a final, more complex example. Note that the antecedent makes an explicit assumption about the existence of function values for the function (denoted by) f.

EXAMPLE 10.6.13. The following is a varying domain tableau proof of $\{(\forall x)\Box E(f(x)) \wedge (\forall x)\Diamond\langle\lambda y.R(x, y)\rangle(f(x))\} \supset (\forall x)\Diamond(\exists y)R(x, y)$.

$$1 \quad \neg\big[\{(\forall x)\Box E(f(x)) \wedge (\forall x)\Diamond\langle\lambda y.R(x, y)\rangle(f(x))\}$$
$$\qquad \supset (\forall x)\Diamond(\exists y)R(x, y)\big] \quad 1.$$
$$1 \quad (\forall x)\Box E(f(x)) \wedge (\forall x)\Diamond\langle\lambda y.R(x, y)\rangle(f(x)) \quad 2.$$
$$1 \quad \neg(\forall x)\Diamond(\exists y)R(x, y) \quad 3.$$
$$1 \quad (\forall x)\Box E(f(x)) \quad 4.$$
$$1 \quad (\forall x)\Diamond\langle\lambda y.R(x, y)\rangle(f(x)) \quad 5.$$
$$1 \quad \neg\Diamond(\exists y)R(p_1, y) \quad 6.$$
$$1 \quad \Diamond\langle\lambda y.R(p_1, y)\rangle(f(p_1)) \quad 7.$$
$$1.1 \quad \langle\lambda y.R(p_1, y)\rangle(f(p_1)) \quad 8.$$
$$1.1 \quad R(p_1, f_{1.1}(p_1)) \quad 9.$$
$$1.1 \, \neg(\exists y)(R(p_1, y)) \quad 10.$$
$$1 \quad \Box E(f(p_1)) \quad 11.$$
$$1.1 \quad E(f(p_1)) \quad 12.$$
$$1.1 \, \neg R(p_1, f_{1.1}(p_1)) \quad 13.$$

Items 2 and 3 are from 1 by a Conjunction Rule, as are 4 and 5 from 2; 6 is from 3 by an Existential Rule; 7 is from 5 by a Universal Rule; 8 is from 7 by a Possibility Rule; 9 is from 8 by a Predicate Abstraction Rule; 10 is from 6 by a Necessitation Rule; 11 is from 4 by a Universal Rule.; 12 is from 11 by a Necessitation Rule; 13 is from 10 and 12 using the Object Existence Derived Rule. Now this branch closes because of 9 and 13.

Here is another derived rule that is often useful.

DEFINITION 10.6.14. [Parameter Existence Derived Rule] A branch of a tableau that contains $\sigma \neg E(p_\sigma)$ closes.

EXERCISES

EXERCISE 10.6.1. Justify the Parameter Existence derived rule.

EXERCISE 10.6.2. Give a varying domain **K** tableau proof of $\langle \lambda x, y. \Diamond (x = y) \rangle (a, b) \supset \langle \lambda x, y.(x = y) \rangle (a, b)$.

EXERCISE 10.6.3. Discuss what goes wrong with an attempt to provide a varying domain **K** proof of $\langle \lambda x, y.(x = y) \rangle (a, b) \supset \Box \langle \lambda x, y.(x = y) \rangle (a, b)$. Contrast this with the provability of $\langle \lambda x, y.(x = y) \rangle (a, b) \supset \langle \lambda x, y.\Box (x = y) \rangle (a, b)$.

EXERCISE 10.6.4. Give a varying domain **K** tableau proof of the following.

$$\{ \Box (\forall x)[A(x) \supset \langle \lambda y.B(y) \rangle (f(x))] \wedge \Box (\forall x) E(f(x)) \} \supset$$
$$\{ \Diamond \langle \lambda x.A(x) \rangle (c) \supset \Diamond (\exists x) B(x) \}$$

EXERCISE 10.6.5. Consider the following three formulas:

1. $\langle \lambda x. \Box \langle \lambda y. P(y) \rangle (x) \rangle (c)$
2. $\langle \lambda x. \langle \lambda y. \Box P(y) \rangle (x) \rangle (c)$
3. $\Box \langle \lambda y. P(y) \rangle (c)$

where P is a one-place relation symbol and c is a constant symbol. Use varying domain **K** tableaus to show there are two of them that imply each other (you figure out which two). Then give models to show the third neither implies nor is implied by the others.

EXERCISE 10.6.6. This exercise fills in missing work from Section 10.2. We define four formulas as follows.

$$A_3 = \Box\langle\lambda y.\langle\lambda x.\Phi(y) \supset \Phi(c)\rangle(c)\rangle(c) \supset$$
$$\langle\lambda y.\Box\langle\lambda x.\Phi(y) \supset \Phi(x)\rangle(c)\rangle(c)$$
$$A_4 = \langle\lambda x.\Box\Phi(x)\rangle(c) \supset \Box\langle\lambda x.\Phi(x)\rangle(c)$$
$$B_4 = \langle\lambda y.\langle\lambda x.\Box(\Phi(x) \supset \Phi(y))\rangle(c)\rangle(c) \supset$$
$$\langle\lambda y.\Box\langle\lambda x.\Phi(x) \supset \Phi(y)\rangle(c)\rangle(c)$$
$$B_3 = \Box\langle\lambda x.\Phi(x)\rangle(c) \supset \langle\lambda x.\Box\Phi(x)\rangle(c)$$

Give tableau proofs, using varying domain **K** rules, of the following.

1. $A_3 \supset A_4$.
2. $B_4 \supset B_3$.

EXERCISE 10.6.7. This finishes the proof of Proposition 10.2.5. Prove each of the following using varying domain **K** rules.

1. $\langle\lambda y.\Box\langle\lambda x.x = y\rangle(c)\rangle(c) \supset [\Box\langle\lambda x.P(x)\rangle(c) \supset \langle\lambda x.\Box P(x)\rangle(c)]$
2. $\langle\lambda y.\Box\langle\lambda x.x = y\rangle(c)\rangle(c) \supset [\langle\lambda x.\Box P(x)\rangle(c) \supset \Box\langle\lambda x.P(x)\rangle(c)]$

EXERCISE 10.6.8. This completes a missing piece from Section 10.3. Give a varying domain **D** proof that the conjunction of items 1, 2, and 3 below implies item 4.

1. $\langle\lambda x.E(x)\rangle(c)$
2. $(\forall x)\Box\neg(x \approx x + 1)$
3. $\langle\lambda x.\Box(c \approx x + 1)\rangle(c)$
4. $\neg\langle\lambda x.\Box\langle\lambda y.x = y\rangle(c)\rangle(c)$

EXERCISE 10.6.9. Give a varying domain **K** tableau proof of $(A \wedge B) \supset C$, where A, B, and C are defined as follows.

$$A = (\forall x)\Diamond\langle\lambda y.(\forall z)\Box\langle\lambda w.R(x, y, z, w)\rangle(g(x, z))\rangle(f(x))$$
$$B = (\forall x)\Box E(f(x)) \wedge (\forall x)\Box(\forall y)\Box E(g(x, y))$$
$$C = (\forall x)\Diamond(\exists y)(\forall z)\Box(\exists w) R(x, y, z, w)$$

10.7. TABLEAU RULES, CONSTANT DOMAIN

As we remarked several times, as far as terms are concerned the essential difference between constant and varying domain semantics is that, for constant domains terms always designate existent objects, while for varying domains this is not so. It follows that the earlier constant domain quantifier rules are now too narrow. Here are the replacements.

DEFINITION 10.7.1. [Universal Rules—Constant Domain] In the following we take t to be any *grounded term* whatsoever, provided that all subscripts
in t already occur as prefixes on the tableau branch.

$$\frac{\sigma\,(\forall x)\Phi(x)}{\sigma\,\Phi(t)} \qquad \frac{\sigma\,\neg(\exists x)\Phi(x)}{\sigma\,\neg\Phi(t)}$$

DEFINITION 10.7.2. [Existential Rules—Constant Domain] In the following, p is a parameter that is *new to the tableau branch.*

$$\frac{\sigma\,(\exists x)\Phi(x)}{\sigma\,\Phi(p)} \qquad \frac{\sigma\,\neg(\forall x)\Phi(x)}{\sigma\,\neg\Phi(p)}$$

The existential rules are exactly as they were in Section 5.1. The universal
rules have been broadened to allow grounded terms, and not just parameters. Note that this is essentially equivalent to taking $E(x)$ as an open global
assumption, and using the Object Existence Derived Rule (10.6.12).

The remaining rules are *exactly* as they were in the previous section:
Abstraction, Reflexivity, Substitutivity.

EXAMPLE 10.7.3. Here is a constant domain **K** tableau proof of the sen-
tence $\Diamond\langle\lambda x.P(x)\rangle(c) \supset (\exists x)\Diamond P(x)$, where c is a constant symbol.

$$
\begin{array}{ll}
1 & \neg[\Diamond\langle\lambda x.P(x)\rangle(c) \supset (\exists x)\Diamond P(x)] \quad 1. \\
1 & \Diamond\langle\lambda x.P(x)\rangle(c) \quad 2. \\
1 & \neg(\exists x)\Diamond P(x) \quad 3. \\
1.1 & \langle\lambda x.P(x)\rangle(c) \quad 4. \\
1.1 & P(c_{1.1}) \quad 5. \\
1 & \neg\Diamond P(c_{1.1}) \quad 6. \\
1.1 & \neg P(c_{1.1}) \quad 7.
\end{array}
$$

In this, 2 and 3 are from 1 by a Conjunction Rule; 4 is from 2 by a Possibility
Rule; 5 is from 4 by an Abstraction Rule; 6 is from 3 by a Universal Quantifier
Rule; and 7 is from 6 by a Necessity Rrule.

EXAMPLE 10.7.4. Here is a constant domain **K** tableau proof of

$$(\forall x)\Diamond\langle\lambda y.R(x, y)\rangle(f(x)) \supset (\forall x)\Diamond(\exists y)R(x, y).$$

We proved a sentence very much like this using varying domain rules, as Example 10.6.13. But there we needed an explicit existence assumption, which

is unnecessary now.

$$1 \quad \neg\big[(\forall x)\Diamond\langle\lambda y.R(x, y)\rangle(f(x)) \supset (\forall x)\Diamond(\exists y)R(x, y)\big] \quad 1.$$
$$1 \quad (\forall x)\Diamond\langle\lambda y.R(x, y)\rangle(f(x)) \quad 2.$$
$$1 \quad \neg(\forall x)\Diamond(\exists y)R(x, y) \quad 3.$$
$$1 \quad \neg\Diamond(\exists y)R(p, y) \quad 4.$$
$$1 \quad \Diamond\langle\lambda y.R(p, y)\rangle(f(p)) \quad 5.$$
$$1.1 \quad \langle\lambda y.R(p, y)\rangle(f(p)) \quad 6.$$
$$1.1 \,\neg(\exists y)R(p, y) \quad 7.$$
$$1.1 \quad R(p, f_{1.1}(p)) \quad 8.$$
$$1.1 \,\neg R(p, f_{1.1}(p)) \quad 9.$$

In this, 2 and 3 are from 1 by a Conjunction Rule; 4 is from 3 by an Existential Rule; 5 is from 2 by a Universal Rule; 6 is from 5 by a Possibility Rule; 7 is from 4 by a Necessity Rule; 8 is from 6 by a Predicate Abstraction Rule; and 9 is from 7 by a Universal Rule.

EXERCISES

EXERCISE 10.7.1.   Give a constant domain **K** tableau proof of the sentence $\langle\lambda x.\Diamond P(x)\rangle(c) \supset (\exists x)\Diamond P(x)$, where c is a constant symbol.

EXERCISE 10.7.2.   Give a constant domain **D** proof of Exercise 10.6.4 without using item 1, the existence clause.

EXERCISE 10.7.3.   In Exercise 10.6.9 A, B, and C were defined, and you were asked to provide a varying domain proof of $(A \wedge B) \supset C$. Now give a constant domain proof of $A \supset C$.

CHAPTER ELEVEN

DESIGNATION

There is a natural confusion between existence and designation, but these are really orthogonal issues. Terms designate; objects exist. For instance the phrase "the first President of the United States" designates George Washington, though thinking temporally, the person being designated is no longer with us—the person designated does not exist, though he once did. The nonexistent George Washington is designated by the phrase *now*. On the other hand the phrase, "the present King of France," does not designate anybody now, living or dead, though at certain past instances it did designate.

In our formal treatment we have allowed terms to designate nonexistent objects at worlds of varying domain models. More properly, a term may designate, at a world, an object not in the domain of that world, though it must be in the domain of *some* world. But we have adopted the convenient fiction that terms always do designate. The problem for this chapter, then, is to modify the formal machinery to allow for non-designating terms.

While definite descriptions are the classic examples of things that may fail to designate, we do not introduce them formally until Chapter 12. Nonetheless, we can consider them as motivational examples, as we did above, and they serve to guide us in our revised treatment of terms. We used the phrase "the first President of the United States" as an example of a term that designates, though the object designated does not now exist—this suggests we should allow constant symbols to lack values at some worlds. But also, consider the closely related (no pun intended) phrase, "the eldest son of the first President of the United States." George Washington had no children, so this phrase does not designate anybody, now or at any time in the past. (Of course one can imagine alternate worlds in which George Washington had a son. We are informally using a temporal model in which possible worlds are the current and the previous states of the actual world, and in this model the phrase never designates.) It is "the eldest son of" that gives rise to the problem. This phrase can be thought of as a natural language specification of a function, mapping objects to their eldest sons. As such, it is a *partial* function. The second President of the United States had an eldest son, the first President did not. For that matter, the Taj Mahal and Niagara Falls also do not have eldest sons.

Our treatment of function symbols must be modified, so they can stand for partial functions. As noted above, our treatment of constant symbols, which can be thought of as 0-place function symbols, must also be modified—they must be allowed to be undefined at some possible worlds. *Partiality* is the theme for now.

Suppose we are at a possible world and the term t fails to designate any object at that world. What should we take as the truth value of $\langle\lambda x.\Phi\rangle(t)$ at that world? There are, basically, only two reasonable approaches. One possibility is to allow the sentence itself to fail to designate, that is, $\langle\lambda x.\Phi\rangle(t)$ is neither true nor false under these circumstances—its truth value is undefined. This moves us beyond the classical two truth values, to a three-valued, or a partial logic. (That is, we can say there is a third truth value of undefined, or equivalently we can say the assignment of truth values is a partially defined function.) There are good arguments one can make for such an approach. However, we wish to stay as close to classical logic as we can, with every sentence having a classical truth value at each world.

So we take the straightforward approach that the application of any predicate abstract to a non-designating term yields a false sentence. A term that designates can designate an object that does not exist, or it can designate one that does. But since we can ascribe no properties to what is designated by a term that does not designate, we will see that we cannot assert either the existence or the nonexistence of an object designated by a term that does not designate. We are thinking here of nonexistence as a positive property in its own right, rather than as the lack of the existence property. Careful attention to such matters will allow us to avoid the usual paradoxes. Now the details.

11.1. THE FORMAL MACHINERY

The notion of a non-rigid interpretation was given in Definition 9.4.3. We now modify this to allow for *partiality*.

DEFINITION 11.1.1. [General Non-Rigid Interpretation]
Let $\mathcal{F} = \langle \mathcal{G}, \mathcal{R}, \mathcal{D} \rangle$ be an augmented frame. $\mathcal{I}$ is a *general non-rigid interpretation* in $\mathcal{F}$ if $\mathcal{I}$ assigns relations to relation symbols (these are completely defined, of course), as in Definition 4.6.2, and in addition:

1. To each constant symbol c, and to *some* (possibly no) members $\Gamma \in \mathcal{G}$, $\mathcal{I}$ assigns a member of the domain of $\mathcal{F}$. That is, $\mathcal{I}(c, \Gamma)$ may not be defined, but if it is then $\mathcal{I}(c, \Gamma) \in \mathcal{D}(\mathcal{F})$.
2. To each n-place function symbol f, and to each $\Gamma \in \mathcal{G}$, $\mathcal{I}$ assigns some *partial* n-ary function from the domain of $\mathcal{F}$ to itself. That is, $\mathcal{I}(f, \Gamma)$ is a function from a subset, not necessarily all, of $\mathcal{D}(\mathcal{F})^n$ to $\mathcal{D}(\mathcal{F})$.

DEFINITION 11.1.2. [General Non-Rigid Model]
A *general non-rigid model* is a structure $\mathcal{M} = \langle \mathcal{G}, \mathcal{R}, \mathcal{D}, \mathcal{I} \rangle$ where, as before,

$\langle \mathcal{G}, \mathcal{R}, \mathcal{D} \rangle$ is an augmented frame, and $\mathcal{I}$ is a general non-rigid interpretation in it.

Next, we must modify Definition 9.4.5 to allow for terms to fail to designate.

DEFINITION 11.1.3. [Term Evaluation] Let $\mathcal{M} = \langle \mathcal{G}, \mathcal{R}, \mathcal{D}, \mathcal{I} \rangle$ be a general non-rigid model, let $\Gamma \in \mathcal{G}$, and let v be a valuation. This time we associate, at Γ, values in $\mathcal{D}(\mathcal{F})$ with *some* terms t, not necessarily all of them. We continue to use the notation $(v \star \mathcal{I})(t, \Gamma)$ for the object designated by t, if t designates at Γ (in $\mathcal{M}$, with respect to v). It is defined as follows.

1. If x is a free variable (including parameters), x designates at Γ, and $(v \star \mathcal{I})(x, \Gamma) = v(x)$.
2. If c is a constant symbol, c designates at Γ provided $\mathcal{I}(c, \Gamma)$ is defined, and if it is, $(v \star \mathcal{I})(c, \Gamma) = \mathcal{I}(c, \Gamma)$.
3. If f is an n-place function symbol, and each of $t_1, \ldots, t_n$ designates at Γ, and $\langle (v \star \mathcal{I})(t_1, \Gamma), \ldots, (v \star \mathcal{I})(t_n, \Gamma) \rangle$ is in the domain of the function $\mathcal{I}(f, \Gamma)$, then $f(t_1, \ldots, t_n)$ designates at Γ, and

$$(v \star \mathcal{I})(f(t_1, \ldots, t_n), \Gamma) = \mathcal{I}(f, \Gamma)((v \star \mathcal{I})(t_1, \Gamma), \ldots, (v \star \mathcal{I})(t_n, \Gamma))$$

If t does not designate at Γ, $v(t, \Gamma)$ is undefined.

Finally, we define when formulas are true at possible worlds of models. The following is to replace Definition 9.4.6, which assumed that terms always designated. Incidentally, we continue to treat equality as we did in Chapter 7—*throughout this chapter, all models are normal.* We will not say, every time, that the equality relation symbol is always interpreted by the equality relation at each possible world.

DEFINITION 11.1.4. [Truth in a General Non-Rigid Model]
Let $\mathcal{M} = \langle \mathcal{G}, \mathcal{R}, \mathcal{D}, \mathcal{I} \rangle$ be a general non-rigid model. The definition of $\mathcal{M}, \Gamma \Vdash_v X$ remains the same as in Definition 4.6.7, with the following additions:

9. If t designates at Γ in $\mathcal{M}$ with respect to v, $\mathcal{M}, \Gamma \Vdash_v \langle \lambda x.\Phi \rangle(t) \iff \mathcal{M}, \Gamma \Vdash_w \Phi$, where w is the x-variant of v such that $w(x) = (v \star \mathcal{I})(t, \Gamma)$.
10. If t fails to designate at Γ in $\mathcal{M}$ with respect to v, $\mathcal{M}, \Gamma \not\Vdash_v \langle \lambda x.\Phi \rangle(t)$.

Thus we have a formal counterpart to the idea that no property can be correctly ascribed to what is designated by a non-designating term. The formal version is that the application of any predicate abstract to a non-designating term is always false. Note that Definition 11.1.4 reduces to Definition 9.4.6 if all terms designate.

11.2. DESIGNATION AND EXISTENCE

Designation is an inherently semantic notion. Nonetheless, a syntactic counterpart is available for it. We begin this section by introducing a "designation property," then we investigate its relationships with the existence property from Chapter 8.

DEFINITION 11.2.1. [**D**(t)] **D** abbreviates $\langle \lambda x.x = x \rangle$, and so **D**(t) abbreviates $\langle \lambda x.x = x \rangle(t)$.

Think of **D**(t) as intended to assert that t designates. The choice of the formula $x = x$ is a little arbitrary. All we really want is a formula that cannot be falsified; $P(x) \vee \neg P(x)$ would do as well, though it requires us to choose a relation symbol P, and this has an even greater element of arbitrariness about it. At any rate, the key reason we introduce **D**(t) is that it allows us to move a meta-level notion into the object language. Here is the formal statement of this fact.

PROPOSITION 11.2.2.   Let $\mathcal{M} = \langle \mathcal{G}, \mathcal{R}, \mathcal{D}, \mathcal{I} \rangle$ be a general non-rigid model, and let t be a term. Then $\mathcal{M}, \Gamma \Vdash_v \mathbf{D}(t)$ if and only if t designates at Γ in $\mathcal{M}$ with respect to v.

Proof This is quite simple, and follows directly from our definitions. If t designates at Γ in $\mathcal{M}$, $(v \star \mathcal{I})(t, \Gamma)$ is defined, and since we are assuming $\mathcal{M}$ is a normal model, it follows from Definition 11.1.4 that $\mathcal{M}, \Gamma \Vdash_v \langle \lambda x.x = x \rangle(t)$. On the other hand if t does not designate at Γ, by Definition 11.1.4 again, $\mathcal{M}, \Gamma \Vdash_v \langle \lambda x.\Phi \rangle(t)$ is false no matter what Φ may be, so in particular, $\mathcal{M}, \Gamma \Vdash_v \langle \lambda x.x = x \rangle(t)$ is false. ■

Earlier we introduced an *existence* abstract, **E**, for use in varying domain models. If you recall,

$$\mathbf{E}(t) = \langle \lambda x.E(x) \rangle(t) = \langle \lambda x.(\exists y)(y = x) \rangle(t).$$

In a similar way we can define a *nonexistence* predicate:

$$\overline{\mathbf{E}}(t) = \langle \lambda x.\neg E(x) \rangle(t) = \langle \lambda x.\neg(\exists y)(y = x) \rangle(t).$$

It might seem that this new predicate has no uses. After all, everything exists, and nothing has the nonexistence property. Formally, both $(\forall x)E(x)$ and $\neg(\exists x)\overline{E}(x)$ are valid in varying domain models. But this validity is nothing profound—quantifiers only quantify over things that exist. The *open* formula, $E(x)$, is not valid, since we might assign to the variable x a value that does not exist at some given world. This suggests that if we want to look at the relationships between existence and designation, we should avoid the use of quantifiers and their built-in existence presuppositions.

PROPOSITION 11.2.3. For each term t, the following is valid in varying domain models:

$$\mathbf{D}(t) \equiv [\mathbf{E}(t) \vee \overline{\mathbf{E}}(t)].$$

Proof One direction is quite simple. If t does not designate (at some particular world), no predicate abstracts correctly apply to t; and thus in particular, both $\mathbf{E}(t)$ and $\overline{\mathbf{E}}(t)$ are false.

The other direction is only a little more work. Suppose t does designate at a world of the model $\mathcal{M} = \langle \mathcal{G}, \mathcal{R}, \mathcal{D}, \mathcal{I} \rangle$, say $\mathcal{M}, \Gamma \Vdash_v \mathbf{D}(t)$. Then $(v * \mathcal{I})(t)$ is defined. Now, either $(v * \mathcal{I})(t)$ is in $\mathcal{D}(\Gamma)$ or it is not. If it is, $\mathcal{M}, \Gamma \Vdash_v \mathbf{E}(t)$ is true, and if it is not, $\mathcal{M}, \Gamma \Vdash_v \overline{\mathbf{E}}(t)$ is true. Either way, we have $\mathbf{E}(t) \vee \overline{\mathbf{E}}(t)$. ∎

If the term t designates at a world, it follows from this Proposition that $\mathbf{E}(t) \vee \overline{\mathbf{E}}(t)$, and hence $\overline{\mathbf{E}}(t) \equiv \neg\mathbf{E}(t)$. Since in earlier chapters all terms always designated, a special nonexistence predicate abstract was redundant— negating the existence predicate abstract sufficed. This accounts for why we did not introduce a nonexistence predicate abstract earlier. Things are different now, however. If t does not designate at a particular world, *both* $\mathbf{E}(t)$ and $\overline{\mathbf{E}}(t)$ must be false there, and we do not have $\overline{\mathbf{E}}(t) \equiv \neg\mathbf{E}(t)$!

Let us state this fact in an unabbreviated version. If t does not designate at a world, $\langle \lambda x.\neg(\exists y)(y = x) \rangle(t)$ and $\neg\langle \lambda x.(\exists y)(y = x) \rangle(t)$ are not equivalent at that world. In earlier chapters the machinery of predicate abstraction was significant when dealing with modal notions, but its effects were invisible where the classical connectives and quantifiers were concerned. Now that non-designation is allowed, predicate abstraction is visible even at the propositional level: as we have just seen, negation inside and negation outside a predicate abstract can be different. This fact was noted long ago by Russell (1905) in his treatment of definite descriptions, which are the classic examples of possible non-designation.

EXERCISES

EXERCISE 11.2.1. Give formal details of the proof of Proposition 11.2.3, which we omitted.

EXERCISE 11.2.2. Explain why $\neg\langle \lambda x.\varphi(x) \rangle(y) \equiv \langle \lambda x.\neg\varphi(x) \rangle(y)$ is valid, where y is a variable.

EXERCISE 11.2.3. Analogous to the nonexistence predicate introduced above, we could also introduce a non-designation predicate. Do so, then explain why the result is useless.

EXERCISE 11.2.4.   Which of the following are valid:

1. $\langle \lambda x.\neg\varphi \rangle(t) \supset \neg\langle \lambda x.\varphi \rangle(t)$.
2. $\neg\langle \lambda x.\varphi \rangle(t) \supset \langle \lambda x.\neg\varphi \rangle(t)$.
3. $\langle \lambda x.\varphi \supset \psi \rangle(t) \supset [\langle \lambda x.\varphi \rangle(t) \supset \langle \lambda x.\psi \rangle(t)]$.
4. $[\langle \lambda x.\varphi \rangle(t) \supset \langle \lambda x.\psi \rangle(t)] \supset \langle \lambda x.\varphi \supset \psi \rangle(t)$.

11.3. EXISTENCE AND DESIGNATION

Throughout this section only varying domain models will be considered. The use of actualist quantification is critical for the points we wish to make.

One of the interesting results of our treatment of varying domain semantics is that everything exists; however, as Quine put it, we quibble over cases. The "x exists" formula, defined as

(11.1) $E(x)$ or $(\exists y)(y = x)$,

is an open sentence with a free variable. It is important to distinguish (11.1) from

(11.2) $(\forall x)E(x)$ or $(\forall x)(\exists y)(y = x)$

which is a (closed) sentence and, importantly, a logical truth: Everything exists. The open formula (11.1) is far from a logical truth. To the contrary, it is (speaking loosely) true of some things and false of others. Bear in mind our use of free variables. An object that is not in the domain of the actual world can be the value of a free variable. Such an object is a nonexistent object in the actual world. And if such an object is taken to be the value of the free variable x, then (11.1) is not true of it: there is nothing in the domain of this world with which it is identical.

By contrast with (11.1), the formula

(11.3) $x = x$,

which is also an open sentence with one free variable, is true of every existent and every nonexistent. Formula (11.3) is true for any object that exists in any possible world. We might, if we wanted to keep the Meinongian terminology alive, take the predicate abstracted from (11.3), $\langle \lambda x.x = x \rangle$, to be the "has being" predicate. But we prefer to think of it as saying that a term designates.

Although everything exists, we can say, of a given object, that *it* exists, and speak either truly or falsely; and we can say, of a given object, that *it* does not exist, and speak either truly or falsely. There are no paradoxical consequences to be drawn from this. We cannot say everything has being: if we attempt to do so, we come out with the slightly different

(11.4) $(\forall x)(x = x)$,

which is true of everything in the range of the quantifier, i.e., everything that exists. What we can say is that every instance of (11.3) is true, and so, we cannot truthfully deny *of* a given object that *it* has being.

Does this mean that we are saddled with the Meinongian thesis that the round-square has being or that the least prime number between 19 and 23 has being? No. We are not committed to Meinong's comprehension schema. The meaning of a singular term need not be identified with the thing it designates, and so, a singular term can be meaningful even if it doesn't designate anything. Furthermore, there is no need to suppose that inconsistent singular terms need to designate anything to be meaningful. The singular term "the round-square" does not designate in any possible world. And in the case of "the round-square," we can coherently deny that the round-square exists, even that the round-square is identical with itself; what we cannot do, since the term fails to designate, is to deny *of* the round-square that it exists, or again, deny *of* the round-square that it has being.

In Chapter 12 we give a formal version of these ideas, but until definite descriptions have been discussed in detail, such a treatment must wait. In the meantime, we continue with our informal exposition, preparing the way for the formalizations that follow.

Let us give a more concrete example of the distinctions we have been drawing. Suppose we are working within a temporal framework, so that our frame contains a set of possible worlds with a temporal accessibility relation. In the present temporal world, George Washington doesn't exist. This is not to say that *something* fails to exist, for the point of saying that he doesn't exist is that there is nothing identical with him—so, of course, it could not be that something fails to exist. The expression "George Washington" *designates*, but fails to *designate an existent*. It designates an object that exists in an earlier temporal world. Recall our temporal operators **P** and **F** (*It was the case that, It will be the case that*). Now, George Washington purportedly had wooden teeth. We use the following abbreviations: g for "George Washington" and $T(x)$ for "x has wooden teeth." So, we have the following truth:

(11.5) $\mathbf{P}\langle \lambda x.T(x)\rangle(g)$

Kripke has argued that a proper name like "George Washington" is a rigid designator when we interpret $\square$ as metaphysical necessity. There is no *a priori* reason to think that because an expression is rigid under one interpretation of $\square$ that it is therefore rigid under all; but the sort of arguments put forward for the claim that proper names are rigid in the context of metaphysical necessity appear to be equally appropriate for the case of temporal necessity. The salient difference between the alethic and the temporal case is that George Washington does not exist *now*. As a result, we have a rigid designator, "George Washington", that rigidly designates a nonexistent and further, one whose baptismal ceremony took place in another (earlier) world in the model, but not the actual (present) one. This further underscores the correctness of separating out the issues of rigidity and existence.

It is reasonable to argue that "George Washington" is temporally a rigid designator, and this implies that (11.5) is logically indistinguishable from

(11.6) $\langle \lambda x.PT(x)\rangle(g)$

George Washington doesn't exist—certainly not in this temporal world, although he did exist in a previous temporal world. So, the upshot is that rigid designators can be introduced for nonexistents. In another temporal world, the man exists and is so baptized; we intend to maintain that reference even though the man no longer exists. We can speak about him even though he doesn't exist.

So we have the proper name "George Washington" and even though we have true things to say about George Washington, he doesn't exist. In particular, even if it turns out to be true that he, i.e., George Washington, had wooden teeth, it does not follow that *something* or *someone*—where these are understood as existential quantifications—had wooden teeth. From (11.6), then, we cannot infer

(11.7) $(\exists y)\langle \lambda x.PT(x)\rangle(y)$.

And the reason is quite clear: George Washington does not exist now, and so existential generalization does not apply.

Contrast the expressions "George Washington" and "George Washington's eldest son." Not only does this latter expression fail to designate in this temporal world (as does "George Washington"), but it fails to designate in every temporal world. There is no temporal world in which George Washington has an eldest son. From a temporal world perspective, then, there is no "possible world" in which George Washington's eldest son exists. We have here a somewhat different situation from that of George Washington. George Washington, we say, doesn't exist but did. There is some possible world in which we find him in the domain. George Washington's eldest son doesn't exist and can't. His existence is impossible because of the semantics chosen in the model, so "George Washington's eldest son" fails to designate. (This is the formal way of saying that he lacks being, in Meinongian terms.) We can, of course, deny his existence. But here we must be careful. Suppose we let w be "George Washington's eldest son." This is a constant which fails to designate anything that exists in the current (or any other) temporal world. Since it fails to designate in any world, it fails to designate anything that has a property; so, in particular, it fails to designate anything that has the nonexistence property. This means that we cannot express the claim

(11.8) George Washington's eldest son does not exist

as

(11.9) $\langle \lambda x. \neg (\exists y)(y = x) \rangle (w)$

Rather, we must do so like this:

(11.10) $\neg \langle \lambda x. (\exists y)(y = x) \rangle (w)$

It is not, then, that w fails to exist; the term fails to designate! We could just as well have used our existence formula $E(x)$ to represent (11.9) and (11.10), respectively, as

(11.11) $\langle \lambda x. \neg E(x) \rangle (w)$, or equivalently, $\overline{E}(w)$

which is false, since it asserts that w has a property, nonexistence, and

(11.12) $\neg \langle \lambda x. E(x) \rangle (w)$, or equivalently, $\neg E(w)$

which is true. Predicate abstraction notation is crucial here. Without it, the symbolization of (11.8) would be the ambiguous

(11.13) $\neg E(w)$.

The most radical type of designation failure is an inconsistent description, e.g., "the round square" or "the smallest prime number between 19 and 23." These are meaningful phrases, but we don't have to suppose that their meaning consists in what they denote, as Russell did. They don't denote at all. We can, however, deny their existence in the manner of (11.10). And, if we use the E predicate, we can even make it look like a subject/predicate proposition, as in (11.12). The strongest type of designation is a term that designates an existent. In this temporal world, Margaret Thatcher exists: the term "Margaret Thatcher" is meaningful, and it designates an existent. The middle case is a term that designates something in another possible world but not this one, like "George Washington" or "Lord Wellington." Although they don't exist, i.e., nothing *is* identical with them, they *did* exist. Finally, we consider expressions like "the present king of France." This designates something in an earlier temporal world, e.g., Louis XIV. We can say not that Louis XIV is the present King of France, but rather, that it was the case that

Louis XIV is the present King of France. The expression "the present King of France" fails to designate in this temporal world, but it does in an earlier one. "The present King of France" is a nonrigid designator. So, to deny the existence of the present King of France, we must understand the denial as:

(11.14) $\neg\langle\lambda x.E(x)\rangle$(the present King of France) or

(11.15) $\qquad\qquad\neg E$(the present King of France)

not as

(11.16) $\langle\lambda x.\neg E(x)\rangle$(the present King of France) or

(11.17) $\qquad\qquad\overline{E}$(the present King of France).

George Washington, even though he fails to exist, does have properties. George Washington, for example, has the property of nonexistence: *He* does not exist. On the other hand, he has the property of *having existed*. Recall our use of **P** and **F** to abbreviate *It was the case that* and *It will be the case that*, respectively. Then, the following are both true:

(11.18) $\langle\lambda x.\neg E(x)\rangle(w)$

(11.19) $\langle\lambda x.PE(x)\rangle(w)$

We are saying *of* George Washington that he does not exist but he did. In each case, we are ascribing properties to him: the property of nonexistence, the property of past existence. And since "George Washington" is a rigid designator, w is invisible to the scope distinction, i.e., the following is also true:

(11.20) $P\langle\lambda x.E(x)\rangle(w)$

George Washington has various positive properties, for example, the property of *having been* the president of the United States and the property of *having had* wooden teeth. It has sometimes been argued that nonexistent objects can have no positive properties because there is nothing to hang them on. They can, so the view has it, possess only negative properties. There seems to be a confusion in this reasoning that is nicely brought out by predicate abstraction notation. There are two places one can bring in negation: (*a*) one can ascribe happiness or *un*happiness, existence or *non*existence to an individual; (*b*) one can ascribe or *not* ascribe happiness, existence, etc. to an individual. It is nigh impossible to distinguish positive and negative properties, as any number of philosophers and logicians have pointed out. But positive and negative *ascriptions* of properties can be distinguished on purely syntactic grounds. For while (11.18) is an instance of the form $F(w)$ the following,

(11.21) $\neg\langle\lambda x.E(x)\rangle(w)$

is not. We might say, then, *that* nonexistent objects have positive or negative properties as it seems reasonable; but for a term *t* that does not designate, there are no true sentences that say *of t* that it has a positive property.

How far can we ascribe properties to nonexistents? Is George Washington a President? Does George Washington have wooden teeth? These are all questions that demand an answer, in the sense that in the indicative forms, each of these has a truth value *in this world*. What is the truth value? It is not certain how we are to answer in each case. If we think that these predicates are true only of things in the domain of this world, then we will take each to be false. But we don't have to think this way. It is possible to take a more relaxed attitude toward these predicates and suppose that some will be true of George Washington and others false. It seems to us reasonable, for example, to say that George Washington is a President, though not the current one, for it seems right to suppose that anyone who was a President is a President. It also seems reasonable to suppose that George Washington is a man, even though he does not exist. It is not so clear what is to be said about his having wooden teeth. Essentially, these are not issues of what is true in a formal model, but rather, which formal model best reflects the way we use natural language. Reasonable people can debate this, but within each formal model, truth and falsity for sentences is clearly ascribed.

Finally, we mention one nice outcome of our way of dealing with rigidity, namely, we can extend the notion of rigidity to general terms in a natural way. Kripke's original definition does not extend readily to general terms. With his notion of rigidity we cannot say, for example, that "tiger" designates the same objects in every possible world because the extension of "tiger" in one world need not be the same as the extension of "tiger" in another if, as we suppose, we are dealing with a variable domain model. But once rigidity and existence have been separated, we can say that a general expression $F(x)$ is rigid (or, alternatively, a *natural kind term*) if it is an essential property of anything that possesses it, i.e.,

$$(11.22) \ \langle \lambda x . \Diamond F(x) \rangle (y) \equiv \langle \lambda x . \Box F(x) \rangle (y).$$

Our use of free variables enables us to speak generally over the domain of the frame.

EXERCISES

EXERCISE 11.3.1. Let *t* be a term that does not designate. Are there any true sentences that say of *t* that it has a negative property?

11.4. FICTION

In a series of unpublished lectures, Kripke has suggested that names that are introduced in fiction, like "Sherlock Holmes," are introduced, *a fortiori* for things that fail to exist. If it should turn out that in fact there was an individual who lived on Baker Street, smoked a pipe and solved crimes with a close friend Watson, it would be entirely accidental (as the usual disclaimers in movies and books say). On some readings, Kripke appears to say that Holmes is a necessary nonexistent. Some flavor of this view can be found in the *Addenda* to Kripke (1980). He says there:

Similarly, I hold the metaphysical view that, granted that there is no Sherlock Holmes, one cannot say of any possible person that he *would have been* Sherlock Homes, had he existed. Several distinct possible people, and even actual ones such as Darwin or Jack the Ripper, might have performed the exploits of Holmes, but there is none of whom we can say that he would have *been* Holmes had he performed these exploits. For if so, which ones? I thus could no longer write, as I once did, that 'Holmes does not exist, but in other states of affairs, he would have existed.' ... The quoted assertion gives the erroneous impression that a fictional name such as 'Holmes' names a particular possible-but-not-actual individual. The substantive point I was trying to make, however, remains and is independent of any linguistic theory of the status of names in fiction. The point was that, in other possible worlds 'some actually existing individuals may be absent while new individuals ... may appear' ... and if in an open formula $A(x)$ the free variable is assigned a given individual as value, a problem arises as to whether (in a model-theoretic treatment of modal logic) a truth-value is to be assigned to the formula in worlds in which the individual in question does not exist. (p. 80)

Kripke's discussion here is far from clear. It is true, as he mentions at the very end, that there is a question about assigning a truth value to a formula containing a free variable whose value is taken to be an individual that does not exist (at that world). We have taken a stand on this: we do assign it a truth value. For us, $F(x) \vee \neg F(x)$ is true whatever object in the frame is assigned to x. But the semantic treatment of free variables does not by itself determine the semantic treatment of constants or more complex terms. And we have taken a very tolerant attitude about constants, admitting quite different types of constants and quite different semantic treatments for these constants. Some constants don't designate; some constants are rigid designators; some rigid designators designate existents and some don't; some constants are nonrigid; some of the constants that are nonrigid designate nonexistents; some of the constants that are nonrigid designate in other worlds but not this one. The richness of our treatment of singular terms enables us to make many distinctions that comfortably mirror our intuitions. We have urged that "George Washington" is a rigid designator of a nonexistent. Here we should like to make more plausible the view of severing rigidity from existence by extending the idea of rigid designation of nonexistents to such names as "Holmes."

Kripke appears to believe that proper names can only be introduced for existents. Whether or not this is his view, however, the view is not quite right. From the perspective of temporal modalities, George Washington fails to exist, and yet the name "George Washington" is a rigid designator. The term designates a nonexistent object. The term was introduced in an earlier temporal world in which George Washington existed. Our problem is to generalize this so that proper names can be introduced in other possible worlds and used in this one. In another possible world there is a flying horse that is named "Pegasus"; we in this world use the name to refer to this nonexistent.

Kripke has urged that we hold language constant across possible worlds. So, he says, when we use the name "Nixon" to refer to an individual in another possible world, we do not look to see whom they call "Nixon" in that world. The designation of the name has been fixed here, and we use it to refer in another world to whoever has been so named here. As he sometimes puts it, we don't suppose the language to vary from world to world. But, if a name can have its designation fixed in the actual world, it could also have its designation fixed in another possible world. We do not have to suppose that the language is varying from world to world. It is the same language we use here, for, of course, we use that very same name to refer to that object: one that does not exist here. The name is rigid, i.e., it designates the same thing in every possible world; it is just that the designation has been fixed in another world. And this is not to change the language in any fundamental way. We are supposing domains vary, and objects exist in other worlds that do not exist in ours. We don't change the language (in this world) when we baptize a new baby; *a fortiori* the language is not changed when a new baby is baptized in another world. It might very well be that in another possible world, someone other than Nixon is baptized "Nixon." Which one are we referring to when we use the name "Nixon?" But, in this very world, more than one person has been baptized "Aristotle." Nonetheless, when I speak of Book Γ of Aristotle's *Metaphysics*, it is the Greek philosopher I speak of, not the Greek shipping magnate. However the details of the ambiguity of "Aristotle" are to be worked out, they will serve also for the case of "Nixon."

It might be objected that Kripke has a specific view of how possible worlds are to be understood. They are to be understood counterfactually. We don't look at another possible world through our Jules-Verne-O-Scope and see who it is that is the closest counterpart to our own Nixon. We wonder what the world would be like if *this* man, Nixon, were such-and-such. True, but this has very little impact on the matter of rigid designation. After all, just as we can suppose there are worlds in which Nixon doesn't exist, we can suppose there are worlds containing things that do not exist in this world. This is fundamental to our understanding of the varying domain model. If this way of looking at possible worlds is workable, then we should be able to make sense of the counterfactual story in which things other than those in this world exist. And if Kripke's counterfactual story is to remain intact, then our speaking of a counterfactual world is to be understood in terms other than the Jules-Verne-O-Scope model. This means that the idea of rigidly designating a nonexistent cannot be ruled out *a priori*.

It might also be objected that from the temporal world perspective, there is a causal/historical story to be told about the name "George Washington." None, however, appears to be available for "Pegasus." Again, this is not quite right. The name "Pegasus" is in our language. We use it with the intention of referring to whatever or whomever was originally so baptized. But nothing was so baptized; there is no such thing as a flying horse. Nothing, that is, that exists in this world. But in another possible world (one of our imagination), there is such a thing; in that world, that thing was named "Pegasus," and we use "Pegasus" to refer to that thing. Pegasus does not exist; but that does not mean that "Pegasus" does not designate. Pegasus is a pretend object; but that does not mean that "Pegasus" is a pretend name. On the contrary, "Pegasus" is a proper name, serves that way in the language, and should play the same role as other proper names. Just as "Nixon" is not short for a description, so "Pegasus" isn't. The argument originally made for names like "Richard Nixon" and "Ben Franklin" are easily made for "Pegasus." Ben Franklin is the inventor of bifocals, but it is not essential to him that he invented bifocals; in another possible world, someone other than Ben Franklin invented bifocals. "Pegasus" is a rigid designator: it designates the same thing in every possible world. Consider, however, the description "the winged horse captured by Bellerophon." Is this property, viz., *having been captured by Bellerophon*, an essential property of Pegasus? Surely not. This is only a contingent property. In some possible world, he is never captured at all. That is, it seems as much a mistake to suppose that the story told about Pegasus is essential to him as it is, now shifting to another famous example Kripke discusses, to suppose that the story told about Moses is essential to him.

So, we can regard a myth or a story as describing a collection of possible worlds. And if names are introduced for characters in the myth or the story, these are names in the language in which the story is framed, and their role is the very same in the story as is the role of names in nonfiction.

Now, it is rather clear that Pegasus and Holmes fail to exist; indeed, if anything actual were to turn up having exactly the properties of Pegasus (or of Holmes), we would not suppose that it is Pegasus (or Holmes). Let *h* abbreviate "Holmes." The following is thus true:

(11.23) $\neg \mathbf{E}(h)$.

For the following argument, assume *h* designates. We leave it to you to check the validity, in varying domain models, of the following open formula.

(11.24) $[\langle \lambda z.x = z \rangle(h) \wedge \mathsf{E}(x)] \supset \mathbf{E}(h)$

From this and (11.23) we have, by elementary logic,

(11.25) $\mathsf{E}(x) \supset \neg \langle \lambda z.x = z \rangle(h)$.

But since *h* designates, we also have the following.

(11.26) $\neg \langle \lambda z.x = z \rangle(h) \supset \langle \lambda z.\neg(x = z) \rangle(h)$
$$\supset \langle \lambda z.\Box \neg(x = z) \rangle(h)$$

Then by transitivity,

(11.27) $\mathsf{E}(x) \supset \langle \lambda z.\Box \neg(x = z) \rangle(h)$.

That is, each thing that exists is necessarily distinct from Holmes. This seems to capture Kripke's intuition about the fictional status of Holmes. There is no need for anything stronger, and in particular, for anything like

(11.28) $\Box \neg (\exists x)(x = h)$

to which Kripke appears to be committing himself when he says that "granted that there *is* no Sherlock Holmes, one cannot say of any possible person that he *would have been* Sherlock Holmes, if he existed." If he means to deny that Sherlock Holmes does not exist in some possible world, then Kripke's view is somewhat different from the one being put forward here.

EXERCISES

EXERCISE 11.4.1.   Kripke (1979a) speaks of a young Frenchman, Pierre, who is unaware that the city he knows as *Londres* is the very same city as the one he knows as *London*. The example is supposed to expose difficulties about substituting the names "London" and "Londres" one for the other. However, our discussion focuses on a different aspect of the example. The exercise is to provide a plausible answer (story) for the following question: "Is 'Pierre' a rigid designator?"

EXERCISE 11.4.2.   We claim, in the text, that Pegasus is a pretend object, but this does not mean that "Pegasus" is a pretend name. Write an essay either defending or criticizing this claim.

EXERCISE 11.4.3.   In the text, we claim about proper names that "their role is the very same in the story as is the role of names in non-fiction." This is controversial. Write an essay either defending or criticizing this claim.

11.5. Tableau Rules

Almost all of the tableau rules presented up to this point remain as they have been; a few are modified. But the modifications are quite straightforward. The idea is to arrange things so that a grounded term can be used on a branch only if we are sure it designates. To make it easy to state the modifications, a little terminology is introduced.

DEFINITION 11.5.1.   [Designates on a Branch] We say a grounded term *designates on a branch* if it is a parameter, or else it if already occurs somewhere on the branch.

Informally parameters, being free variables, always denote. For grounded terms involving constant or function symbols, we will arrange things so that if they are introduced onto a branch, we have sufficient guarantee that they have designations.

In Section 10.6 we gave two Abstraction Rules. One of these remains unchanged, the other undergoes a modification. We repeat the first for convenience, then give the revised version of the second.

DEFINITION 11.5.2.   [Positive Abstraction Rule] With no special assumptions concerning $t@\sigma$,

$$\frac{\sigma \ \langle \lambda x.\Phi(x)\rangle(t)}{\sigma \ \Phi(t@\sigma)}$$

DEFINITION 11.5.3.   [Negative Abstraction Rule] If $t@\sigma$ designates on the tableau branch, then

$$\frac{\sigma \ \neg\langle \lambda x.\Phi(x)\rangle(t)}{\sigma \ \neg\Phi(t@\sigma)}$$

Here are the informal ideas behind these rules. To apply the first rule we must have $\sigma \langle \lambda x.\Phi(x) \rangle (t)$ on a tableau branch. Intuitively, this asserts that a certain abstract, $\langle \lambda x.\Phi(x) \rangle$, correctly applies to t at world σ so t must, in fact, designate at σ. Consequently we are permitted to introduce $t@\sigma$ into the tableau, which we can think of as the object designated by t. With the second rule, on the other hand, we have $\sigma \neg \langle \lambda x.\Phi(x) \rangle (t)$ on a tableau branch and, intuitively, this asserts that property $\langle \lambda x.\Phi(x) \rangle$ does not apply to t. From this we can conclude nothing about whether or not t designates at σ, and so we can use $t@\sigma$ only if we have information from elsewhere on the tableau branch that t designates.

There were two tableau rules for equality in Chapter 7. Of these, the Substitutivity Rule rule remains unchanged. The other rule undergoes a modification exactly like that above.

DEFINITION 11.5.4. [Reflexivity Rule] If t is a grounded term that designates on a tableau branch and σ is a prefix that occurs on the branch, then $\sigma (t = t)$ can be added to the end of the branch.

This completes the tableau rule modifications and the intent should be clear. We conclude with an example of their use.

EXAMPLE 11.5.5. Although $\neg \langle \lambda x.\varphi(x) \rangle (t) \supset \langle \lambda x.\neg\varphi(x) \rangle (t)$ is not valid if t does not designate, we do have the validity of the following.

$$\mathbf{D}(t) \supset [\neg \langle \lambda x.\varphi(x) \rangle (t) \supset \langle \lambda x.\neg\varphi(x) \rangle (t)]$$

Here is a tableau proof using varying domain **K** rules.

$$
\begin{aligned}
&1 \; \neg\{\mathbf{D}(t) \supset [\neg \langle \lambda x.\varphi(x) \rangle (t) \supset \langle \lambda x.\neg\varphi(x) \rangle (t)]\} \quad 1. \\
&1 \quad \mathbf{D}(t) \quad 2. \\
&1 \; \neg [\neg \langle \lambda x.\varphi(x) \rangle (t) \supset \langle \lambda x.\neg\varphi(x) \rangle (t)] \quad 3. \\
&1 \; \neg \langle \lambda x.\varphi(x) \rangle (t) \quad 4. \\
&1 \; \neg \langle \lambda x.\neg\varphi(x) \rangle (t) \quad 5. \\
&1 \quad \langle \lambda x.x = x \rangle (t) \quad 6. \\
&1 \quad t_1 = t_1 \quad 7. \\
&1 \; \neg\varphi(t_1) \quad 8. \\
&1 \; \neg\neg\varphi(t_1) \quad 9.
\end{aligned}
$$

Items 2 and 3 are from 1 by a Conjunctive Rule, as are 4 and 5 from 3; 6 is simply 2 rewritten, from which 7 follows by the Positive Abstraction Rule; now 8 is from 4 and 9 is from 5 by the Negative Abstraction Rule, and we have closure at this point.

A look at items 2, 6 and 7 in the tableau of the example above shows we have the following derived rule concerning designation.

DEFINITION 11.5.6. [Derived Designation Rule] If σ $\mathbf{D}(t)$ occurs on a tableau branch, then $t@\sigma$ designates on that branch.

EXERCISES

EXERCISE 11.5.1.   Give a tableau proof of the following (which is the converse of the Example above):

$$[\neg\langle\lambda x.\varphi(x)\rangle(t) \supset \langle\lambda x.\neg\varphi(x)\rangle(t)] \supset \mathbf{D}(t)$$

EXERCISE 11.5.2.   Give a tableau proof of $\mathbf{D}(t) \equiv [\mathbf{E}(t) \vee \overline{\mathbf{E}}(t)]$.

CHAPTER TWELVE

DEFINITE DESCRIPTIONS

We have used phrases like "the King of France" several times, though we always treated them like non-rigid constant symbols. But such phrases have more structure than constant symbols—they do not arbitrarily designate. The King of France, for instance, has the property of being King of France, provided there is one, and the phrase "the King of France" designates him because he alone has that property. Phrases of the form "the so-and-so" are called *definite descriptions*. In this chapter we examine the behavior of definite descriptions in modal contexts.

12.1. NOTATION

Let Φ be a formula. The notation $\imath x.\Phi$ is read "the x such that Φ," and is called a *definite description*. For example, if Φ is a full description of what it means to be the King of France, then $\imath x.\Phi$ is read as "the x such that x is the King of France," or more briefly just "the King of France." The definite description $\imath x.\Phi$ is used syntactically just as if it were a term in the sense of earlier chapters.

However, the syntactical situation is more complex than it appears at first glance. Definite descriptions contain formulas. But definite descriptions are treated like terms, so they can occur in formulas. Consequently it is no longer possible to define terms first, and then formulas, as we did in earlier chapters. The two must be defined together. Still, adjusting for this new complexity actually involves only making a few changes to earlier definitions.

First of all, *atomic formulas* are defined as before: an atomic formula is an expression of the form $R(x_1, x_2, \ldots, x_n)$, where R is an n-place relation symbol and $x_1, x_2, \ldots, x_n$ are variables. All of $x_1, x_2, \ldots, x_n$ occur free in $R(x_1, x_2, \ldots, x_n)$. Now for the build-up part.

DEFINITION 12.1.1. [Formulas and Terms Together] The notions of formula, term, and free variable occurrence are defined as follows.

1. A variable x is a term in which x occurs free.
2. A constant symbol is a term with no free variable occurrences.
3. If f is an n-place function symbol, and $t_1, \ldots, t_n$ are terms, $f(t_1, \ldots, t_n)$ is a term; the free variable occurrences of $f(t_1, \ldots, t_n)$ are those of t_1 together with those of $t_2, \ldots$, together with those of t_n.

4. If x is a variable and Φ is a formula then $\imath x.\Phi$ is a term; its free variable occurrences are those of Φ except for occurrences of x.

5. An atomic formula is a formula; every variable occurrence in an atomic formula is a free variable occurrence.

6. If X is a formula, so is $\neg X$; the free variable occurrences of $\neg X$ are those of X.

7. If X and Y are formulas, and $\circ$ is a binary connective, $(X \circ Y)$ is a formula; the free variable occurrences of $(X \circ Y)$ are those of X together with those of Y.

8. If X is a formula, so are $\Box X$ and $\Diamond X$; the free variable occurrences of $\Box X$ and of $\Diamond X$ are those of X.

9. If X is a formula and v is a variable, both $(\forall v)X$ and $(\exists v)X$ are formulas; the free variable occurrences of $(\forall v)X$ and of $(\exists v)X$ are those of X, except for occurrences of v.

10. If Φ is a formula, v is a variable, and t is a term, $\langle \lambda v.\Phi \rangle (t)$ is a formula; the free variable occurrences of $\langle \lambda v.\Phi \rangle (t)$ are those of Φ except for occurrences of v, together with all free variable occurrences in t.

The key thing to notice is that part of the definition of term—item 4—involves the notion of formula, while part of the definition of formula—item 10—involves the notion of term. Don't be put off by all this. In what follows, an intuitive feel for what is a formula and what is a term should be sufficient.

EXERCISES

EXERCISE 12.1.1.   Show, using Definition 12.1.1, that the following is a term with no free variable occurrences (P is a one-place relation symbol):

$$\imath x.\langle \lambda y.x = y \rangle (\imath z.P(z)).$$

This term can be read: "the x that is equal to the z such that P."

12.2. TWO THEORIES OF DESCRIPTIONS

Two types of theories of descriptions are well-known in the literature: (a) those based on Frege's treatment; (b) those based on Russell's treatment. The account of descriptions presented in this text falls into neither of these two categories neatly, as we shall see in the next section.

Frege (1893) considered a definite description to be a genuine singular term (as do we), so that a sentence like

(12.1)   The present King of France is bald

would have the same logical form as "Harry Truman is bald." What happens if there is no present King of France? To avoid truth value gaps (or the introduction of a third truth value), Frege arbitrarily assigned a default designatum to the description, namely, the expression itself, "the present King of France." Carnap, working within Frege's paradigm, took a somewhat narrower approach, having a single default designatum rather than a family of them. On his approach, the present King of France turns out to be identical with the present Queen of France, because France has neither a King nor Queen, and so both terms designate the same default object. Apart from the arbitrariness of the default designatum, there remains something unsatisfying about the idea that the phrase "the King of France" designates, but not a King of France. One also wonders how denials of existence are to be treated.

Russell (1905), on the other hand, denied that definite descriptions were genuine singular terms. Descriptions were, on his view, "incomplete expressions," and he provided a *contextual* account of descriptions. Sentence (12.1) is rephrased so that the apparent singular term is eliminated:

>   At least one thing presently kings France, and
>
>   At most one thing presently kings France, and
>
>   Whatever presently kings France is bald.

The first two conjuncts express existence and uniqueness conditions, respectively; the third conjunct expresses that the uniquely existent individual has a certain property. Let $K(x)$ abbreviate "x presently kings France" and $B(x)$ abbreviate "x is bald." Then (12.1) can be symbolized as

(12.2) $(\exists x)((K(x) \wedge (\forall y)(K(y) \supset y = x)) \wedge B(x))$

But this paraphrase of sentences involving descriptions is only one part of Russell's celebrated theory. We must not overlook the scope distinction, only dimly understood in Russell (1905), and presented there hurriedly as the much weaker distinction between a *primary* and *secondary* occurrence of a description. (A full treatment is presented in (Whitehead and Russell, 1925).) It turns out that it is this scope distinction that is doing most of the work for Russell. And we can see this in his solutions to the three puzzles he sets forward to test theories of descriptions.

EXAMPLE 12.2.1. [The Law of Excluded Middle] Here is Russell's account:

By the law of excluded middle, either 'A is B' or 'A is not B' must be true. Hence either 'the present King of France is bald' or 'the present King of France is not bald' must be true. Yet if we enumerated the things that are bald, and then the things that are not bald, we should not find the present King of France in either list. Hegelians, who love a synthesis, will probably conclude that he wears a wig. (Russell, 1905, p. 48)

Russell's solution makes explicit appeal to the scope distinction. Using our predicate abstraction notation rather than his notational devices, (12.1) is symbolized as

(12.3) $\langle \lambda x.B(x) \rangle (\imath x.K(x))$

The sentence

(12.4) The present King of France is *not* bald,

can be symbolized in two distinct ways:

(12.5) $\langle \lambda x.\neg B(x) \rangle (\imath x.K(x))$
(12.6) $\neg \langle \lambda x.B(x) \rangle (\imath x.K(x))$

(12.3) says *of* $\imath x.K(x)$ that he is bald; (12.5) says *of* $\imath x.K(x)$ that he is non-bald; (12.6) says *that* $\imath x.K(x)$ is not bald. (12.3) and (12.5) are both false because the term is non-designating; but (12.6) is true. Russell preserves the Law of Excluded Middle by distinguishing (12.5) from (12.6). (12.5) is not the contradictory of (12.3), but (12.6) is; and (12.3) and (12.6) in fact differ in truth value.

EXAMPLE 12.2.2. [The Paradox of NonBeing] Russell's version is a bit baroque:

Consider the proposition 'A differs from B'. If this is true, there is a difference between A and B, which fact may be expressed in the form 'the difference between A and B subsists'. But if it is false that A differs from B, then there is no difference between A and B, which fact may be expressed in the form 'the difference between A and B does not subsist'. But how can a non-entity be the subject of a proposition? 'I think, therefore I am' is no more evident than 'I am the subject of a proposition, therefore I am', provided 'I am' is taken to assert subsistence or being, not existence. Hence, it would appear, it must always be self-contradictory to deny the being of anything; but we have seen, in connexion with Meinong, that to admit being also sometimes leads to contradictions. Thus if A and B do not differ, to suppose either that there is, or that there is not, such an object as 'the difference between A and B' seems equally impossible. (Russell, 1905, p. 48)

Russell's is the classic Deflationist answer to the Paradox. But his solution is misleading, for he assumes in his discussion that "x exists" is not a predicate. On Russell's view,

(12.7) The present King of France exists

is a conjunction of the two claims:

At least one thing presently kings France, and

At most one thing presently kings France.

To deny (12.7) is to deny the conjunction, i.e., to deny that anything uniquely kings France at present:

(12.8) $\neg(\exists x)(K(x) \wedge (\forall y)(K(y) \supset x = y))$

Contrast (12.7) with (12.1). In the latter case we have three conjuncts; in the former case we have only two. There are two ways of negating (12.1): on the *narrow scope* reading of the definite description, one negates the conjunction of the three clauses; on the *wide scope* reading, one negates only the last of the three clauses. Since, in the case of (12.7), there are only two clauses, the negation of (12.7) engenders no scope ambiguity.

There is no third conjunct in the expansion of (12.7) because Russell simply assumes "x exists" is not a predicate. This, as we have argued, is prejudicial. If we had a predicate "x exists," then (12.7) would be understood to be expanded to three conjuncts

At least one thing presently kings France, and

At most one thing presently kings France, and

Whatever presently kings France exists.

And, in this case, there would be two ways of denying (12.7): either deny the conjunction of all three conjuncts, or simply deny the last conjunct. If we simply denied the last, we should be denying *of* the present King of France *that* he exists, and it is precisely this reading Russell sought to rule out.

EXAMPLE 12.2.3. [The Puzzle About Identity] Russell, once more:

If a is identical with b, whatever is true of the one is true of the other, and either may be substituted for the other in any proposition without altering the truth or falsehood of that proposition. Now George IV wished to know whether Scott was the author of *Waverley*; and in fact Scott *was* the author of *Waverley*. Hence we may substitute *Scott* for *the author of 'Waverley'*, and thereby prove that George IV wished to know whether Scott was Scott. Yet an interest in the law of identity can hardly be attributed to the first gentleman of Europe. (Russell, 1905, pp. 47-8)

Despite the truth of the identity,

   (12.9)   Scott was the author of Waverley,

substitution of one expression for the other in

   (12.10) George IV wished to know whether Scott was the author of Waverley,

turns the true sentence into a false one:

   (12.11) George IV wished to know whether Scott was Scott.

But, substituting coextensional terms in non-truth-functional contexts is not truth preserving. With a small scope reading, we cannot substitute for the description and preserve truth value: (12.11) does not follow from (12.9) and (12.10).

   This last example shows very clearly that Russell had recognized as early as (Russell, 1905) that descriptions fail to act as proper names not just when they fail to designate, but when they occur inside non-truth-functional contexts (like *A believes that*). Indeed, it is just this character of Russell's treatment of descriptions that enabled Smullyan to seize upon the scope distinction as a structural way of marking the *de re*/*de dicto* distinction.[23]

12.3. THE SEMANTICS OF DEFINITE DESCRIPTIONS

How should definite descriptions behave semantically? The choice of possibilist or actualist quantification is not critical—either will serve for this

---

[23] Smullyan (1948) calls our attention to the fact that theorem 14.18 of Whitehead and Russell (1925),

$$E!\imath x.\Phi x \supset [(\forall x)\Psi x \supset \Psi\imath x.\Phi x]$$

which says that if the Φ exists, then we can substitute "the Φ" for a variable, is further constrained by theorem 14.3,

$$((\forall p)(\forall q)(p \equiv q \supset f(p) \equiv f(q)) \land E!\imath x.\phi x) \supset$$
$$f\{[\imath x.\phi x].\xi\imath x.\phi x\} \equiv [\imath x.\phi x].f\{\xi\imath x.\phi x\}$$

which says says that scope does not matter if the Φ exists *and* f is truth-functional. Quine was apparently unaware of this constraint on substitution, for as late as (Quine, 1961b, p. 154), he accuses Smullyan of "altering" Russell's theory with this additional constraint, a charge excised in later editions.

section. The central issue is: what do we do when there is nothing for a definite description to describe?

One approach that is attractive at first glance is to say, "The King of France is bald" is neither true nor false since it refers to something that does not exist, namely the King of France. This leads to a partial, or three-valued logic, which is something we wish to avoid here—see (Blamey, 1986) for a presentation of such an approach. It is not so simple to say that we will allow some formulas to lack a truth value; the question is "which ones?" and this is not always obvious. For instance, consider "Either the King of France is bald, or it is not the case that the King of France is bald." We might want to take this as lacking a truth value because it refers to something that does not exist, or we might want to take this as true, since it has the form $X \vee \neg X$. Even more strongly, what about, "The King of France does not exist;" if it is not to be taken as true, how would we tell a Frenchman he is kingless? Fortunately it is not necessary to move to a partial logic in order to deal with the problem satisfactorily.

We can, in fact, avoid the issue. Unlike Frege and Russell, we have all along allowed terms to designate nonexistent objects, and we have provided evaluation rules to assign truth values to sentences containing such terms. And, starting with Chapter 11 we have provided evaluation rules to assign truth values to sentences containing terms that lack designation altogether. We simply extend these rules to definite descriptions. Here are the details.

In Definition 11.1.3 of Chapter 11 we defined the notion of term evaluation, allowing the possibility of non-designation for a term. We now add one more case to this definition.

DEFINITION 12.3.1. [Extending Definition 11.1.3] Let $\mathcal{M} = \langle \mathcal{G}, \mathcal{R}, \mathcal{D}, \mathcal{I} \rangle$ be a general non-rigid model, let $\Gamma \in \mathcal{G}$, and let v be a valuation.

4. If $\mathcal{M}, \Gamma \Vdash_{v'} \Phi$ for exactly one x-variant v' of v, then $\imath x.\Phi$ designates at Γ, with respect to v, and $(v \star \mathcal{I})(\imath x.\Phi, \Gamma) = v'(x)$.

We will soon see that this does the job nicely. But first, honor requires us to note that by adding this new part to Definition 11.1.3 we have complicated the technical details considerably. In order to determine whether $\imath x.\Phi$ designates at a possible world Γ, and to determine its denotation if it does, we need to check whether Φ is true at Γ under various valuations. But to check the truth of a formula at a possible world we need to know about the denotations of terms that appear in the formula, and Φ could contain additional definite descriptions. The situation is circular in the sense that we can no longer define designations for terms first, and then truth at worlds afterwards. Both must be defined together.

From now on, Definition 11.1.4, of truth in a general model, and Definition 11.1.3, of term evaluation (with the extra item above), should be thought of as a single definition. Of course, what makes all this work is that each step of term evaluation or of truth-value determination involves a reduction in syntactic complexity, and so the circular references "bottom out" eventually. We skip the formal details and rely on your intuition that this is so.

We have sidestepped entirely the existence issue, which historically has been pretty much at the forefront of the discussion of definite descriptions. In doing so, we have adopted Frege's semantic idea that definite descriptions are singular terms, and combined it with Russell's scope distinction which enables us to capture the variability of designation—and even the effect of their nondesignation. Frege's intuitions are central in our model theoretic account above of the behavior of descriptions, but Russell's ideas will resurface. In Section 12.7, we will show how Russell's account can be embedded in a Fregean account.

12.4. Some Examples

Is the present King of France King of France? Is the flying horse of Greek mythology a horse? And is the round square a square? We find ourselves in agreement with Whitehead and Russell (1925), who observed:

> ... such a proposition as "the man who wrote Waverley wrote Waverley" does not embody a logically necessary truth, since it would be false if Waverley had not been written, or had been written by two men in collaboration. For example, "the man who squared the circle squared the circle" is a false proposition.

The sentence "The present King of France is king of France" is not true in this world, because there is no French king now. The round square case is different. The defining property of round squares is contradictory, and so round squares can exist in no possible world of any model. Consequently it is false at every world that the round square is square, since the definite description can never designate.

For this section, assume semantics is varying domain, because we wish to consider the roles of existence and nonexistence. Our distinction between designation and existence is of great importance in understanding the semantics of definite descriptions. The description "the first President of the United States" designates George Washington. The definite description designates at the present time, but it does not designate something that exists at the present time. By contrast, the definite description, "the present King of France," does not even designate now. It is easy to overlook this distinction because existence assumptions, as opposed to designation assumptions, are frequently tacitly built into a definite description. For example, "the tallest person in the world," takes for granted that we mean to designate an existent person. By way of contrast, "the tallest person in the world in 1995" presumably also designates, but it is contingent whether the person designated currently exists or not—it is not an assumed condition.

In the present section we will explain how definite descriptions operate by looking at examples, beginning more formally, and ending up, hopefully, in accordance with colloquial usage.

EXAMPLE 12.4.1. Suppose we have a varying domain model $\mathcal{M} = \langle \mathcal{G}, \mathcal{R}, \mathcal{D}, \mathcal{I} \rangle$ with $\mathcal{G}$ containing exactly four possible worlds, with all worlds accessible from each (thus it is an **S5** model, though this plays no special role). Suppose further that the domains associated with the four possible worlds, and the truth-value at worlds for atomic formulas involving the one-place relation symbol P are given by the following schematic diagram.

$$\Gamma_1 \quad \boxed{a, b} \quad \Vdash P(c)$$

$$\Gamma_2 \quad \boxed{b, c} \qquad\qquad \Gamma_3 \quad \boxed{d, e} \quad \begin{array}{l} \Vdash P(c) \\ \Vdash P(e) \end{array}$$

$$\Gamma_4 \quad \boxed{f, g} \quad \Vdash P(f)$$

In this model, $\imath x.P(x)$ designates at Γ_4, and in fact it designates f there. We check this claim in detail, and skip the formalities for later assertions.

Let v be any valuation, and consider $(v * \mathcal{I})(\imath x.P(x), \Gamma_4)$. Obviously if v' is the x-variant of v such that $v'(x) = f$, we have that $\mathcal{M}, \Gamma_4 \Vdash_{v'} P(x)$. But also, if v'' is any x-variant of v besides v', so that $v''(x) \neq f$, then $\mathcal{M}, \Gamma_4 \nVdash_{v''} P(x)$. Then according to Definition 11.1.3, $\imath x.P(x)$ designates at Γ_4, and $(v * \mathcal{I})(\imath x.P(x), \Gamma_4) = v'(x) = f$.

Since $(v * \mathcal{I})(\imath x.P(x), \Gamma_4)$ exists, we have that $\mathbf{D}(\imath x.P(x))$ is true at Γ_4, where we have used the designation predicate abstract of Chapter 11. And further, since f is in the domain of Γ_4 we also have that $\mathbf{E}(\imath x.P(x))$ is true at Γ_4.

At Γ_1, $\imath x.P(x)$ also designates—here it designates c. Thus $\mathbf{D}(\imath x.P(x))$ is true at Γ_1. But since c is not in the domain of Γ_1 we have $\overline{\mathbf{E}}(\imath x.P(x))$ at Γ_1.

At Γ_2, $\imath x.P(x)$ does not designate, because P is true of nothing at Γ_2. Likewise at Γ_3, $\imath x.P(x)$ does not designate, but now because there is more than one thing that P is true of at Γ_3.

EXAMPLE 12.4.2. The formula $\langle \lambda x.\Phi \rangle (\imath x.\Phi)$ is not valid, because it is false if $\imath x.\Phi$ does not designate, though things change if designation is ensured. The following *is* valid.

$$\mathbf{D}(\imath x.\Phi) \supset \langle \lambda x.\Phi \rangle (\imath x.\Phi)$$

It is not difficult to check the validity of this. Suppose we have a model $\mathcal{M} = \langle \mathcal{G}, \mathcal{R}, \mathcal{D}, \mathcal{I} \rangle$ and, for $\Gamma \in \mathcal{G}$ and valuation v, we have

$$\mathcal{M}, \Gamma \Vdash_v \mathbf{D}(\imath x.\Phi)$$

or equivalently,

$$\mathcal{M}, \Gamma \Vdash_v \langle \lambda x.x = x \rangle (\imath x.\Phi).$$

Then $\imath x.\Phi$ must designate at Γ with respect to v, so according to Definition 11.1.3 there is exactly one x-variant, v' of v such that $\mathcal{M}, \Gamma \Vdash_{v'} \Phi$, and $v'(x) = (v * \mathcal{I})(\imath x.\Phi)$. But then by definition of satisfiability for predicate abstracts,

$$\mathcal{M}, \Gamma \Vdash_v \langle \lambda x.\Phi \rangle (\imath x.\Phi).$$

As a matter of fact, the converse implication is also valid. We thus have a central fact—the following is valid:

$$\mathbf{D}(\imath x.\varphi(x)) \equiv \langle \lambda x.\varphi(x) \rangle (\imath x.\varphi(x)) \qquad (12.12)$$

EXAMPLE 12.4.3. Think of the real world and its conceivable alternatives, including the world of Greek mythology, as constituting a modal model. Take the model as varying domain—quantifiers are actualist. By the equivalence (12.12) above, the flying horse of Greek mythology is a horse provided the term designates. In fact, Pegasus does not exist, but in the world of Greek mythology it does. Thus, if $P(x)$ is a one-place relation symbol intended to characterize the flying horse the Greeks spoke of, the following is true at the real world of our informal model:

$$\overline{\mathbf{E}}(\imath x.P(x)) \wedge \Diamond \mathbf{E}(\imath x.P(x)) \wedge \mathbf{D}(\imath x.P(x))$$

One might be uneasy with this modeling of our intuition. For if Pegasus is a horse one might feel obligated to include treatment of Pegasus in the study of biology. An alternative modeling takes the description to be:

the unique x such that

it is said that (x flies and x is a horse)

The description designates, for the x such that it is said to fly and be a horse *is* said to fly and be a horse. But we don't have to worry about the biology of such a creature.

EXAMPLE 12.4.4. This example involves modal notions. Consider the sentence, "Someday, somebody could be taller than the tallest person in the world." Suppose we formulate this in a modal logic in which possible worlds are intended to be temporal states of the world, reading $\Box X$ as asserting that X is and always will be the case. Take the model to be varying domain. Let $T(x)$ be a one-place relation symbol that characterizes the tallest person in the world—the intention is that $T(x)$ should be true of an object precisely when that object is a person, that person currently exists, and all other currently existing people are shorter. Now, here is a formalization of "Someday, somebody could be taller than the tallest person in the world."

$$\langle \lambda x.\Diamond \neg \langle \lambda y.x = y\rangle(\imath z.T(z))\rangle(\imath z.T(z)) \tag{12.13}$$

In order to read this properly, first note that $\langle \lambda y.x = y\rangle$ is the "same as x" predicate, so the subformula $\neg \langle \lambda y.x = y\rangle(\imath z.T(z))$ essentially says the tallest person in the world is not the same as x. Combining this with the rest of the formula, then, we have: "It is true of the tallest person in the world (now) that, someday, the tallest person in the world will not be that person." Sentence (12.13) is satisfiable, but we can do better than that. Let us make some reasonable assumptions about our notions. First, existence is an inherent assumption about "the tallest person in the world." Let us make this explicit with $\mathbf{E}(\imath x.T(x))$. Second, let us postulate that any given person might not be the tallest in the world someday, whether or not they are now: $(\forall x)\Diamond \neg T(x)$. From these formula (12.13) follows. That is, the following is valid in the simplest modal logic we have considered, varying domain **K**:

$$[\mathbf{E}(\imath x.T(x)) \wedge (\forall x)\Diamond \neg T(x)] \supset$$
$$\langle \lambda x.\Diamond \neg \langle \lambda y.x = y\rangle(\imath z.T(z))\rangle(\imath z.T(z)) \tag{12.14}$$

We postpone showing the validity of this until tableau machinery becomes available.

There is a little more of interest that can be extracted from this example. Sentence (12.13) is trivially equivalent to

$$\langle \lambda x.\neg\Box \langle \lambda y.x = y\rangle(\imath z.T(z))\rangle(\imath z.T(z)).$$

We leave it to you to check that, even for terms t that can fail to designate, $\langle \lambda x.\neg\Phi\rangle(t) \supset \neg\langle \lambda x.\Phi\rangle(t)$ is valid. So a consequence of (12.13) is

$$\neg\langle \lambda x.\Box \langle \lambda y.x = y\rangle(\imath z.T(z))\rangle(\imath z.T(z)).$$

Now, Proposition 10.2.5 (extended from constant symbols to definite descriptions) says this is the negation of the assertion that $\imath z.T(z)$ is rigid. (Actually, Proposition 10.2.5 was proved under the assumption that terms always designated, but designation is a consequence of the existence assumption we made above about $\imath z.T(z)$.) Non-rigidity should be an "obvious" consequence, since if $\imath z.T(z)$ could not vary its designation, the tallest person in the world could never change.

EXAMPLE 12.4.5.   Consider the description "the present King of France." There is no present King of France (in this world) and so, if the term designated, it would not designate an existing object. But it does not even designate, because the predicate "x presently kings France" is true uniquely for no value of the variable. The present King of France fails to exist; and the term "the present King of France" fails to designate. Of course, there are other worlds in which it does designate, and indeed, is an object existing in that world. We might put the point this way:

$$\langle \lambda x.\Diamond Ex\rangle(\imath x.K(x))$$

is false; but

$$\Diamond\langle \lambda x.Ex\rangle(\imath x.K(x))$$

is true.

EXAMPLE 12.4.6.
    Here are a few more Russell-like examples, in addition to "George IV didn't know that Scott was the author of Waverley."

    Let $W(z)$ be a formula asserting that z is the author of Waverley. Consider the sentence: "George IV didn't know the author of Waverley was the author of Waverley." More precisely worded, it is, "George IV didn't know, of the author of Waverley, that he was the author of Waverley." Suppose we read $\Box$ epistemically as "George IV knows that . . . ." Then this formalizes directly as the following.

$$\langle \lambda x.\neg\Box\langle\lambda y.x = y\rangle(\imath z.W(z))\rangle(\imath z.W(z))$$

In addition, we could also represent it as the following less complex formula.

$$\langle \lambda x.\neg\Box W(x)\rangle(\imath z.W(z))$$

We give as exercises below to show that these are satisfiable.

Next sentence: "George IV knew there was an author of Waverley, but he didn't know who it was." This formalizes as follows.

$$\Box \mathbf{E}(\imath z.W(z)) \wedge (\forall w)\neg\Box\langle\lambda y.w = y\rangle(\imath z.W(z))$$

We also give the satisfiability of this as an exercise.

Finally, to continue with the previous sentence, the following is not an adequate formalization of "George IV knew there was an author of Waverley, but he didn't know who it was."

$$\Box \mathbf{E}(\imath z.W(z)) \wedge (\forall w)\langle\lambda y.\neg\Box w = y\rangle(\imath z.W(z))$$

This is *not* satisfiable—in fact, its negation is valid. We give this as an exercise once tableau rules have been introduced.

EXERCISES

EXERCISE 12.4.1.   Let P be a one-place relation symbol. Give an explicit model showing that $\langle\lambda x.P(x)\rangle(\imath x.P(x))$ is not valid.

EXERCISE 12.4.2.   Show the validity of the following:

$$\langle\lambda x.\psi(x)\rangle(\imath x.\varphi(x)) \supset \mathbf{D}(\imath x.\varphi(x)).$$

EXERCISE 12.4.3.   Show the satisfiability of the following sentences, all taken from Example 12.4.6.
1. $\langle\lambda x.\neg\Box\langle\lambda y.x = y\rangle(\imath z.W(z))\rangle(\imath z.W(z))$
2. $\langle\lambda x.\neg\Box W(x)\rangle(\imath z.W(z))$
3. $\Box\mathbf{E}(\imath z.W(z)) \wedge (\forall z)\neg\Box\langle\lambda y.z = y\rangle(\imath z.W(z))$

EXERCISE 12.4.4.   Just as we defined existence and nonexistence properties, we can define possible-existence and necessary-existence properties.

$$\mathbf{E}_\Box(t) = \langle\lambda x.\Box\mathbf{E}(x)\rangle(t) = \langle\lambda x.\Box(\exists y)(y = x)\rangle(t)$$
$$\mathbf{E}_\Diamond(t) = \langle\lambda x.\Diamond\mathbf{E}(x)\rangle(t) = \langle\lambda x.\Diamond(\exists y)(y = x)\rangle(t)$$

Use the model of Example 12.4.1 and determine at which worlds of it the sentence $\mathbf{E}_\Diamond(\imath x.P(x))$ is true; and similarly for $\Diamond\mathbf{E}(\imath x.P(x))$ is true. Do the same with $\mathbf{E}_\Box(\imath x.P(x))$ and $\Box\mathbf{E}(\imath x.P(x))$.

EXERCISE 12.4.5. Using the notation of Exercise 12.4.4, determine the relationship between $\mathbf{E}_\Diamond(t)$ and $\Diamond\mathbf{E}(t)$, where t is a definite description. That is, does the first always imply the second, or not, and similarly the other way around. Also do the same for the pair $\mathbf{E}_\Box(t)$ and $\Box\mathbf{E}(t)$.

EXERCISE 12.4.6. One ontological argument for the existence of God hinges on making necessary existence part of the defining condition. Consider the necessarily existent being, $\imath x.\mathbf{E}_\Box(x)$. Discuss the relationships between assertions that this term denotes, that the being denoted has existence, and that the being has necessary existence.

EXERCISE 12.4.7. Consider the description "the possible fat man in the doorway." Construct three distinct formal representations of this description and discuss how they differ.

12.5. HINTIKKA'S SCHEMA AND VARIATIONS

Definition 11.1.3, specifying the semantic behavior of definite descriptions, works whether constant or varying domains semantics is used. But tableau rules for the two versions differ, and for these we must be more careful.

If we use possibilist quantifiers—constant domains—Russell's treatment of definite descriptions is applicable. In this approach definite descriptions are translated away, using quantifiers and equality. But it is important that quantifiers be possibilist so that everything, and not just everything exist-ent, is in quantifier range. Then if we follow Russell, we need no additional tableau proof machinery to deal with definite descriptions—we just translate all occurrences of them away and use the machinery we already have.

A possibilist approach hides issues of existence. Precisely because the distinction between designation and existence is of interest, we begin with an approach that uses actualist quantification. We will consider Russell's treatment in Section 12.7, but until then, our models are varying domain.

Hintikka (1959) characterizes definite descriptions in classical logic axio-matically, taking as axioms universal closures of all formulas of the following form, where y does not occur in $\varphi(x)$:

$$\{x = \imath x.\varphi(x)\} \equiv \{\varphi(x) \wedge (\forall y)[\varphi(y) \supset y = x]\}.$$

Hintikka's axiom scheme involves universal quantifiers. There is an explicit occurrence of $(\forall y)$, and there are also the quantifiers involved in forming the universal closure. Unfortunately, these must all be understood as *possibilist*, not actualist. We can do away with the need for closure quantifiers by working with instances of open formulas. Of course predicate abstraction machinery must come in too. This produces the following tentative version, where t is a term. (We also assume x and y are different—something we generally don't bother stating after this.)

$$(t \approx \imath x.\varphi(x)) \equiv \{\langle\lambda x.\varphi(x)\rangle(t) \wedge (\forall y)(\varphi(y) \supset \langle\lambda x.y = x\rangle(t))\}$$

But we still have the explicit $(\forall y)$, and it is harder to deal with. Fortunately there is a way out, and it will lead us to reasonable actualist tableau rules in the next section. It involves replacing the quantifier above with a free variable, breaking the Hintikka equivalence into two implications, and treating each implication separately. To motivate what follows, suppose the sentence $A \equiv (\forall x)\varphi(x)$ is *classically* valid. Then each of the implications $A \supset (\forall x)\varphi(x)$ and $(\forall x)\varphi(x) \supset A$ is valid. Since classically we have universal instantiation, $(\forall x)\varphi(x) \supset \varphi(x)$ is valid, so from the first of our implications we get the classical *validity* of $A \supset \varphi(x)$. But using standard quantifier manipulations, the second implication is equivalent to $(\exists x)[\varphi(x) \supset A]$, and from this follows the classical *satisfiability* of $\varphi(x) \supset A$, and not its validity.

The following two propositions show that modally the Hintikka equivalence can be replaced with two implications, one valid, one satisfiable. The arguments are more complex than they were classically. When dealing with actualist quantification we don't have familiar things like universal instantiation available. Since the proofs that follow are rather baroque, you may decide to skip them. The tableau rules based on them are much simpler to apply than their justifications are to understand.

PROPOSITION 12.5.1. Let t be a term, and suppose the variable y does not occur in $\varphi(x)$ or in t. The following are valid in varying domain models.

1. $\langle\lambda y, x.y = x\rangle(t, \imath x.\varphi(x)) \supset \{\langle\lambda x.\varphi(x)\rangle(t) \wedge (\varphi(y) \supset \langle\lambda x.y = x\rangle(t))\}$
2. $\langle\lambda y, x.y = x\rangle(t, \imath x.\varphi(x)) \supset \langle\lambda x.\varphi(x)\rangle(t)$
3. $\langle\lambda y, x.y = x\rangle(t, \imath x.\varphi(x)) \supset (\varphi(y) \supset \langle\lambda x.y = x\rangle(t))$

Proof Item 1 is equivalent to items 2 and 3 together; we show 2 and 3. For the rest of this proof, let $\mathcal{M} = \langle\mathcal{G}, \mathcal{R}, \mathcal{D}, \mathcal{I}\rangle$ be a varying domain model, let $\Gamma \in \mathcal{G}$, and let v be a valuation. We assume $\mathcal{M}, \Gamma \Vdash_v \langle\lambda y, x.y = x\rangle(t, \imath x.\varphi(x))$.

Part One First we show that $\mathcal{M}, \Gamma \Vdash_v \langle\lambda x.\varphi(x)\rangle(t)$, which will establish item 2.

Recall that $\langle\lambda y, x.y = x\rangle(t, \imath x.\varphi(x))$ is an abbreviation for the formula $\langle\lambda y.\langle\lambda x.y = x\rangle(\imath x.\varphi(x))\rangle(t)$. Since we assume this is true at Γ with respect to v, t designates at Γ with respect to v. But then, if v' is the y-variant of v

such that $v'(y) = (v * \mathit{l})(t, \Gamma)$, we have that $\mathcal{M}, \Gamma \Vdash_{v'} \langle \lambda x.y = x \rangle (\imath x.\varphi(x))$. It follows that $\imath x.\varphi(x)$ also designates at Γ, with respect to v', so if v'' is the x-variant of v' such that $v''(x) = (v' * \mathit{l})(\imath x.\varphi(x), \Gamma)$, then $\mathcal{M}, \Gamma \Vdash_{v''} y = x$. It follows that $v''(y) = v''(x)$.

By definition, $v'(y) = (v * \mathit{l})(t, \Gamma)$, and since v' and v'' are x-variants they agree on y, so $v''(y) = (v * \mathit{l})(t, \Gamma)$. Also by definition, $v''(x) = (v' * \mathit{l})(\imath x.\varphi(x), \Gamma)$. Now, v' and v are y-variants, and y does not occur in $\imath x.\varphi(x)$, so $(v' * \mathit{l})(\imath x.\varphi(x), \Gamma) = (v * \mathit{l})(\imath x.\varphi(x), \Gamma)$, and hence $v''(x) = (v * \mathit{l})(\imath x.\varphi(x), \Gamma)$. Since we know $v''(y) = v''(x)$, we have that $(v * \mathit{l})(t, \Gamma) = (v * \mathit{l})(\imath x.\varphi(x), \Gamma)$. Note, in particular, that $\imath x.\varphi(x)$ designates at Γ with respect to v.

Since $\imath x.\varphi(x)$ designates at Γ with respect to v, by formula (12.12) we have $\mathcal{M}, \Gamma \Vdash_v \langle \lambda x.\varphi(x) \rangle (\imath x.\varphi(x))$. Also $(v * \mathit{l})(t, \Gamma) = (v * \mathit{l})(\imath x.\varphi(x), \Gamma)$, so it follows that $\mathcal{M}, \Gamma \Vdash_v \langle \lambda x.\varphi(x) \rangle (t)$, which is what we wanted to show.

Part Two Next we show that $\mathcal{M}, \Gamma \Vdash_v \varphi(y) \supset \langle \lambda x.y = x \rangle (t)$, which gives us item 3.

Assume $\mathcal{M}, \Gamma \Vdash_v \varphi(y)$. We must show $\mathcal{M}, \Gamma \Vdash_v \langle \lambda x.y = x \rangle (t)$. Let w be the x-variant of v such that $w(x) = (v * \mathit{l})(t, \Gamma)$. We must show $\mathcal{M}, \Gamma \Vdash_w y = x$. To show this, we must show $w(y) = w(x)$. By definition of w, this means we must show $w(y) = (v * \mathit{l})(t, \Gamma)$. And since w is an x-variant of v, the two agree on y, so finally, what we must show is $v(y) = (v * \mathit{l})(t, \Gamma)$.

We saw in Part One of this proof that $\imath x.\varphi(x)$ does, in fact, designate at Γ with respect to v, and that $(v * \mathit{l})(t, \Gamma) = (v * \mathit{l})(\imath x.\varphi(x), \Gamma)$. Thus it is enough for us to show that $v(y) = (v * \mathit{l})(\imath x.\varphi(x), \Gamma)$. We are assuming that $\mathcal{M}, \Gamma \Vdash_v \varphi(y)$. Then if we let v' be the x-variant of v such that $v'(x) = v(y)$, it follows that $\mathcal{M}, \Gamma \Vdash_{v'} \varphi(x)$. Since $\imath x.\varphi(x)$ designates at Γ with respect to v, by Definition 12.3.1 there is *exactly one* x-variant of v such that $\varphi(x)$ is true at Γ with respect to it. Then v' must be this x-variant, and so by Definition 12.3.1 again, $v'(x) = (v * \mathit{l})(\imath x.\varphi(x), \Gamma)$. It follows that $v(y) = (v * \mathit{l})(\imath x.\varphi(x), \Gamma)$, and we are done.

■

Now we treat the other, not valid but satisfiable, direction.

PROPOSITION 12.5.2. Let t be a term, suppose y does not occur in $\varphi(x)$, and neither x nor y occurs in t. Also, let $\mathcal{M} = \langle \mathcal{G}, \mathcal{R}, \mathcal{D}, \mathit{l} \rangle$ be a varying domain model with Γ a world of it and v a valuation in it. Then, there is a y-variant w of v such that the following is true at Γ with respect to valuation w:

(12.15) $\{ \langle \langle \lambda x.\varphi(x) \rangle (t) \wedge (\varphi(y) \supset \langle \lambda x.y = x \rangle (t)) \} \supset$
$$\langle \lambda y, x.y = x \rangle (t, \imath x.\varphi(x)).$$

What this says, intuitively, is that in any model, given any valuation, we can make (12.15) be true at any world we want, provided we assign the "right" value to y, and change nothing else. In particular, doing so will not affect the truth value of any formulas that do not contain free occurrences of y.

Proof While the proof of this Proposition is not deep it has numerous cases, and in different cases we need different y-variants of v to make (12.15) be true at Γ. Most of the difficulty in what follows is keeping track of the cases; each particular case is not hard.

1. If t does not designate at Γ with respect to v, $\mathcal{M}, \Gamma \not\Vdash_v \langle \lambda x.\varphi(x) \rangle(t)$, and so (12.15) is true at Γ with respect to v since part of the antecedent fails. In this situation, we simply take w to be v itself and we are done. Consequently *for the rest of this proof, we can assume t designates at Γ with respect to v.*

2. If we have that $\mathcal{M}, \Gamma \Vdash_v \langle \lambda y, x.y = x \rangle(t, \imath x.\varphi(x))$, (12.15) is again true at Γ with respect to v, since we have the consequent. In this case too, we take w to be v. So *for the rest of this proof, we can assume $\mathcal{M}, \Gamma \not\Vdash_v \langle \lambda y, x.y = x \rangle(t, \imath x.\varphi(x))$, which unabbreviated says $\mathcal{M}, \Gamma \not\Vdash_v \langle \lambda y.\langle \lambda x.y = x \rangle(\imath x.\varphi(x)) \rangle(t)$.*

3. Since we assume t designates at Γ with respect to v, by 1, we can define a y-variant of v by setting $v'(y) = (v * \mathit{l})(t, \Gamma)$. By the assumption of 2, we have $\mathcal{M}, \Gamma \not\Vdash_{v'} \langle \lambda x.y = x \rangle(\imath x.\varphi(x))$. Also, since v and v' are y-variants, and y does not occur in t, then t designates at Γ with respect to v' too, and $(v * \mathit{l})(t, \Gamma) = (v' * \mathit{l})(t, \Gamma)$.

4. By 3, t designates at Γ with respect to v'. Let v'' be the x-variant of v' defined by $v''(x) = (v' * \mathit{l})(t, \Gamma)$.

5. If $\mathcal{M}, \Gamma \not\Vdash_{v''} \varphi(x)$, by definition, $\mathcal{M}, \Gamma \not\Vdash_{v'} \langle \lambda x.\varphi(x) \rangle(t)$, and (12.15) holds with respect to v', since part of the antecedent is false at Γ. In this case take w to be v', and we have satisfiability. Consequently, *for the rest of this proof we can assume $\mathcal{M}, \Gamma \Vdash_{v''} \varphi(x)$.*

6. Suppose $\mathcal{M}, \Gamma \Vdash_{w_1} \varphi(x)$, where w_1 is an x-variant of v', distinct from v''. Since w_1 and v'' are distinct they must differ at x, since they agree on all other variables. Define a valuation w_2 to be like w_1 except for y, and set $w_2(y) = w_1(x)$. We show $\mathcal{M}, \Gamma \not\Vdash_{w_2} (\varphi(y) \supset \langle \lambda x.y = x \rangle(t))$.

   Since y does not occur in $\varphi(x)$, and we have $\mathcal{M}, \Gamma \Vdash_{w_1} \varphi(x)$, and w_2 "behaves on" y the way w_1 does on x, we immediately have $\mathcal{M}, \Gamma \Vdash_{w_2} \varphi(y)$ (technically, this is by Proposition 4.6.9).

   To show $\mathcal{M}, \Gamma \not\Vdash_{w_2} \langle \lambda x.y = x \rangle(t)$ we must show $w_2(y) \neq (w_2 * \mathit{l})(t, \Gamma)$. Now, $w_2(y) = w_1(x)$ by definition, so we must show $w_1(x) \neq (w_2 * \mathit{l})(t, \Gamma)$. But also, since y does not occur in t, and w_2 and w_1 differ only on y, $(w_2 * \mathit{l})(t, \Gamma) = (w_1 * \mathit{l})(t, \Gamma)$, so we must show $w_1(x) \neq$

$(w_1 * \mathcal{I})(t, \Gamma)$. Further, since x is not in t, and w_1 and v' differ only on x, this reduces to showing $w_1(x) \neq (v' * \mathcal{I})(t, \Gamma)$. But $(v' * \mathcal{I})(t, \Gamma)$ is $v''(x)$ by 4, and we know w_1 and v'' differ on x.

We now have $\mathcal{M}, \Gamma \not\Vdash_{w_2} (\varphi(y) \supset \langle \lambda x.y = x \rangle(t))$ and consequently (12.15) is true at Γ with respect to w_2, since part of the antecedent fails. Unfortunately we cannot simply take w to be w_2 since w_2 may not be a y-variant of v. Following through the definitions above, w_2 is a y-variant of w_1, which is an x-variant of v', which is a y-variant of v, so w_2 and v might differ on both x and y. Fortunately, this does not really matter since x does not occur free in (12.15). So, set w to be like w_2, except that $w(x) = v(x)$. Then w is a y-variant of v, and (12.15) will be true at Γ with respect to w as well, completing this case.

Consequently, *for the rest of this proof we can assume there is no x-variant of v', distinct from v'', such that $\varphi(x)$ is true at Γ with respect to it.*

7. By our assumption at the end of 5, $\varphi(x)$ is true at Γ with respect to v'', and by our assumption at the end of 6, v'' is the only x-variant of v' for which this is the case. It follows that $\imath x.\varphi(x)$ designates at Γ with respect to v', and $v'(\imath x.\varphi(x), \Gamma) = v''(x)$.

In this case we show we can take the valuation w to be v', by showing $\mathcal{M}, \Gamma \Vdash_{v'} \langle \lambda y, x.y = x \rangle(t, \imath x.\varphi(x))$, which means we have (12.15) because we have the consequent. To show this we need $(v' * \mathcal{I})(t, \Gamma) = (v' * \mathcal{I})(\imath x.\varphi(x))$, but this is true because both sides are, simply, $v''(x)$.

We have now considered all the cases, and the proof is complete. ■

The two propositions above form the basis of our formal proof methods for definite descriptions, in the next section.

12.6. VARYING DOMAIN TABLEAUS

We have seen that a modified version of Hintikka's approach to definite descriptions is available to us in modal logic, allowing for actualist quantification and varying domain models. Our varying domain proof rules, then, are quite straightforward—we essentially use the formulas presented in the previous section as global assumptions. Since definite descriptions are complex constructs, it should not be expected that tableau proofs of sentences in which they appear are simple things. Indeed, they are not. But generally, a tableau proof is an easier thing to construct than a corresponding direct argument about models. And we will shortly introduce some derived rules that simplify proofs considerably.

To apply the tableau rules for predicate abstracts, Section 10.6, we must
be able to form $t@\sigma$, where t is a term and σ is a prefix. For terms other
than definite descriptions we have already said what this means—it amounts
to the addition of σ as a subscript to those constant and function symbols in
the term that do not already have a subscript. We treat definite descriptions in
a similar way.

DEFINITION 12.6.1.   Let $\imath x.\Phi$ be a definite description and σ be a prefix.
By $(\imath x.\Phi)@\sigma$ we mean the formal expression $[\imath x.\Phi]_\sigma$.

In t, if every constant symbol, function symbol, *and definite description*
has a subscript, t is *grounded*. (This extends Definition 10.6.1.)

As usual, think of $[\imath x.\Phi]_\sigma$ as the object that $\imath x.\Phi$ designates at world σ, if
there is such an object. The Positive and the Negative Predicate Abstraction
Rules, from Chapter 11 still apply, but with a broader notion of term, since
definite descriptions are now included.

In addition to broadening the abstraction rules, we add some new tableau
rules specifically for definite descriptions. The rules, very simply, allow the
formulas of Propositions 12.5.1 and 12.5.2 to be used as global assumptions,
essentially as in Definition 8.9.3.

DEFINITION 12.6.2.   [Branch Sentence] Let $\mathcal{B}$ be a tableau branch. A $\mathcal{B}$-
sentence is a formula X that may contain grounded terms that designate on
$\mathcal{B}$ (including parameters), but otherwise contains no free variables.

The first of our new tableau rules, based on Proposition 12.5.1, says we
can add to a tableau branch a formula asserting that a definite description
satisfies its defining condition.

DEFINITION 12.6.3.   [Description Satisfaction Rule] For any prefix σ occurring on a tableau branch $\mathcal{B}$, σX can be added to the end of the branch,
where X is a $\mathcal{B}$-sentence of the following form:

$$\langle \lambda y, x.y = x \rangle (t, \imath x.\varphi(x)) \supset \langle \lambda x.\varphi(x) \rangle (t)$$

provided y does not occur in $\varphi(x)$ or in t.

The next tableau rule, also based on Proposition 12.5.1 says we can add to
a tableau branch a formula asserting that a designating definite description is
the *only* thing satisfying its defining condition.

DEFINITION 12.6.4.   [Description Uniqueness Rule] For any prefix σ occurring on a tableau branch $\mathcal{B}$, σX can be added to the end of the branch,
where X is a $\mathcal{B}$-sentence of the following form:

$$\langle \lambda y, x.y = x \rangle (t, \imath x.\varphi(x)) \supset (\varphi(y) \supset \langle \lambda x.y = x \rangle (t))$$

provided y does not occur in $\varphi(x)$ or in t.

The third tableau rule provides us with machinery that can be used to conclude a definite description actually designates on a branch. It comes directly from Proposition 12.5.2. However, that Proposition gives us a formula that is not valid, but *satisfiable*—roughly, an appropriate value for a certain free variable can be chosen so as to make the formula true. The counterpart of this, for tableaus, is the introduction of a new parameter.

DEFINITION 12.6.5. [Description Designation Rule] For any prefix σ occurring on a tableau branch $\mathcal{B}$, σX can be added to the end of the branch, where X is a $\mathcal{B}$-sentence of the following form:

$$\{\langle \lambda x.\varphi(x)\rangle(t) \wedge (\varphi(p_\sigma) \supset \langle \lambda x.p_\sigma = x\rangle(t))\} \supset \langle \lambda y, x.y = x\rangle(t, \imath x.\varphi(x))$$

provided p_σ is a new parameter on the branch, and x does not occur in t.

EXAMPLE 12.6.6. We give a varying domain tableau proof of $(\forall z)\langle \lambda x.z = x\rangle(\imath x.z = x)$. Note that $\langle \lambda x.z = x\rangle$ is essentially the property of being z, so the sentence asserts that, for all objects z, the thing that is z has the property of being z. We give the proof for varying domain **K**, though as usual the actual choice of modal logic does not really matter. Since the tableau as a whole is somewhat large, we present it in pieces, one branch at a time. It begins as follows.

$$1 \neg(\forall z)\langle \lambda x.z = x\rangle(\imath x.z = x) \quad 1.$$
$$1 \neg\langle \lambda x.p_1 = x\rangle(\imath x.p_1 = x) \quad 2.$$
$$1 \quad [\langle \lambda x.p_1 = x\rangle(p_1) \wedge (p_1 = q_1 \supset \langle \lambda x.q_1 = x\rangle(p_1))] \supset$$
$$\langle \lambda y, x.y = x\rangle(p_1, \imath x.p_1 = x) \quad 3.$$

Item 2 is from 1 using an Existential Rule, so p_1 is a new parameter. Note that p_1 designates on the branch, since it is a parameter. Item 3 is added using the Description Designation Rule, in which q_1 is a new parameter.

Because of item 3 the tableau branches at this point. At this point we continue *with the right branch only*.

$$1 \neg(\forall z)\langle \lambda x.z = x\rangle(\imath x.z = x) \quad 1.$$
$$1 \neg\langle \lambda x.p_1 = x\rangle(\imath x.p_1 = x) \quad 2.$$
$$1 \quad [\langle \lambda x.p_1 = x\rangle(p_1) \wedge (p_1 = q_1 \supset \langle \lambda x.q_1 = x\rangle(p_1))] \supset$$
$$\langle \lambda y, x.y = x\rangle(p_1, \imath x.p_1 = x) \quad 3.$$
$$1 \quad \langle \lambda y, x.y = x\rangle(p_1, \imath x.p_1 = x) \quad 4.$$
$$1 \quad \langle \lambda x.p_1 = x\rangle(\imath x.p_1 = x) \quad 5.$$
$$1 \quad p_1 = [\imath x.p_1 = x]_1 \quad 6.$$
$$1 \neg(p_1 = [\imath x.p_1 = x]_1) \quad 7.$$

Item 4 is the first formula on the right-hand branch that results when a Disjunctive Rule is applied to item 3. Item 5 is from item 4 by application of the Positive Predicate Abstraction Rule. Note that item 4 is an abbreviation that expands to nested predicate abstracts, with that involving y being outermost. Item 6 is from item 5 by the Positive Predicate Abstraction Rule. At this point $[\imath x.p_1 = x]_1$ designates on the branch, so the Negative Predicate Abstraction Rule can be applied to item 2, yielding item 7. The branch is closed at this point.

There is also a left-hand branch, resulting from the Disjunctive Rule applied to item 3. It remains to be dealt with. It begins as follows.

$$1 \; \neg(\forall z)\langle \lambda x.z = x\rangle(\imath x.z = x) \quad 1.$$
$$1 \; \neg\langle \lambda x.p_1 = x\rangle(\imath x.p_1 = x) \quad 2.$$
$$1 \quad \left[\langle \lambda x.p_1 = x\rangle(p_1) \wedge (p_1 = c_1 \supset \langle \lambda x.c_1 = x\rangle(p_1))\right] \supset$$
$$\langle \lambda y, x.y = x\rangle(p_1, \imath x.p_1 = x) \quad 3.$$
$$1 \; \neg\left[\langle \lambda x.p_1 = x\rangle(p_1) \wedge (p_1 = c_1 \supset \langle \lambda x.c_1 = x\rangle(p_1))\right] \quad 8.$$

We leave it to you to show this branch also closes.

The tableau rules for definite descriptions are rather complex. In the proof above, the key item was picking the right formula to use in an application of the Description Designation Rule. This choice of an appropriate formula is at the heart of all the proofs in this section. Fortunately, there are some derived rules that can simplify things a bit. Here are three that we have found to be useful.

PROPOSITION 12.6.7. [Description Satisfaction Derived Rule]
If $[\imath x.\varphi(x)]_\sigma$ is an extended term that designates on a tableau branch, then $\sigma \, \varphi([\imath x.\varphi(x)]_\sigma)$ may be added to the end of the branch. Schematically,

$$\frac{}{\sigma \, \varphi([\imath x.\varphi(x)]_\sigma)}$$

Proof Suppose at some point of a tableau construction we know $[\imath x.\varphi(x)]_\sigma$ designates on a tableau branch. Using the Description Satisfaction Rule we can add the following to the end of the branch.

$$\sigma \, \langle \lambda y, x.y = x\rangle(\imath x.\varphi(x), \imath x.\varphi(x)) \supset \langle \lambda x.\varphi(x)\rangle(\imath x.\varphi(x))$$

Now using a Disjunctive Rule, we have a branching. The left branch begins as follows.

$$\sigma \, \neg\langle \lambda y, x.y = x\rangle(\imath x.\varphi(x), \imath x.\varphi(x))$$

Since we are assuming that $[\imath x.\varphi(x)]_\sigma$ designates on the branch, using the Negative Abstraction Rule twice we can add the following to the left branch.

$$\sigma \neg([\imath x.\varphi(x)]_\sigma = [\imath x.\varphi(x)]_\sigma)$$

Then this branch can be closed using the Reflexivity Rule for equality.

Now that the left branch has closed, we can continue with the right branch which starts as follows.

$$\sigma \langle \lambda x.\varphi(x) \rangle (\imath x.\varphi(x))$$

An application of the Positive Predicate Abstraction Rule allows us to add the desired formula:

$$\sigma \varphi([\imath x.\varphi(x)]_\sigma)$$

thus accomplishing what the derived rule asserted could be done. ∎

Here is the second of our derived rules. We leave its proof to you as an exercise.

PROPOSITION 12.6.8. [Description Uniqueness Derived Rule]
Suppose $[\imath x.\varphi(x)]_\sigma$ and t are both extended terms that designate on a tableau branch. A branch containing $\sigma \varphi(t)$ may be extended with $\sigma t = [\imath x.\varphi(x)]_\sigma$. Schematically,

$$\frac{\sigma \varphi(t)}{\sigma t = [\imath x.\varphi(x)]_\sigma}$$

The third of our derived rules says a change of variable in a definite description can be carried out provided there is no variable conflict. In proving it we make use of the two derived rules above.

PROPOSITION 12.6.9. [Change of Variables Derived Rule]
Suppose x does not occur in $\psi(y)$, and $[\imath y.\psi(y)]_\sigma$ designates on a tableau branch. Then $\sigma [\imath y.\psi(y)]_\sigma = [\imath x.\psi(x)]_\sigma$ can be added to the branch (after which $[\imath x.\psi(x)]_\sigma$ also designates on the branch).

Proof We begin by using the Description Designation Rule, taking $\varphi(x)$ to be $\psi(x)$, t to be $\imath y.\psi(y)$, and p_σ to be a new parameter. Note that because of our assumption about x not occurring in $\psi(y)$, if we substitute an expression for free x in $\psi(x)$ and for free y in $\psi(y)$ we get the same thing. Here are the tableau steps we add in order to accomplish the effect of using the derived rule. First we add the following, using the Description Designation Rule.

$$\sigma \{\langle \lambda x.\psi(x)\rangle(\imath y.\psi(y)) \wedge (\psi(p_\sigma) \supset \langle \lambda x.p_\sigma = x\rangle(\imath y.\psi(y)))\}$$
$$\supset \langle \lambda y, x.y = x\rangle(\imath y.\psi(y), \imath x.\psi(x))$$

Next we apply a Disjunctive Rule. The right branch begins with

$$\sigma \langle \lambda y, x.y = x\rangle(\imath y.\psi(y), \imath x.\psi(x))$$

and using the Positive Predicate Abstraction Rule twice, we can add

$$\sigma [\imath y.\psi(y)]_\sigma = [\imath x.\psi(x)]_\sigma$$

as desired. It remains to demonstrate that the left fork of the branch closes. The following does this.

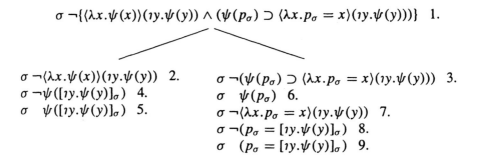

$$\sigma \neg\{\langle \lambda x.\psi(x)\rangle(\imath y.\psi(y)) \wedge (\psi(p_\sigma) \supset \langle \lambda x.p_\sigma = x\rangle(\imath y.\psi(y)))\} \quad 1.$$

$\sigma \neg\langle \lambda x.\psi(x)\rangle(\imath y.\psi(y)) \quad 2.$
$\sigma \neg\psi([\imath y.\psi(y)]_\sigma) \quad 4.$
$\sigma \ \ \psi([\imath y.\psi(y)]_\sigma) \quad 5.$

$\sigma \neg(\psi(p_\sigma) \supset \langle \lambda x.p_\sigma = x\rangle(\imath y.\psi(y))) \quad 3.$
$\sigma \ \ \psi(p_\sigma) \quad 6.$
$\sigma \neg\langle \lambda x.p_\sigma = x\rangle(\imath y.\psi(y)) \quad 7.$
$\sigma \neg(p_\sigma = [\imath y.\psi(y)]_\sigma) \quad 8.$
$\sigma \ \ (p_\sigma = [\imath y.\psi(y)]_\sigma) \quad 9.$

Item 1, of course, is how the left fork of the disjunctive split begins; items 2 and 3 are from 1 by a Disjunctive Rule; 4 is from 2 by the Negative Predicate Abstraction Rule (note that $[\imath y.\psi(y)]_\sigma$ designates on the branch by assumption); 5 is by the Description Satisfaction Derived Rule; 6 and 7 are from 3 by a Conjunctive Rule; 8 is from 7 by the Negative Predicate Abstraction Rule; 9 is by the Description Uniqueness Derived Rule. ∎

Finally, here an example that uses derived rules, and also involves modal notions. It continues our earlier Example 12.4.4.

EXAMPLE 12.6.10.   We give a tableau proof of formula (12.14) which asserts, under reasonable conditions, that someday the tallest person in the world might not be the tallest person in the world. The formula to be proved is:

$$[\mathbf{E}(\imath x.T(x)) \wedge (\forall x)\Diamond\neg T(x)] \supset$$
$$\langle \lambda x.\Diamond\neg\langle \lambda y.x = y\rangle(\imath z.T(z))\rangle(\imath z.T(z))$$

Here is the tableau establishing it, using varying domain **K**.

$$1 \quad \neg\{[\mathbf{E}(\imath x.T(x)) \wedge (\forall x)\Diamond\neg T(x)] \supset$$
$$\langle\lambda x.\Diamond\neg\langle\lambda y.x = y\rangle(\imath z.T(z))\rangle(\imath z.T(z))\} \quad 1.$$
$$1 \quad \mathbf{E}(\imath x.T(x)) \wedge (\forall x)\Diamond\neg T(x) \quad 2.$$
$$1 \quad \neg\langle\lambda x.\Diamond\neg\langle\lambda y.x = y\rangle(\imath z.T(z))\rangle(\imath z.T(z)) \quad 3.$$
$$1 \quad \mathbf{E}(\imath x.T(x)) \quad 4.$$
$$1 \quad (\forall x)\Diamond\neg T(x) \quad 5.$$
$$1 \quad \Diamond\neg T([\imath x.T(x)]_1) \quad 6.$$
$$1.1 \;\neg T([\imath x.T(x)]_1) \quad 7.$$
$$1 \quad [\imath x.T(x)]_1 = [\imath z.T(z)]_1 \quad 8.$$
$$1.1 \;\neg T([\imath z.T(z)]_1) \quad 9.$$
$$1 \quad \neg\Diamond\neg\langle\lambda y.[\imath z.T(z)]_1 = y\rangle(\imath z.T(z)) \quad 10.$$
$$1.1 \;\neg\neg\langle\lambda y.[\imath z.T(z)]_1 = y\rangle(\imath z.T(z)) \quad 11.$$
$$1.1 \quad \langle\lambda y.[\imath z.T(z)]_1 = y\rangle(\imath z.T(z)) \quad 12.$$
$$1.1 \quad [\imath z.T(z)]_1 = [\imath z.T(z)]_{1.1} \quad 13.$$
$$1.1 \;\neg T([\imath z.T(z)]_{1.1}) \quad 14.$$
$$1.1 \quad T([\imath z.T(z)]_{1.1}) \quad 15.$$

Items 2 and 3 are from 1, and 4 and 5 are from 2, by a Conjunctive Rule; 6 is from 4 and 5 by the Object Existence Derived Rule; 7 is from 6 by a Possibility Rule; 8 is by the Change of Variables Derived Rule; 9 is from 7 and 8 by Substitutivity; 10 is from 3 by the Negative Predicate Abstraction Rule; 11 is from 10 by a Necessity Rule; 12 is from 11 by the Double Negation Rule; 13 is from 12 by the Positive Predicate Abstraction Rule; 14 is from 9 and 13 by Substitutivity; and 15 is by the Description Satisfaction Derived Rule.

EXERCISES

EXERCISE 12.6.1.   Complete the tableau of Example 12.6.6.

EXERCISE 12.6.2.   Supply the proof for Proposition 12.6.8.

EXERCISE 12.6.3.   Two of our basic tableau rules, Description Satisfaction and Description Uniqueness, can in turn be proved using the simpler Derived Description Satisfaction and Derived Description Uniqueness rules. Suppose, to keep things simple, that c is a constant symbol, the only free variable in $\varphi(x)$ is x, and z does not occur in $\varphi(x)$.

1. Give a tableau proof of

$$\langle \lambda y, x.y = x \rangle (c, \imath x.\varphi(x)) \supset \langle \lambda x.\varphi(x) \rangle (c)$$

that uses the Description Satisfaction Derived Rule, but not the Description Satisfaction Rule.

2. Give a tableau proof of

$$(\forall z)\{[\langle \lambda y, x.y = x \rangle (c, \imath x.\varphi(x)) \supset \varphi(z)] \supset \langle \lambda x.z = x \rangle (c)\}$$

that uses the Description Uniqueness Derived Rule, but not the Description Uniqueness Rule.

EXERCISE 12.6.4.   Give varying domain tableau proofs of the following.

1. $\mathbf{D}(\imath x.\varphi(x)) \supset \mathbf{D}(\imath y.\varphi(y))$. where y does not occur in $\varphi(x)$.
2. $(\forall x)\mathbf{E}(\imath y.y = x)$.
3. $\mathbf{D}(\imath x.\varphi(x)) \equiv \langle \lambda x.\varphi(x) \rangle (\imath x.\varphi(x))$.
4. $\mathbf{D}(\imath x.\varphi(x)) \supset [(\exists x)\varphi(x) \supset \mathbf{E}(\imath x.\varphi(x))]$.

EXERCISE 12.6.5.   Give a varying domain proof of the following (see Example 12.4.6)

$$\neg\{\Box\mathbf{E}(\imath z.W(z)) \wedge (\forall w)\langle \lambda y.\neg\Box w = y \rangle (\imath z.W(z))\}$$

EXERCISE 12.6.6.   In Exercise 12.4.4 we defined an $\mathbf{E}_\Box$ predicate abstract, and we noted in subsequent exercises that it played a role in some versions of the ontological argument. Now give a varying domain tableau proof, in the logic $\mathbf{T}$, of $\neg\overline{\mathbf{E}}(\imath x.\mathbf{E}_\Box(x))$. Discuss the differences between this and the sentence $\mathbf{E}(\imath x.\mathbf{E}_\Box(x))$.

EXERCISE 12.6.7.   Suppose all P's are Q's. Then *the* P and *the* Q will be the same thing, provided both definite descriptions make sense. The following sentence expresses this more precisely.

$$[\mathbf{E}(\imath x.P(x)) \wedge \mathbf{D}(\imath x.Q(x))] \supset$$
$$[(\forall x)(P(x) \supset Q(x)) \supset \langle \lambda x, y.x = y \rangle (\imath x.P(x), \imath x.Q(x))]$$

Give a varying domain tableau proof of this sentence. Also, what goes wrong if $\mathbf{E}(\imath x.P(x))$ is weakened to $\mathbf{D}(\imath x.P(x))$?

EXERCISE 12.6.8.   The following sentence says that if *the* P designates rigidly (Proposition 10.2.5), and *the possible* P designates at all, then the two definite descriptions designate the same thing.

$$[\langle \lambda x.\Box\langle \lambda y.x = y \rangle (\imath x.P(x)) \rangle (\imath x.P(x)) \wedge \mathbf{D}(\imath x.\Diamond P(x))] \supset$$
$$\langle \lambda x, y.x = y \rangle (\imath x.P(x), \imath x.\Diamond P(x))$$

Give a varying domain tableau proof of this sentence.

EXERCISE 12.6.9. Give a varying domain proof of the following.

$$[\mathbf{D}(\imath z.\Diamond P(z)) \wedge \Box \mathbf{D}(\imath z.P(z))] \supset \langle \lambda x.\Diamond \langle \lambda y.x = y\rangle(\imath z.P(z))\rangle(\imath z.\Diamond P(z))$$

EXERCISE 12.6.10. Give a varying domain proof of the following.

$$\Box \mathbf{D}(\imath z.P(z)) \supset (\forall x)[\langle \lambda y.y = x\rangle(\imath z.\Diamond P(z)) \supset \Diamond \langle \lambda y.y = x\rangle(\imath z.P(z))]$$

12.7. RUSSELL'S APPROACH

We have set things up so that definite descriptions can designate nonexistent objects. This is essential for descriptions like "the chief assassin of Julius Caesar" or "the first President of the United States," neither of whom exist in a temporal sense. But for the similar sounding "the current President of the United States," current existence of the object designated is assumed to be part of the criteria for designation. We would be very surprised if "the current President of the United States" designated somebody who did not exist. (It describes a state of affairs that has not occurred at least since Warren Harding's term.) In this section we see that, even with actualist quantifiers, Russell's treatment of definite descriptions applies very nicely to the special class of definite descriptions for which existence of the designated object is an inherent part. This leads to a simple tableau approach based on possibilist quantifiers.

As we said earlier, unlike Frege and Carnap, Russell held that though definite descriptions function syntactically like terms, semantically they are myths. He shifted the question from what "the present King of France" designates to how it behaves in context. According to his theory, to say "The present King of France is bald" is to say: "there is one and only one object that meets the conditions for being the King of France, and that object is also bald." Definite descriptions simply disappear.

Russell's version, developed in the framework of classical logic, extends to modal logic quite well. We can think of it as a translation device, paraphrasing away all definite description occurrences, in context.

DEFINITION 12.7.1. [Russell Paraphrase] Let $\langle \lambda x.\varphi(x)\rangle(\imath y.\psi(y))$ be a formula. Its *Russell paraphrase* is

$$(\exists x)\{ \; \varphi(x)\wedge$$
$$(\forall z)\,[\varphi(z) \supset z = x]\wedge$$
$$\psi(x)$$
$$\}$$

In this, z is a variable that does not occur in $\varphi(x)$.

In a Russell paraphrase the separate parts each play a well-defined role: there is something such that φ; nothing else is such that φ; and that thing is also such that ψ. It is sometimes convenient to give this paraphrase in a more compact form, which we might also call Russell's paraphrase:

$$(\exists x)\{(\forall z)[\varphi(z) \equiv z = x] \wedge \psi(x)\}.$$

A treatment of definite descriptions based on Russell's ideas is simple. Before evaluating a formula at a world of a model, or attempting to prove it using tableaus, use Russell's paraphrase to remove all definite description occurrences, and evaluate the result instead.

It should not be surprising that Russell's approach gives different results from ours when actualist quantifiers are employed. Russell's explicit introduction of the existential quantifier imposes existence assumptions, as opposed to designation assumptions. Nonetheless, his approach and ours coincide for those definite descriptions that make an existence assumption.

DEFINITION 12.7.2. [Existence Supposition] We say a definite description *makes an existence supposition* if it is of the form $\imath x.(\mathrm{E}(x) \wedge \Phi)$.

PROPOSITION 12.7.3. The following are equivalent formulas in varying domain models:

1. $\langle \lambda x.\varphi(x) \rangle (\imath y.\mathrm{E}(y) \wedge \psi(y))$.
2. The Russell paraphrase of $\langle \lambda x.\varphi(x) \rangle (\imath y.\psi(y))$.
3. The Russell paraphrase of $\langle \lambda x.\varphi(x) \rangle (\imath y.\mathrm{E}(y) \wedge \psi(y))$.

We leave verification of this to you as an exercise. Note that if y is a variable, $\mathrm{E}(y)$ and $\mathbf{E}(y)$ are equivalent formulas. We have opted for the simpler version in the Proposition above.

EXERCISES

EXERCISE 12.7.1. Give the proof of Proposition 12.7.3. The easiest way is to use varying domain tableaus to show the equivalence of the three versions.

EXERCISE 12.7.2. Show the equivalence of the two versions of Russell's paraphrase:

1. $(\exists x)\{(\forall y)[\varphi(y) \equiv y = x] \wedge \psi(x)\}$;
2. $(\exists x)\{\varphi(x) \wedge (\forall y)[\varphi(y) \supset y = x] \wedge \psi(x)\}$.

EXERCISE 12.7.3.   Example 12.4.2 has the following more specialized version. Suppose definite descriptions are treated using Russell's paraphrase (or equivalently, that they make an existential assumption). Under this assumption, show the validity of

$$\mathbf{E}(\imath x.\varphi(x)) \equiv \langle \lambda x.\varphi(x) \rangle (\imath x.\varphi(x)).$$

12.8. POSSIBILIST QUANTIFIERS

We have been using actualist quantification because we wanted to investigate the interplay between existence and designation where definite descriptions are concerned. But if we use possibilist quantifiers throughout, both the theory and the tableau rules become much simpler since with possibilist quantifiers explicit existence assumptions are unnecessary. More precisely, in constant domain models each formula of the form $E(y) \wedge \psi(y)$ is equivalent to the simpler formula $\psi(y)$. It follows from Proposition 12.7.3 that for constant domain models $\langle \lambda y.\varphi(x) \rangle (\imath y.\psi(y))$ is equivalent to the Russell paraphrase of $\langle \lambda x.\varphi(x) \rangle (\imath y.\psi(y))$. Then for a possibilist approach, we need no additional machinery to treat definite descriptions via tableaus. We can use Russell's paraphrase to *eliminate* them before starting a tableau proof.

Perhaps the best approach in the long run is to use possibilist quantifiers exclusively, and incorporate an explicit existence predicate into the language as we did in Section 4.8, thus retaining the ability to express what actualist quantifiers express. As we said, we have not done this here because we wanted to emphasize the differing roles of existence and designation, and we did not want part of the problem hidden behind quantifiers. Once the issues are understood, simplicity should rule.

EXERCISES

EXERCISE 12.8.1.   Give a constant domain model in which the sentence $\lozenge\mathbf{D}(\imath x.P(x)) \supset \mathbf{D}(\imath x.\lozenge P(x))$ is not valid.

EXERCISE 12.8.2.   Consider the sentence $\mathbf{D}(\imath x.\lozenge P(x)) \supset \lozenge\mathbf{D}(\imath x.P(x))$.

1. Give a constant domain proof by first using Russell's paraphrase to eliminate the definite descriptions, then using constant domain tableau rules.
2. Give a varying domain model in which the sentence is not valid.
3. Provide natural language examples illustrating parts 1 and 2.

References

Ackrill, J. K. (ed.): 1963, *Aristotle's* Categories *and* De Interpretatione. Oxford: Clarendon Press.

Angle, P. M. (ed.): 1991, *The Complete Lincoln-Douglas Debates of 1858*. Chicago: University of Chicago Press.

Bencivenga, E.: 1986, *Free Logics*, Chapt. III-6, pp. 373–426. In (Gabbay and Guenthner, 1989).

Berlin, I.: 1949-50, 'Logical Translation'. *Proceedings of the Aristotelian Society* NS, **50**, 157–188.

Blamey, S.: 1986, *Partial Logic*, Chapt. III-1, pp. 1–70. In (Gabbay and Guenthner, 1989).

Bressan, A.: 1972, *A General Interpreted Modal Calculus*. Yale University Press.

Bull, R. A. and K. Segerberg: 1984, *Basic Modal Logic*, Chapt. II-1, pp. 1–88. In (Gabbay and Guenthner, 1989).

Carnap, R.: 1947, *Meaning and Necessity: A Study in Semantics and Modal Logic*. Chicago: University of Chicago Press.

Cartwright, R.: 1960, 'Negative Existentials'. *The Journal of Philosophy* **57**, 629–639. Reprinted in R. L. Cartwright, *Philosophical Essays*, MIT Press, Cambridge, 1987.

Chellas, B.: 1980, *Modal Logic*. Cambridge: Cambridge University Press.

Chisholm, R. M. (ed.): 1960, *Realism and the Background of Phenomenology*. New York: Free Press.

Cornford, F.: 1957, *Plato and Parmenides*. Indianapolis: Bobbs-Merrill.

D'Agostino, M., D. Gabbay, R. Hähnle, and J. Posegga (eds.): 1998, *Handbook of Tableau Methods*. Dordrecht: Kluwer.

Fitting, M. C.: 1972a, 'An Epsilon-calculus system for first-order S4'. In: W. Hodges (ed.): *Conference in Mathematical Logic, London '70*. pp. 103–110. *Springer Lecture Notes in Mathematics, No. 255*.

Fitting, M. C.: 1972b, 'Tableau methods of proof for modal logics'. *Notre Dame Journal of Formal Logic* **13**, 237–247.

Fitting, M. C.: 1973, 'A Modal logic analog of Smullyan's fundamental theorem'. *Zeitschrift für mathematische Logik und Gründlagen der Mathematik* **19**, 1–16.

Fitting, M. C.: 1975, 'A Modal logic epsilon-calculus'. *Notre Dame Journal of Formal Logic* **16**, 1–16.

Fitting, M. C.: 1983, *Proof Methods for Modal and Intuitionistic Logics*. Dordrecht: D. Reidel Publishing Co.

Fitting, M. C.: 1991, 'Modal logic should say more than it does'. In: J.-L. Lassez and G. Plotkin (eds.): *Computational Logic, Essays in Honor of Alan Robinson*. Cambridge, MA: MIT Press, pp. 113–135.

Fitting, M. C.: 1993, 'Basic modal logic'. In: D. M. Gabbay, C. J. Hogger, and

J. A. Robinson (eds.): *Handbook of Logic in Artificial Intelligence and Logic Programming*, Vol. 1. Oxford University Press, pp. 368–448.

Fitting, M. C.: 1996a, *First-Order Logic and Automated Theorem Proving*. Springer-Verlag. First edition, 1990.

Fitting, M. C.: 1996b, 'A Modal Herbrand Theorem'. *Fundamenta Informaticae* **28**, 101–122.

Fitting, M. C.: 1998, 'Introduction'. In (D'Agostino et al., 1998).

Fraassen, B. C. V.: 1966, 'Singular Terms, Truth-Value Gaps, and Free Logic'. *The Journal of Philosophy* **63**, 481–495.

Frege, G.: 1879, *Begriffsschrift, eine der arithmetischen nachgebildete Formelsprache des reinen Denkens*. Halle: Verlag von L. Nebert. Partial translation in (Frege, 1952).

Frege, G.: 1884, 'Dialogue with Pünjer on Existence'. In: F. K. Hans Hermes and F. Kaulbach (eds.): *Gottlob Frege, Posthumous Writings*. Oxford, 1979: Basil Blackwell, pp. 53–67. tr. Peter Long and Roger White.

Frege, G.: 1892, 'Uber Sinn und Bedeutung'. *Zeitschrift fur Philosophie und philosophische Kritik* **100**, 25–50. "On Sense and Reference" translated in (Frege, 1952).

Frege, G.: 1893, *Grundgesetze der Arithmetic*, Vol. I. Jena: Pohle. reprinted in 1962 Olms: Hildesheim; partial translation in (Furth, 1967).

Frege, G.: 1952, *Translations from the Philosophical Writings of Gottlob Frege*. Oxford: Basil Blackwell. P. Geach and M. Black editors.

Furth, M. (ed.): 1967, *The Basic Laws of Arithmetic: Exposition of the System*. Berkeley and Los Angeles: University of California Press. Furth, Montgomery (tr.).

Gabbay, D. and F. Guenthner: 1983–1989, *Handbook of Philosophical Logic*, Synthese Library. Dordrecht: Kluwer. Four volumes.

Garson, J. W.: 1984, *Quantification in Modal Logic*, Chapt. II–5, pp. 249–307. In (Gabbay and Guenthner, 1989).

Gödel, K.: 1933, 'Eine Interpretation Des Intuitionistischen Aussagenkalküls'. *Ergebnisse eines mathematischen Kolloquiums* **4**, 39–40. 'An Interpretation of the Intuitionistic Propositional Calculus" is translated in S. Feferman, ed., *Kurt Gödel, Collected Works, Volume One*, Oxford, pp. 300-303, 1986.

Goré, R.: 1998, 'Tableau methods for modal and temporal logics'. In (D'Agostino et al., 1998).

Harel, D.: 1984, *Dynamic Logic*, Chapt. II–10, pp. 497–604. in (Gabbay and Guenthner, 1989).

Hintikka, J.: 1959, 'Towards a theory of definite descriptions'. *Analysis* **19**, 79–85.

Hintikka, J.: 1961, 'Modality and Quantification'. *Theoria* **27**, 119–128.

Hintikka, J.: 1962, *Knowledge and Belief*. Ithaca: Cornell University Press.

Hughes, G. E. and M. J. Cresswell: 1968, *An Introduction to Modal Logic*. London: Methuen.

Hughes, G. E. and M. J. Cresswell: 1984, *A Companion to Modal Logic*. London: Methuen.

Hughes, G. E. and M. J. Cresswell: 1996, *A New Introduction to Modal Logic*. London: Routledge.

Hume, D.: 1888, *A Treatise of Human Nature*. Oxford: Clarendon Press.

Kant, I.: 1781, *Immanuel Kant's Critique of Pure Reason*. London: Macmillan & Co. trans. Norman Kemp Smith, 1964.

Kneale, W. and M. Kneale: 1962, *The Development of Logic*. Oxford: Clarendon Press.

Kneale, W. C.: 1962, 'Modality *De Dicto* and *De Re*'. In: *Logic, Methodology and Philosophy of Science*. Stanford: North-Holland Publishing Company, pp. 622–633.

Koslow, A.: 1992, *A Structuralist Theory of Logic*. Cambridge: Cambridge University Press.

Kripke, S.: 1963a, 'Semantical analysis of modal logic I, normal propositional calculi'. *Zeitschrift für mathematische Logik und Grundlagen der Mathematik* **9**, 67–96.

Kripke, S.: 1963b, 'Semantical Considerations on Modal Logic'. *Acta Philosophica Fennica* **16**, 83–94.

Kripke, S.: 1965, 'Semantical analysis of modal logic II, non-normal modal propositional calculi'. In: J. W. Addison, L. Henkin, and A. Tarski (eds.): *The Theory of Models*. Amsterdam: North-Holland, pp. 206–220.

Kripke, S.: 1979a, 'A Puzzle About Belief'. In: A. Margalit (ed.): *Meaning and Use*. Dordrecht: D. Reidel, pp. 239–283.

Kripke, S.: 1979b, 'Speaker's Reference and Semantic Reference'. In: P. French, T. Uehling, and H. Wettstein (eds.): *Contemporary Perspectives in the Philosophy of Language*. Minneapolis: University of Minnesota Press, pp. 6–27.

Kripke, S.: 1980, *Naming and Necessity*. Cambridge: Harvard University Press.

Lewis, C. and C. Langford: 1959, *Symbolic Logic*. Dover. Second edition. First edition 1932.

Lewis, C. I.: 1918, *A Survey of Symbolic Logic*. New York: Dover.

Lewis, D.: 1968, 'Counterpart Theory and Quantified Modal Logic'. *The Journal of Philosophy* **65**, 113–126.

Lukasiewicz, J.: 1953, 'A System of Modal Logic'. *The Journal of Computing systems* **1**, 111–149.

Marcus, R. B.: 1946, 'A functional calculus of first order based on strict implication'. *Journal of Symbolic Logic* **11**, 1–16.

Marcus, R. B.: 1992, *Modalities*. New York: Oxford University Press.

Massacci, F.: 1994, 'Strongly analytic tableaux for normal modal logics'. In: A. Bundy (ed.): *Proceedings of CADE 12*, Vol. 814 of *Lecture Notes in Artificial Intelligence*. Berlin, pp. 723–737.

Massacci, F.: 1998, 'Single Step Tableaux for Modal Logics: methodology, computations, algorithms'. Technical Report TR-04, Dipartimento di Informatica e Sistemistica, Università di Roma "La Sapienza". http://www.dis.uniroma1.it/PUB/AI/papers/mass-98-c.ps.gz.

Mates, B.: 1961, *Stoic Logic*. Berkeley and Los Angeles: University of California Press.

Meinong, A.: 1889, 'On The Theory of Objects'. Reprinted in (Chisholm, 1960).

Mendelsohn, R. L.: 1982, 'Frege's *Begriffsschrift* Theory of Identity'. *Journal of the History of Philosophy* xx, 279–299.

Nozick, R.: 1981, *Philosophical Explanations*. Cambridge: Harvard University Press.

Parsons, T.: 1985, *Nonexistent Objects*. New Haven: Yale University Press.

Prior, A. N.: 1955, *Formal Logic*. Oxford: Clarendon Press.

Prior, A. N.: 1957, *Time and Modality*. Oxford: Clarendon Press.

Prior, A. N.: 1967, *Past, Present and Future*. Oxford: Clarendon Press.

Quine, W. V. O.: 1943, 'Notes on Existence and Necessity'. *The Journal of Philosophy* 40, pp. 113–127.

Quine, W. V. O.: 1948, 'On What There Is'. *Review of Metaphysics* 2. Reprinted in (Quine, 1961a).

Quine, W. V. O.: 1953, 'Three Grades of Modal Involvement'. In: *The Ways of Paradox and Other Essays*. New York: Random House, pp. 156–174.

Quine, W. V. O.: 1961a, *From a Logical Point of View*. New York: Harper & Row, 2nd. rev. edition.

Quine, W. V. O.: 1961b, 'Reference and Modality'. pp. 139–159. In (Quine, 1961a).

Ross, W. D. (ed.): 1928, *The Works of Aristotle*, Vol. I. Oxford: Oxford University Press.

Routley, R.: 1980, *Exploring Meinong's Jungle and Beyond: An Investigation of Noneism and the Theory of Items*, Department of Philosophy, Monograph No. 3. Canberra: Australian National University.

Russell, B.: 1905, 'On Denoting'. *Mind* 14, 479–493. Reprinted in Robert C. Marsh, ed., *Logic and Knowledge: Essays 1901-1950, by Bertrand Russell*, Allen & Unwin, London, 1956.

Russell, B.: 1938, *The Principles of Mathematics*. New York: Norton.

Ryle, G.: 1932, 'Systematically Misleading Expressions'. *Proceedings of the Aristotelian Society* 32. Reprinted in G. Ryle, *Collected Papers, Vol. II*, pp. 39-62, Barnes & Noble, New York, 1971.

Searle, J.: 1968, *Speech Acts*. Cambridge: Cambridge University Press.

Smullyan, A. F.: 1948, 'Modality and Description'. *The Journal of Symbolic Logic* 13, 31–37.

Smullyan, R. M.: 1968, *First-Order Logic*. Berlin: Springer-Verlag. Revised Edition, Dover Press, New York, 1994.

Smullyan, R. M. and M. C. Fitting: 1996, *Set Theory and the Continuum Problem*. Oxford: Oxford University Press.

Sorabji, R.: 1980, *Necessity, Cause, and Blame: Perspectives on Aristotle's Theory*. Ithaca: Cornell University Press.

Stalnaker, R. and R. Thomason: 1968, 'Abstraction in first-order modal logic'. *Theoria* 34, 203–207.

Strawson, P. F.: 1950, 'On Referring'. *Mind* pp. 320–344.

Thomason, R. and R. Stalnaker: 1968, 'Modality and reference'. *Nous* 2, 359–372.

Whitehead, A. N. and B. Russell: 1925, *Principia Mathematica*. Cambridge, England: Cambridge University Press, second edition. Three volumes.

Wittgenstein, L.: 1922, *Tractatus Logico-Philosophicus*. Annalen der Naturphilo-sophie. tr. by D. F. Pears and B. F. McGuinness, Routledge & Kegan Paul, London, 1961.

INDEX

189. H. Siegel, *Relativism Refuted*. A Critique of Contemporary Epistemological Relativism. 1987
     ISBN 90-277-2469-5
190. W. Callebaut and R. Pinxten, *Evolutionary Epistemology*. A Multiparadigm Program, with a Complete Evolutionary Epistemology Bibliograph. 1987   ISBN 90-277-2582-9
191. J. Kmita, *Problems in Historical Epistemology*. 1988   ISBN 90-277-2199-8
192. J. H. Fetzer (ed.), *Probability and Causality*. Essays in Honor of Wesley C. Salmon, with an Annotated Bibliography. 1988   ISBN 90-277-2607-8; Pb 1-5560-8052-2
193. A. Donovan, L. Laudan and R. Laudan (eds.), *Scrutinizing Science*. Empirical Studies of Scientific Change. 1988   ISBN 90-277-2608-6
194. H.R. Otto and J.A. Tuedio (eds.), *Perspectives on Mind*. 1988   ISBN 90-277-2640-X
195. D. Batens and J.P. van Bendegem (eds.), *Theory and Experiment*. Recent Insights and New Perspectives on Their Relation. 1988   ISBN 90-277-2645-0
196. J. Österberg, *Self and Others*. A Study of Ethical Egoism. 1988   ISBN 90-277-2648-5
197. D.H. Helman (ed.), *Analogical Reasoning*. Perspectives of Artificial Intelligence, Cognitive Science, and Philosophy. 1988   ISBN 90-277-2711-2
198. J. Wolenski, *Logic and Philosophy in the Lvov-Warsaw School*. 1989   ISBN 90-277-2749-X
199. R. Wójcicki, *Theory of Logical Calculi*. Basic Theory of Consequence Operations. 1988
     ISBN 90-277-2785-6
200. J. Hintikka and M.B. Hintikka, *The Logic of Epistemology and the Epistemology of Logic*. Selected Essays. 1989   ISBN 0-7923-0040-8; Pb 0-7923-0041-6
201. E. Agazzi (ed.), *Probability in the Sciences*. 1988   ISBN 90-277-2808-9
202. M. Meyer (ed.), *From Metaphysics to Rhetoric*. 1989   ISBN 90-277-2814-3
203. R.L. Tieszen, *Mathematical Intuition*. Phenomenology and Mathematical Knowledge. 1989
     ISBN 0-7923-0131-5
204. A. Melnick, *Space, Time, and Thought in Kant*. 1989   ISBN 0-7923-0135-8
205. D.W. Smith, *The Circle of Acquaintance*. Perception, Consciousness, and Empathy. 1989
     ISBN 0-7923-0252-4
206. M.H. Salmon (ed.), *The Philosophy of Logical Mechanism*. Essays in Honor of Arthur W. Burks. With his Responses, and with a Bibliography of Burk's Work. 1990
     ISBN 0-7923-0325-3
207. M. Kusch, *Language as Calculus vs. Language as Universal Medium*. A Study in Husserl, Heidegger, and Gadamer. 1989   ISBN 0-7923-0333-4
208. T.C. Meyering, *Historical Roots of Cognitive Science*. The Rise of a Cognitive Theory of Perception from Antiquity to the Nineteenth Century. 1989   ISBN 0-7923-0349-0
209. P. Kosso, *Observability and Observation in Physical Science*. 1989   ISBN 0-7923-0389-X
210. J. Kmita, *Essays on the Theory of Scientific Cognition*. 1990   ISBN 0-7923-0441-1
211. W. Sieg (ed.), *Acting and Reflecting*. The Interdisciplinary Turn in Philosophy. 1990
     ISBN 0-7923-0512-4
212. J. Karpiński, *Causality in Sociological Research*. 1990   ISBN 0-7923-0546-9
213. H.A. Lewis (ed.), *Peter Geach: Philosophical Encounters*. 1991   ISBN 0-7923-0823-9
214. M. Ter Hark, *Beyond the Inner and the Outer*. Wittgenstein's Philosophy of Psychology. 1990
     ISBN 0-7923-0850-6
215. M. Gosselin, *Nominalism and Contemporary Nominalism*. Ontological and Epistemological Implications of the Work of W.V.O. Quine and of N. Goodman. 1990   ISBN 0-7923-0904-9
216. J.H. Fetzer, D. Shatz and G. Schlesinger (eds.), *Definitions and Definability*. Philosophical Perspectives. 1991   ISBN 0-7923-1046-2
217. E. Agazzi and A. Cordero (eds.), *Philosophy and the Origin and Evolution of the Universe*. 1991   ISBN 0-7923-1322-4

SYNTHESE LIBRARY

218. M. Kusch, *Foucault's Strata and Fields*. An Investigation into Archaeological and Genealogical Science Studies. 1991   ISBN 0-7923-1462-X

219. C.J. Posy, *Kant's Philosophy of Mathematics*. Modern Essays. 1992   ISBN 0-7923-1495-6

220. G. Van de Vijver, *New Perspectives on Cybernetics*. Self-Organization, Autonomy and Connectionism. 1992   ISBN 0-7923-1519-7

221. J.C. Nyíri, *Tradition and Individuality*. Essays. 1992   ISBN 0-7923-1566-9

222. R. Howell, *Kant's Transcendental Deduction*. An Analysis of Main Themes in His Critical Philosophy. 1992   ISBN 0-7923-1571-5

223. A. García de la Sienra, *The Logical Foundations of the Marxian Theory of Value*. 1992   ISBN 0-7923-1778-5

224. D.S. Shwayder, *Statement and Referent*. An Inquiry into the Foundations of Our Conceptual Order. 1992   ISBN 0-7923-1803-X

225. M. Rosen, *Problems of the Hegelian Dialectic*. Dialectic Reconstructed as a Logic of Human Reality. 1993   ISBN 0-7923-2047-6

226. P. Suppes, *Models and Methods in the Philosophy of Science: Selected Essays*. 1993   ISBN 0-7923-2211-8

227. R. M. Dancy (ed.), *Kant and Critique: New Essays in Honor of W. H. Werkmeister*. 1993   ISBN 0-7923-2244-4

228. J. Woleński (ed.), *Philosophical Logic in Poland*. 1993   ISBN 0-7923-2293-2

229. M. De Rijke (ed.), *Diamonds and Defaults*. Studies in Pure and Applied Intensional Logic. 1993   ISBN 0-7923-2342-4

230. B.K. Matilal and A. Chakrabarti (eds.), *Knowing from Words*. Western and Indian Philosophical Analysis of Understanding and Testimony. 1994   ISBN 0-7923-2345-9

231. S.A. Kleiner, *The Logic of Discovery*. A Theory of the Rationality of Scientific Research. 1993   ISBN 0-7923-2371-8

232. R. Festa, *Optimum Inductive Methods*. A Study in Inductive Probability, Bayesian Statistics, and Verisimilitude. 1993   ISBN 0-7923-2460-9

233. P. Humphreys (ed.), *Patrick Suppes: Scientific Philosopher*. Vol. 1: Probability and Probabilistic Causality. 1994   ISBN 0-7923-2552-4

234. P. Humphreys (ed.), *Patrick Suppes: Scientific Philosopher*. Vol. 2: Philosophy of Physics, Theory Structure, and Measurement Theory. 1994   ISBN 0-7923-2553-2

235. P. Humphreys (ed.), *Patrick Suppes: Scientific Philosopher*. Vol. 3: Language, Logic, and Psychology. 1994   ISBN 0-7923-2862-0
    Set ISBN (Vols 233–235) 0-7923-2554-0

236. D. Prawitz and D. Westerståhl (eds.), *Logic and Philosophy of Science in Uppsala*. Papers from the 9th International Congress of Logic, Methodology, and Philosophy of Science. 1994   ISBN 0-7923-2702-0

237. L. Haaparanta (ed.), *Mind, Meaning and Mathematics*. Essays on the Philosophical Views of Husserl and Frege. 1994   ISBN 0-7923-2703-9

238. J. Hintikka (ed.), *Aspects of Metaphor*. 1994   ISBN 0-7923-2786-1

239. B. McGuinness and G. Oliveri (eds.), *The Philosophy of Michael Dummett*. With Replies from Michael Dummett. 1994   ISBN 0-7923-2804-3

240. D. Jamieson (ed.), *Language, Mind, and Art*. Essays in Appreciation and Analysis, In Honor of Paul Ziff. 1994   ISBN 0-7923-2810-8

241. G. Preyer, F. Siebelt and A. Ulfig (eds.), *Language, Mind and Epistemology*. On Donald Davidson's Philosophy. 1994   ISBN 0-7923-2811-6

242. P. Ehrlich (ed.), *Real Numbers, Generalizations of the Reals, and Theories of Continua*. 1994   ISBN 0-7923-2689-X

243. G. Debrock and M. Hulswit (eds.), *Living Doubt*. Essays concerning the epistemology of Charles Sanders Peirce. 1994　　　　　　　　　　　　ISBN 0-7923-2898-1
244. J. Srzednicki, *To Know or Not to Know*. Beyond Realism and Anti-Realism. 1994
　　　　　　　　　　　　　　　　　　　　　　　　　　　　ISBN 0-7923-2909-0
245. R. Egidi (ed.), *Wittgenstein: Mind and Language*. 1995　　ISBN 0-7923-3171-0
246. A. Hyslop, *Other Minds*. 1995　　　　　　　　　　　　　ISBN 0-7923-3245-8
247. L. Pólos and M. Masuch (eds.), *Applied Logic: How, What and Why*. Logical Approaches to Natural Language. 1995　　　　　　　　　　　　　　　ISBN 0-7923-3432-9
248. M. Krynicki, M. Mostowski and L.M. Szczerba (eds.), *Quantifiers: Logics, Models and Computation*. Volume One: Surveys. 1995　　　　　　　　ISBN 0-7923-3448-5
249. M. Krynicki, M. Mostowski and L.M. Szczerba (eds.), *Quantifiers: Logics, Models and Computation*. Volume Two: Contributions. 1995　　　　　ISBN 0-7923-3449-3
　　　　　　　　　　　　　　　　　Set ISBN (Vols 248 + 249) 0-7923-3450-7
250. R.A. Watson, *Representational Ideas from Plato to Patricia Churchland*. 1995
　　　　　　　　　　　　　　　　　　　　　　　　　　　　ISBN 0-7923-3453-1
251. J. Hintikka (ed.), *From Dedekind to Gödel*. Essays on the Development of the Foundations of Mathematics. 1995　　　　　　　　　　　　　　　　ISBN 0-7923-3484-1
252. A. Wiśniewski, *The Posing of Questions*. Logical Foundations of Erotetic Inferences. 1995
　　　　　　　　　　　　　　　　　　　　　　　　　　　　ISBN 0-7923-3637-2
253. J. Peregrin, *Doing Worlds with Words*. Formal Semantics without Formal Metaphysics. 1995
　　　　　　　　　　　　　　　　　　　　　　　　　　　　ISBN 0-7923-3742-5
254. I.A. Kieseppä, *Truthlikeness for Multidimensional, Quantitative Cognitive Problems*. 1996
　　　　　　　　　　　　　　　　　　　　　　　　　　　　ISBN 0-7923-4005-1
255. P. Hugly and C. Sayward: *Intensionality and Truth*. An Essay on the Philosophy of A.N. Prior. 1996　　　　　　　　　　　　　　　　　　　　ISBN 0-7923-4119-8
256. L. Hankinson Nelson and J. Nelson (eds.): *Feminism, Science, and the Philosophy of Science*. 1997　　　　　　　　　　　　　　　　　　　ISBN 0-7923-4162-7
257. P.I. Bystrov and V.N. Sadovsky (eds.): *Philosophical Logic and Logical Philosophy*. Essays in Honour of Vladimir A. Smirnov. 1996　　　　　　ISBN 0-7923-4270-4
258. Å.E. Andersson and N-E. Sahlin (eds.): *The Complexity of Creativity*. 1996
　　　　　　　　　　　　　　　　　　　　　　　　　　　　ISBN 0-7923-4346-8
259. M.L. Dalla Chiara, K. Doets, D. Mundici and J. van Benthem (eds.): *Logic and Scientific Methods*. Volume One of the Tenth International Congress of Logic, Methodology and Philosophy of Science, Florence, August 1995. 1997　　　　　ISBN 0-7923-4383-2
260. M.L. Dalla Chiara, K. Doets, D. Mundici and J. van Benthem (eds.): *Structures and Norms in Science*. Volume Two of the Tenth International Congress of Logic, Methodology and Philosophy of Science, Florence, August 1995. 1997　　　ISBN 0-7923-4384-0
　　　　　　　　　　　　　　　　　Set ISBN (Vols 259 + 260) 0-7923-4385-9
261. A. Chakrabarti: *Denying Existence*. The Logic, Epistemology and Pragmatics of Negative Existentials and Fictional Discourse. 1997　　　　　ISBN 0-7923-4388-3
262. A. Biletzki: *Talking Wolves*. Thomas Hobbes on the Language of Politics and the Politics of Language. 1997　　　　　　　　　　　　　　　　ISBN 0-7923-4425-1
263. D. Nute (ed.): *Defeasible Deontic Logic*. 1997　　　　　　ISBN 0-7923-4630-0
264. U. Meixner: *Axiomatic Formal Ontology*. 1997　　　　　　ISBN 0-7923-4747-X
265. I. Brinck: *The Indexical 'I'*. The First Person in Thought and Language. 1997
　　　　　　　　　　　　　　　　　　　　　　　　　　　　ISBN 0-7923-4741-2
266. G. Hölmström-Hintikka and R. Tuomela (eds.): *Contemporary Action Theory*. Volume 1: Individual Action. 1997　　　　　　　ISBN 0-7923-4753-6; Set: 0-7923-4754-4

SYNTHESE LIBRARY

Previous volumes are still available.

KLUWER ACADEMIC PUBLISHERS – DORDRECHT / BOSTON / LONDON

Printed in the United States
113019LV00001B/14/A

9 780792 353355